Amarjit S. Basra
Editor

Crop Responses and Adaptations to Temperature Stress

Pre-publication
REVIEWS,
COMMENTARIES,
EVALUATIONS . . .

"*Crop Responses and Adaptations to Temperature Stress* focuses on assisting in the marketing and processing decisions that are based upon an expectation of certain quality characteristics being derived from a given combination of genotype, environment, and crop management. This book provides the crop sector with essential information on the latest technical developments, identified from the newest review.

The text offers a forum for discussion on how increased chilling tolerance will allow for sowing in more northerly areas. Also, soil temperature is critical for radical emergence during seed germination. Root growth, water uptake, and mineral uptake are quickly, and significantly, inhibited by low temperature.

The book updates readers on the latest development in the conservation of genetic resources and the utilization of the genomic tools that are likely to be critical for our future ability to improve thermotolerance and crop productivity in hot environments.

Finally, the text will initiate research projects aimed at reducing the effect of heat stress on yield and quality. This book is an important reference source for anyone with an interest in the crop sector."

Mahamoud T. Nawar, PhD
Chief of Research,
Cotton Research Institute,
Giza, Egypt

Food Products Press
An Imprint of The Haworth Press, Inc.

Crop Responses
and Adaptations
to Temperature Stress

FOOD PRODUCTS PRESS
Crop Science
Amarjit S. Basra, PhD
Senior Editor

New, Recent, and Forthcoming Titles of Related Interest:

Dictionary of Plant Genetics and Molecular Biology by Gurbachan S. Miglani

Advances in Hemp Research by Paolo Ranalli

Wheat: Ecology and Physiology of Yield Determination by Emilio H. Satorre and Gustavo A. Slafer

Mineral Nutrition of Crops: Fundamental Mechanisms and Implications by Zdenko Rengel

Conservation Tillage in U.S. Agriculture: Environmental, Economic, and Policy Issues by Noel D. Uri

Cotton Fibers: Developmental Biology, Quality Improvement, and Textile Processing edited by Amarjit S. Basra

Heterosis and Hybrid Seed Production in Agronomic Crops edited by Amarjit S. Basra

Intensive Cropping: Efficient Use of Water, Nutrients, and Tillage by S. S. Prihar, P. R. Gajri, D. K. Benbi, and V. K. Arora

Physiological Bases for Maize Improvement edited by María E. Otegui and Gustavo A. Slafer

Plant Growth Regulators in Agriculture and Horticulture: Their Role and Commercial Uses edited by Amarjit S. Basra

Crop Responses and Adaptations to Temperature Stress edited by Amarjit S. Basra

Plant Viruses As Molecular Pathogens by Jawid A. A. Khan and Jeanne Dijkstra

Crop Responses and Adaptations to Temperature Stress

Amarjit S. Basra, PhD
Editor

Food Products Press®
An Imprint of The Haworth Press, Inc.
New York • London • Oxford

Published by

Food Products Press®, an imprint of The Haworth Press, Inc., 10 Alice Street, Binghamton, NY 13904-1580

Cover design by Monica L. Seifert.

Library of Congress Cataloging-in-Publication Data

Crop responses and adaptations to temperature stress / Amarjit S. Basra, editor.
     p.   cm.
    Includes bibliographical references and index.
    ISBN 1-56022-890-3 (hardcover : alk. paper) — ISBN 1-56022-906-3 (pbk.)
    1. Crops—Effect of temperature on. 2. Crops—Adaptation. I. Basra, Amarjit S.

S600.7.T45 C76 2000
631.5′233—dc21
                                                        00-039305

CONTENTS

About the Editor ix

Contributors xi

Preface xiii

Chapter 1. Mechanisms of Chilling Injury and Tolerance 1
Tottempudi K. Prasad

Introduction 1
Maize Seedlings Serve As a Model System
   to Study Chilling Effects 2
Chilling Induces Oxidative Stress in the Tissues 5
Mechanisms of Chilling Responses 6
Antioxidant Gene Expression Responsive
   to Chilling Stress 15
Possible Roles of Hydrogen Peroxide
   in Signaling Mechanisms 19
Summary of Chilling Stress and Tolerance Mechanisms
   in Maize 28
Genetic Engineering of Chilling (Cold) Tolerance 30
Concluding Remarks 34

**Chapter 2. Chilling Effects on Active Oxygen Species
and Their Scavenging Systems in Plants** 53
D. Mark Hodges

An Oxygenated Environment 53
Active Oxygen Species 54
Active Oxygen Scavenging Systems 58
Chilling Effects on Plants 60
Chilling Effects on Photosynthesis and Active
   Oxygen Species Production 61
Effects of Chilling on Antioxidant Systems 63
Conclusions 68

Chapter 3. Root System Functions During Chilling Temperatures: Injury and Acclimation   **77**
Pamela L. Sanders
Albert H. Markhart III

Introduction   77
Chilling Inhibition of Root Growth   80
Water Relations During Chilling   82
Mineral Uptake   85
Circadian Rhythms   87
Hormones in Chilled Roots   89
Root System Acclimation   90
Conclusion   98

Chapter 4. Mechanisms of Cold Acclimation   **109**
Jean-Marc Ferullo
Marilyn Griffith

Introduction   109
Control of Water Status   111
Control of Cell Dehydration   114
Membrane Stabilization   121
Induction of Cryoprotective Proteins   122
Molecular Changes Related to Protein Synthesis   130
Cold-Induced Gene Products of Unknown Function   132
Regulation of Molecular Adaptations to Cold   134
Conclusion   136

Chapter 5. Signal Transduction Under Low-Temperature Stress   **151**
Sirpa Nuotio
Pekka Heino
E. Tapio Palva

Introduction   151
Low-Temperature Control of Gene Expression   156
Genetic Dissection of Signal Pathways   165
Engineering Freezing-Tolerant Crops   167

Chapter 6. Mechanisms of Thermotolerance in Crops　　**177**
Natalya Y. Klueva
Elena Maestri
Nelson Marmiroli
Henry T. Nguyen

Introduction　　177
Impact of High Temperature on Crops　　178
Defense Mechanisms Against High-Temperature-
　Induced Damage　　182
Molecular Strategies for Enhancing Thermotolerance
　in Crops　　203
Conclusions and Perspectives　　206

**Chapter 7. Control of the Heat Shock Response
in Crop Plants**　　**219**
Daniel R. Gallie

Introduction　　219
Expression of HSPs Following Stress
　and During Development　　220
The Impact of Heat Shock on Transcription　　222
Effects of Heat Shock on Protein Yield　　224
Heat-Induced Changes in the Translational Machinery
　of Crop Plants　　226
HSP mRNAs Escape Heat Shock-Induced Translational
　Repression　　229
Regulation of RNA Degradatory Machinery Following
　Heat Shock　　230
Conclusions　　232

**Chapter 8. The Effects of Heat Stress on Cereal Yield
and Quality**　　**243**
Peter Stone

Introduction　　243
Effects of High Temperature on Yield　　245
Effects of High Temperature on Quality　　266
Strategies for Coping with Heat Stress　　277
Conclusions　　280

Index　　**293**

ABOUT THE EDITOR

Amarjit S. Basra, PhD, is an eminent plant physiologist at Punjab Agricultural University in Ludhiana, India, and is currently a visiting scientist at the Department of Agronomy and Range Science, University of California, Davis. His outstanding research in the field of plant growth regulation has been published in the world's leading journals and is widely cited.

He has over eighty research papers and eight edited books to his credit. He is a member of the American Association of Plant Physiologists, the Australian Society of Plant Physiologists, the Crop Science Society of America, the American Society of Agronomy, the American Institute of Biological Sciences, the American Society of Horticultural Science, and the International Society of Horticultural Sciences.

Dr. Basra has received coveted scientific awards and honors for his outstanding contributions, including the INSA medal for Young Scientists and the Rafi Ahmad Memorial Prize for Agricultural Research.

He provides leadership in organizing and fostering collaboration in international crop science research and dissemination of information.

CONTRIBUTORS

Jean-Marc Ferullo, PhD, is Researcher, Department of Biotechnologies, Aventis Crop Science, Lyon, Cedex, France.

Daniel R. Gallie, BS, PhD, is Professor, University of California, Riverside, California.

Marilyn Griffith, PhD, is Associate Professor, Department of Biology, University of Waterloo, Waterloo, Ontario, Canada.

Pekka Heino, PhD, is Acting Professor of Genetics, Department of Biosciences, Division of Genetics, University of Helsinki, Helsinki, Finland.

D. Mark Hodges, BSc (Hons.), PhD, is Research Scientist, Atlantic Food and Horticulture Research Centre, Agriculture and Agri-Food Canada, Kentville, Nova Scotia, Canada.

Natalya Y. Klueva, PhD, is Research Associate, Center for Biotechnology and Genomics, Texas Tech University, Lubbock, Texas.

Elena Maestri, PhD, is Senior Research Associate, Department of Environmental Sciences, Division of Genetics and Environmental Biotechnologies, University of Parma, Parma, Italy.

Albert H. Markhart, III, PhD, is Professor in Horticultural Sciences at University of Minnesota, St. Paul, Minnesota.

Nelson Marmiroli, PhD, is Professor, Department of Environmental Sciences, Division of Genetics and Environmental Biotechnologies, University of Parma, Parma, Italy.

Henry T. Nguyen, PhD, is Professor, Plant Molecular Genetics Laboratory, Department of Plant and Soil Sciences, Texas Tech University, Lubbock, Texas.

Sirpa Nuotio, PhD, is Professor of Genetics, Department of Biosciences, Division of Genetics, University of Helsinki, Helsinki, Finland.

E. Tapio Palva, PhD, is Professor of Genetics, Department of Biosciences, Division of Genetics, University of Helsinki, Helsinki, Finland.

Tottempudi K. Prasad, PhD, is Postdoctoral Fellow, University of Colorado Health Sciences Center, Denver, Colorado.

Pamela L. Sanders, PhD, is Assistant Professor of Biology, Southwest State University, Marshall, Minnesota.

Peter Stone, PhD, is Scientist, New Zealand Institute for Crop & Food Research, Ltd., Hasting, New Zealand.

Preface

Crop climate is often unfavorable for optimum expression of growth, productivity, and quality during various stages of plant development, resulting in huge losses of economic yield and quality in various cropping regions of the world. Problems faced by crop plants coping with temperature stress, manifested as chilling, freezing, and high temperatures, and the strategies that are decisive for successful plant functioning and development in a stressful environment are the focus of this volume. Temperature extremes are not only detrimental but also lead to dehydration or osmotic stresses through reduced availability of water, affecting vital cellular functions and maintenance of cellular integrity. A major goal of current crop science is to understand the integrated mechanisms that allow plants to cope with temperature stress constraints and to utilize this knowledge in the genetic modification of plants to achieve improved stress resistance.

Plants have evolved mechanisms to respond to temperature extremes at molecular, cellular, whole-plant, and canopy levels. Some adaptive responses are highly species specific, while others are of ubiquitous nature. This book gives a comprehensive account of what is currently known about these adaptive responses and presents innovative approaches for optimizing yield potential and stabilizing production under varying temperatures, particularly by employing powerful molecular technologies now available in combination with conventional plant breeding.

Tabulation of the selected data along with well-illustrated figures and references makes the contributors' arguments persuasive. This volume brings together internationally acknowledged experts in the discipline and provides new insights and approaches for anyone interested in understanding the mechanisms of stress perception, signal transduction, and responses, with the eventual aim to improve temperature stress tolerance of crops. The book will be extremely useful to advanced students, teachers, and researchers in plant physiology, biochemistry, molecular biology, biotechnology, plant breeding and genetics, ecology, agronomy, horticulture, and forestry, and especially to those working in the areas of "plant stress."

This book would not have been possible without the outstanding cooperation of contributing authors who have shared their specialist knowledge for the benefit of the international scientific community, and the solid support of Bill Cohen, Publisher, and Bill Palmer, Vice President (Book Division), of The Haworth Press/Food Products Press. Melissa Devendorf was always there with efficient administrative support. Amy Rentner, Peg Marr, and Andrew Roy greatly improved the manuscript through their meticulous editorial support.

My wife Ranjit, daughter Sukhmani, son Nishchayjit, and my parents, Sardarni Harbans Kaur Basra and Sardar Joginder Singh Basra, have played an inspirational role that was so vital for the success of this volume!

Amarjit S. Basra
Editor

Chapter 1

Mechanisms of Chilling Injury and Tolerance

Tottempudi K. Prasad

INTRODUCTION

Adverse environmental stress conditions, such as extreme temperatures, are detrimental to plant growth and development and thus affect the productivity of various crops around the world. Many of the important food crops, such as corn, sorghum, tomato, soybean, and rice, originally of tropical and subtropical origins, are now cultivated in areas where the temperatures fall well below the optimum required for their normal growth and development. Therefore, improvements in vegetative growth and yield performance of any of these crops will require selection of hybrids that have more efficient growth habits under unfavorable stress conditions (Greaves, 1996). However, the traditional plant breeding methods used to select for cold-tolerant varieties are cumbersome and sometimes result in a compromise between selection for efficient growth and high yield. Recently developed molecular techniques could be more amenable to isolate such cold-tolerant crop plants. In this regard, it is important first to understand any physiological, biochemical, and genetic changes that may be involved in improving cold tolerance in all of these food crops. Increasing evidence suggests that oxidative stress is involved in mediating chilling injury and tolerance mechanisms. Isolation of genes that respond

The author would like to acknowledge the contributions of Drs. Cecil Stewart, Barry Martin, and Marc Anderson for the progress of this project over the years. Thanks are due to Drs. Mary Reyland and Edward Dempsey for reading the manuscript. This research project was partly funded by grants from the U.S. Department of Agriculture and Pioneer Hi-Bred International.

to cold and/or oxidative stress is a necessary step toward the eventual goal of achieving cold-tolerant plants. Once the genes of interest are identified and cloned, gene transfer technology can be used to introduce foreign genes into crop plants to improve cold tolerance.

This chapter will attempt to identify some of the possible effects and the tolerance mechanisms of chilling-induced stress on crop plants, maize in particular, and to further discuss the progress and the feasibility of developing cold-tolerant crop plants using genetic engineering technology.

MAIZE SEEDLINGS SERVE AS A MODEL SYSTEM TO STUDY CHILLING EFFECTS

Identification of a Developmental Stage with Chilling Sensitivity

Before attempting to isolate the genes that may confer stress protection, it is important first to identify the developmental stage(s) of the plant that is (are) most sensitive to cold stress so that the molecular mechanisms involved in tolerance can be investigated with reference to the developmental stage and sensitivity of the plant. Cold stress can be subdivided into chilling stress, whereby plants are exposed to a low-temperature stress above $0°C$ (maize, tomato, tobacco, etc.), and freezing stress, whereby plants are exposed to temperatures below $0°C$ (soybean, alfalfa, wheat, etc.). Of particular interest are chilling-sensitive crops, such as maize, for studying the effects of chilling stress on early plant growth. Previous studies indicate that preemergent maize seedlings serve as a good model system for this purpose (Anderson et al., 1994; Prasad et al., 1994). Chilling injury is a serious problem during germination and early seedling growth of many plant species, including maize (Bedi and Basra, 1993; Nykiforuk and Johnson-Flanagan, 1998). Planting of maize during early spring is one of the most desired agronomic practices in maize production, to take advantage of more optimal temperatures and summer rainfall, and to avoid hot, dry summer periods during the stages of pollination and fertilization (Stewart et al., 1990a, b). However, early developmental stages are the only ones of the maize life cycle that are often exposed to spring frost in the field and are found to be very sensitive to chilling stress. Such exposure to chilling conditions can severely affect the stand establishment of maize, or any other crop, if the planted variety is not chilling tolerant. Preemergent seedlings are heterotrophic, which means that germination and seedling growth depend on the seed reserves. The first level of chilling damage could involve the inhibition of seed metabolic processes that are responsible for the breakdown of complex carbohydrates into simple sugars to support growth, such as germination and epicotyl elongation. As the seedlings age, they become autotrophic, which means that they synthesize carbohydrates to support their own growth. At this autotrophic stage, the chilling damage could be mostly through photoinhibition, which prevents normal photosynthesis and thus causes growth retardation. Under severe chilling stress conditions, the seedlings (both heterotrophic and autotrophic) may die, although this is typically not due to the effects of direct chilling but rather to their increased susceptibility to diseases during recovery. Thus, the identification of any physiological, biochemical, and genetic characteristics that may be responsible for causing chilling injury and tolerance mechanisms during germination and early seedling growth is of immense interest to maize producers. Therefore, this chapter will focus on describing chilling effects in maize seedlings and attempt to extrapolate the results of this model system to other crops that undergo similar chilling damage.

Chilling Acclimation Phenomenon

To study the mechanisms of chilling injury or tolerance, most studies have utilized comparisons of metabolic differences between chilling-sensitive and chilling-tolerant varieties as model systems (Jhanke, Hull, and Long, 1991; Walker and McKersie, 1993; Pinhero et al., 1997). The disadvantage with such systems is that complex genetic differences exist between sensitive and tolerant plants, and, therefore, it becomes difficult to interpret the limited metabolic differences in terms of chilling injury or tolerance. Working with one variety that can be acclimated but is otherwise chilling sensitive would be advantageous for studying the molecular mechanisms because both sensitivity and resistance to chilling can be investigated in the same genotype, thus eliminating the complexity in genetic differences. Acclimation to chilling results in a lowering of the temperature at which plant is damaged by chilling. Such a chilling acclimation or temperature-conditioning phenomenon has been characterized in a chilling-sensitive maize inbred, G50. Two- to five-day-old dark-grown maize seedlings were used to simulate the preemergence conditions in the field under laboratory conditions. In general, studies with various inbred lines that are chilling sensitive indicate that germination and early seedling growth are much reduced when they are germinated and allowed to grow under suboptimal temperatures (Stewart et al., 1990b). It has been tested whether developing seedlings grown at 27°C, an optimal temperature for maize seedling growth, would survive at below-threshold temperatures and then recover to grow at optimal temperatures (Prasad, 1996). When two- to five-day-old seedlings were directly exposed to 4°C stress for seven days in darkness, followed by recovery for seven days in a greenhouse, only 14 to 23 percent of the seedlings survived chilling stress (see Figure 1.1). Based on visual observations, the symptoms of chilling injury included the inhibition of seedling growth, waterlogged appearance and browning of mesocotyls, and the browning and desiccation of coleoptile and undeveloped leaves. However, when seedlings were preexposed to 14°C for one to three days (an acclimation treatment), 70 to 93 percent of the two- to four-day-old seedlings survived, but only 15 percent of the five-day-old ones survived. These results suggest that although acclimation-induced chilling tolerance was developmentally regulated in two- to four-

FIGURE 1.1. Effects of Chilling and Acclimation on the Survival of Two- to Five-Day-Old Dark-Grown Maize Seedlings

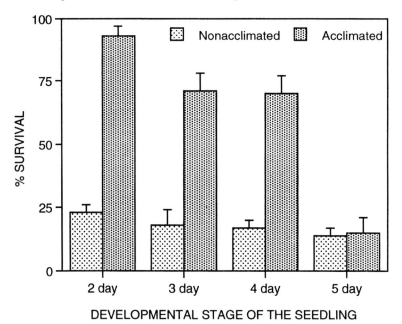

Note: Nonacclimated seedlings were subjected to 4°C stress for seven days, followed by recovery in a greenhouse for seven days. Acclimated seedlings were preexposed to 14°C for three days and then transferred to 4°C for seven days, followed by recovery in a greenhouse for seven days. Values are the means ± SEs of three replicates with fifty seedlings sampled in each replicate.

day-old seedlings, no tolerance was observed in any of the nonacclimated developing seedlings. In addition, chilling tolerance could also be induced either by pretreating the seedlings with heat shock at 37°C for 4 hours (h) (Prasad, unpublished results) or with 1 mM (millimolar) abscisic acid (ABA) for 24 h (Anderson et al., 1994). A similar acclimation phenomenon was demonstrated in other low-temperature-sensitive plant species, such as rice, peas, tomato, and cucumber (Guy, Niemi, and Brambl, 1985; Mohapatra, Poole, and Dhinsda, 1987; Gilmour, Hajela, and Thomashow, 1988; Lin et al., 1990; Guye, Vigh, and Wilson, 1987). This suggests that chilling-sensitive plants have the inherent ability to cope with chilling conditions when preexposed briefly to the appropriate suboptimal temperatures that induce tolerance mechanisms, before being subjected to a prolonged, severe chilling stress.

CHILLING INDUCES OXIDATIVE STRESS IN THE TISSUES

Under aerobic conditions, superoxide radicals and hydrogen peroxide (H_2O_2), the by-products of oxidative metabolism, are the normal metabolites of plant (Elstner, 1991) and animal cells (Halliwell and Gutteridge, 1986). However, superoxides can interact with H_2O_2 to form highly reactive hydroxyl radicals (Haber-Weiss reaction) that are thought to be primarily responsible for oxygen toxicity in the cells. The electron transport chains of mitochondria and chloroplasts are well-documented sources of reactive oxygen species (ROS) in plant systems (Elstner, 1991; Asada, 1992; Bowler, Van Montagu, and Inzé, 1992; Purvis, 1997). Some of the deleterious effects of high levels of ROS include lipid peroxidation, DNA (deoxyribonucleic acid) damage, and protein denaturation (Fridovich, 1978; Halliwell and Gutteridge, 1986). Once formed, ROS can undergo reactions yielding other activated compounds, eventually generating peroxyl radicals (Elstner, 1991). Therefore, these undesired ROS should be kept at low steady-state levels by the action of antioxidant metabolites and enzymes present in the cells. This is accomplished through the action of antioxidant enzymes, such as superoxide dismutase (SOD), catalase (CAT), glutathione reductase (GR), and ascorbate peroxidase (APX), and small-molecule antioxidants, such as ascorbate (AsA), glutathione (GSH), α-tocopherol, and β-carotene, which are located in various compartments of the plant cell (Elstner, 1991; McKersie, 1991; Bowler, Van Montagu, and Inzé, 1992; Scandalios, 1993, 1994).

Low-temperature stress accompanied by light or darkness has been implicated as an inducer of oxidative stress in tissues during chilling (Wise and Naylor, 1987; Bowler, Van Montagu, and Inzé, 1992; Scandalios, 1993, 1994; Prasad et al., 1994; Fadzillah et al., 1996; O'Kane et al., 1996), and exposure to freezing temperatures (Bowler, Van Montagu, and Inzé, 1992; McKersie et al., 1993). Although many of these studies do not report quantitative measurements of superoxide or H_2O_2 levels in their experimental systems, chilling induces oxidative stress conditions, as indicated by the loss or the induction of antioxidant enzymes. However, some studies have reported measurements of the accumulation of H_2O_2 and superoxide in plant tissues subjected to chilling temperatures. Omran (1980) reported a fourfold accumulation of H_2O_2 with a concomitant loss in CAT activity in cucumber seedlings during $5°C$ stress conditions. Chilling also doubled the accumulation of H_2O_2 in winter wheat leaves (Okuda et al., 1991), and in *Arabidopsis* callus tissue (O'Kane et al., 1996). Similarly, a fourfold increase in H_2O_2 was seen in maize seedlings during $4°C$ stress, and in the early stages of acclimation (Prasad et al., 1994; see Figures 1.2 and 1.3). Evidence also indicates that mitochondria isolated from chilling-sensitive green pepper fruit tissue produced increased levels of superoxide in vitro at chilling temperatures (Purvis, Shewfelt, and Gegogeine, 1995).

MECHANISMS OF CHILLING RESPONSES

Numerous mechanisms have been proposed to explain the chilling injury or tolerance in plants. Some of the changes related to low-temperature stress include alterations in gene expression, proteins, lipids, carbohydrate composition, membrane properties, solute leakage, mitochondrial respiration, and photosynthesis (Levitt, 1980; Wang, 1982; Markhart, 1986; Guy, 1990; Thomashow, 1990; McKersie, 1991; Howarth and Ougham, 1993; Smirnoff, 1993; Foyer, Descourvières, and Kunert, 1994). Since low temperatures induce oxidative stress in tissues, chilling damage was partly attributed to the effects of in vivo-generated ROS in the cells. On the contrary, some tolerance mechanisms directed against chilling-induced oxidative stress damage are thought to be induced by in vivo-generated ROS at optimal levels and conditions (Allen et al., 1994; Prasad et al., 1994). Because the effects of chilling damage in various plants were previously reviewed in great detail, this chapter is not intended to furnish a complete overview on chilling damage; rather, it is an attempt to discuss or explore different hypotheses that recently received some increased attention with the hope that this will aid in planning future research. However, one major field of interest that is not dealt with here, photosynthesis, likely requires an independent review on its own merit.

FIGURE 1.2. Levels of Hydrogen Peroxide and Catalase-3 Isozyme Activity in the Mesocotyls of Three-Day-Old Maize Seedlings Subjected to Chilling and Acclimation Treatments

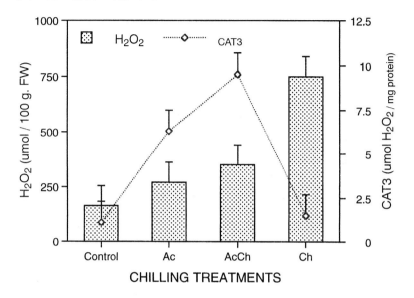

Note: The seedlings were either directly subjected to 4°C stress for four days or acclimated at 14°C for three days before transferring to 4°C stress for four days. Values are the means ± SEs of three replicate samples.Controls, three day-old seedlings; Ac, seedlings acclimated at 14°C for three days; AcCh, Ac seedlings transferred to 4°C stress for four days; and Ch, control seedlings subjected to 4°C stress for four days.

Membrane Alterations

Phase Change Mechanism

Physical transitions of membranes from a liquid-crystalline phase to a solid-gel phase were first suggested to be the primary response of chilling-sensitive plants subjected to low-temperature stress (Lyons and Raison, 1970). Membranes that depend on fluidity begin to solidify at suboptimal temperatures, which severely disrupts the function of membrane-associated proteins (Alonso, Queiroz, and Magalhaes, 1997). Lyons and Raison (1970) showed that the activation energies of mitochondrial enzymes were sharply increased at chilling temperatures, and that this change was attrib-

FIGURE 1.3. Time-Course Study on the Levels of Hydrogen Peroxide and Catalase-3 Isozyme Activity in the Mesocotyls of Maize Seedlings During the Early Stages of Acclimation

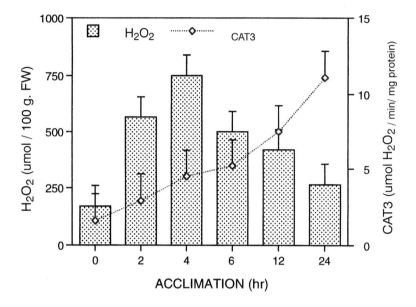

Note: The three-day-old seedlings were subjected to acclimation at 14°C for zero to twenty-four hours. Values are the means ± SEs of three replicate samples.

uted to the membrane phase transition. Using freeze-fracture electron microscopy, Armond and Staehelin (1979) demonstrated that, at solid phase, laterally and vertically oriented proteins were displaced away from the membrane. The temperature at which such phase transition occurs was correlated with chilling sensitivity. Changes in phase transition result in increased leakiness of the membranes that ultimately leads to a loss of compartmentation, collapsed gradients, and disrupted metabolism (Levitt, 1980; Wang, 1982; Markhart, 1986; Collins, Nie, and Saltveit, 1995). Although phase transitions have been correlated with chilling sensitivity in some plants (Armond and Staehelin, 1979; Raison and Orr, 1986), conflicting evidence from various studies makes it difficult to attribute most aspects of chilling injury to the phase transition hypothesis (Leheny and Theg, 1994; Sharon, Willemot, and Thompson, 1994).

Membrane Lipid Composition

Membrane phase changes were proposed to be dependent upon lipid composition, with greater unsaturation promoting the least number of phase transitions (Lyons and Raison, 1970). Because plant plasma membrane has been implicated as the primary site of freezing injury (Steponkus, 1984), the possible role of plant lipids, particularly membrane lipids, in low-temperature susceptibility and tolerance was extensively investigated (Shewfelt, 1992). Chilling tolerance was also linked with the extent of lipid unsaturation in some plant species (Roughan, 1985; Hugly and Somerville, 1992; Riken, Dillwith, and Bergman, 1993), and efforts to alter the degree of unsaturation of membrane lipids have resulted in improved chilling tolerance (Murata et al., 1992; Kodama et al., 1994; Moon et al., 1995). Since alterations in membrane lipid composition are among the major biochemical modifications that occur during chilling treatment, it is reasonable to expect some of the enzymes that catalyze lipid biosynthesis or turnover to be responsive to cold. Hughes and colleagues (1992) isolated a gene, *blt4,* that potentially encodes a phospholipid transfer protein from barley, but the significance of this gene in the regulation of low-temperature tolerance is unknown.

Conflicting evidence suggests that the degree of lipid unsaturation in a membrane is not the sole determinant of the phase transition. Chilling tolerance is not correlated with the extent of membrane lipid unsaturation among many plant species differing in chilling tolerance (Kenrick and Bishop, 1986). Reconstituted thylakoid lipid vesicles did not exhibit a phase change at any temperature above $0°C$, even when the levels of saturated lipids greatly exceeded those found in thylakoids of chilling-sensitive plants (Webb, Lynch, and Green, 1992). Furthermore, *Arabidopsis* mutants *(fab1)* that contain a higher level of saturated fatty acids in phosphatidylglycerols, compared to many chilling-sensitive species, do not show chilling sensitivity (Wu and Browse, 1995).

Changes in Protein Transport

Many of the organelle proteins are nuclear encoded and thus imported from the cytosol. Because organelle membranes may undergo changes in both lipid composition and degree of unsaturation in response to low temperatures, it was suggested that lowered membrane fluidity might be responsible for the cold-induced block of protein import (Nicchitta and Blobel, 1989; Thieringer et al., 1991). However, direct measurement of the envelope phase transition temperature by Fourier Transform Infrared Spectrometry, as well as the activation of the ATP/ADP (adenosine triphosphate/

adenosinediphosphate) translocator in the chloroplast inner membrane at 5 and 25°C, demonstrated that the cold-induced block of protein import into pea chloroplasts observed in vitro was due primarily to energetic considerations (e.g., ATP availability in chloroplasts), and not to decreased membrane fluidity (Leheny and Theg, 1994). It is possible that similar energy requirements may apply for the protein transport into mitochondria. In support of this hypothesis, ATP synthesis was reported to be inhibited during 4°C stress in both nonacclimated and acclimated maize mitochondria (Prasad, Anderson, and Stewart, 1994). Levels of some of the nuclear-encoded proteins, such as F_1-ATPase, Cyt oxidase, and molecular chaperones (HSP70 and HSP60) (Prasad, unpublished results), were also reduced in these mitochondria during chilling stress. These results imply that, at least in part, reduced energy levels in mitochondria may have resulted in reduced protein import from the cytosol during chilling stress. However, it is not known whether differences in phase transitions in the membranes of acclimated and nonacclimated maize mitochondria, if any, also play a role in protein import. Thus, clearly, other biochemical events may contribute to the sensitivity of the membrane to low temperatures, in addition to changes in lipid composition and phase transitions of membranes (Sharon, Willemot, and Thompson, 1994).

Lipid Peroxidation

It is known that ROS trigger lipid peroxidation in cellular membranes (Fridovich, 1978; Halliwell and Gutteridge, 1986). Unsaturated fatty acids are easily peroxidized by hydroxyl radicals that are converted from superoxide (Fridovich, 1978; Halliwell and Gutteridge, 1986; Cadenas, 1989). Reduced cellular and membrane damage (lipid peroxidation) in plants has been linked to increased antioxidant defense system activity against ROS (Dhindsa and Mattowe, 1981; Senaratna, McKersie, and Borochov, 1987; Leprince et al., 1990; Bowler, Van Montagu, and Inzé, 1992; Smirnoff, 1993; Zhang and Kirkham, 1994). Correlative evidence from different studies also suggests that the chilling sensitivity of plants increases with increased lipid peroxidation (Shewfelt and Erickson, 1991; Prasad, 1996). In maize seedlings, lipid peroxidation was increased by about twofold in nonacclimated seedlings, compared to control or acclimated seedlings, during 4°C stress and recovery (Prasad, 1996; see Figure 1.4). Since seedlings pretreated with H_2O_2 and menadione, a superoxide-generating compound, also increased lipid peroxidation (see Figure 1.5), it was suggested that the induced lipid peroxidation was due to the accumulated ROS in nonacclimated seedlings during chilling stress (Prasad, 1996). On the other hand, the reduced lipid peroxidation in acclimated seedlings was

FIGURE 1.4. Changes in Lipid Peroxidation (TBARS), Protein Oxidation (Carbonyl Content), and Endopeptidase Activity in the Mesocotyls of Acclimated and Nonacclimated Three-Day-Old Maize Seedlings

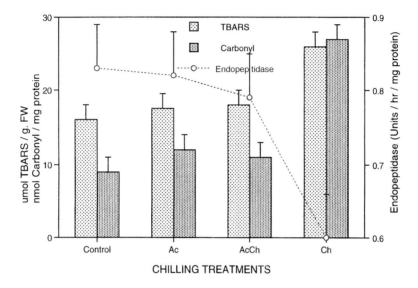

Note: Values are the means ± SEs of three replicate samples. Control, three-day-old seedlings; Ac, acclimated at 14°C for three days; AcCh, Ac seedlings transferred to 4°C stress for seven days; and Ch, control seedlings subjected to 4°C stress for seven days.

due to an increased antioxidant defense system that scavenged ROS during acclimation followed by chilling stress (Prasad, 1996). All of these studies suggest that ROS-induced lipid peroxidation was, at least in part, responsible for increased chilling sensitivities in maize seedlings and, perhaps, all of the chilling-sensitive crop plants.

Mitochondrial Function

The competence and stability of mitochondria are important for plant tissues to survive low temperatures, especially during early seedling growth. Because mitochondria are both the source and a target of oxidative stress in dark-grown seedlings subjected to chilling (Puntarulo et al., 1991; Prasad, Anderson, and Stewart, 1994; Purvis, 1997), the functions of mitochondrial proteins could be affected by chilling-induced oxidative stress. The effects of chilling on respiration are dependent on the severity of the stress and the

FIGURE 1.5. Effect of Hydrogen Peroxide and Menadione on Lipid Peroxidation (TBARS), Protein Oxidation (Carbonyl Content), and Endopeptidase Activity in the Mesocotyls of Three-Day-Old Maize Seedlings

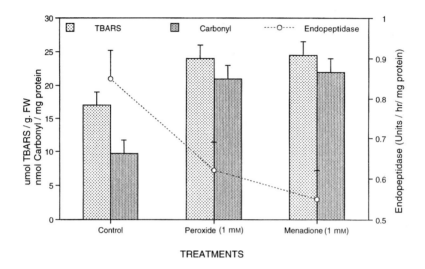

Note: Control, three-day-old dark-grown seedlings; H_2O_2 and menadione, three-day-old seedlings were treated with 1 mM H_2O_2 or menadione for four hours by immersing the seedling roots in treatment solutions at $27°C$ in the darkness. Values are the means $\pm$ SEs of four replicate samples.

sensitivity of the plant species. In general, respiration decreases during stress and sharply increases during recovery. If chilling injury is reversible, respiration declines to prechilled levels after a period of recovery (Lyons, 1973). Chilling causes a decline in the activity of Cyt oxidase and an increase in alternative oxidase activity (Kiener and Bramlage, 1981; Van de Venter, 1985; Elthon et al., 1986; Rybka, 1989; Prasad, Anderson, and Stewart, 1994). Since low-temperature stress is speculated to cause phase transitions, chilling damage of the membrane-associated proteins has been suggested to be the cause of the decline in cyanide-sensitive electron transport (Kiener and Bramlage, 1981; Purvis and Shewfelt, 1993). Consistent with this hypothesis, Cyt oxidase activity and protein level decrease in response to chilling in maize mitochondria (Prasad, Anderson, and Stewart, 1994). This decline may lead to increased generation of superoxide in the mitochondria. Direct measurements of superoxide in chilled green pepper mitochondria also indicate that inhibition of the Cyt pathway leads to superoxide production (Purvis, Shewfelt, and Gegogeine, 1995). Since mitochondria are believed to be the main source of oxidative stress in dark-grown seedlings, a fourfold accumulation of H_2O_2 in nonacclimated chilled maize seedlings could be a result of inhibition of the Cyt pathway in mitochondria (Prasad, Anderson, and Stewart, 1994). Although acclimated seedlings also accumulate H_2O_2 (Prasad, Anderson, and Stewart, 1994), these mitochondria, unlike nonacclimated ones, were protected from chilling injury due to the accumulation of CAT3, a maize mitochondrial catalase isozyme. This is consistent with the assumption that an increase in the mitochondrial antioxidant system is necessary for the cell to develop tolerance against chilling-induced oxidative stress. In a similar analogy, overexpression of plant Mn(manganese)SOD in yeast mitochondria resulted in increased tolerance to oxidative stress conditions in yeast (Zhu and Scandalios, 1992). These observations suggest that protection of mitochondria from oxidative stress is an important component of tolerance mechanism in the cell.

Loss of the Cyt pathway leads to an induction of the alternative pathway in many of the plant tissues subjected to chilling stress. The increase in the alternative oxidase capacity has been shown to be due to an increase in protein synthesis (Stewart et al., 1990a; Vanlerberghe and McIntosh, 1992, 1996). Alternative oxidase is suggested to serve as a nondestructive outlet for electrons when the supply of respiratory electrons exceeds the capacity of the Cyt oxidase (Purvis and Shewfelt, 1993), and thus preventing the formation of superoxides and H_2O_2 (Purvis, 1997). Vanlerberghe and colleagues (1994) reported that *aox1* antisense mutants of tobacco, unlike wild-type plants, could not survive antimycin A, an inhibitor of Cyt oxidase.

In mitochondria of preemergent maize seedlings, the Cyt pathway was inhibited and the alternate pathway was induced in both acclimated and nonacclimated mitochondria at the end of the $4°C$ stress period. However, differences in the selection of a respiratory pathway between acclimated and nonacclimated mitochondria prevailed during recovery. Only the mitochondria of acclimated seedlings that were destined to survive chilling stress switched their respiratory pathway from the alternative to the Cyt pathway. Mitochondria of nonacclimated seedlings that were not going to survive chilling stress remained dependent upon the alternative pathway of respiration (Prasad, Anderson, and Stewart, 1994). Thus, the ability of the maize seedlings and perhaps other plant species to recover from chilling stress seems to depend on restoring electron flow to the normal Cyt pathway during recovery from chilling stress.

In addition, several Krebs cycle enzymes that contribute to normal mitochondrial functions were also shown to be inactivated by chilling

(Lyons and Raison, 1970; Duke, Shrader, and Miller, 1977). Chilling stress also affects the energy-generating capacity of mitochondria. Lyons and colleagues (1964) proposed that mitochondria of chilling-sensitive plants do not possess dynamic properties at low temperatures; thus, phosphorylation is inhibited. Consistent with this hypothesis, ATP synthase activity and its protein levels were inhibited by chilling that resulted in reduced ATP synthesis in both nonacclimated and acclimated maize mitochondria (Prasad, Anderson, and Stewart, 1994). When the seedlings were allowed to recover from chilling stress, only acclimated mitochondria regained the capacity to synthesize ATP. Similar results were reported by Creencia and Bramlage (1971) for maize seedlings. Mitochondria failed to synthesize ATP upon recovery from chilling. Thus, recovery from chilling injury must also depend on whether the mitochondria can quickly and fully regain their capacity to generate ATP. All of these results suggest that alterations in mitochondrial metabolism can affect mitochondrial function during chilling conditions.

Protein Degradation

Proteins damaged by oxidation normally undergo degradation by a number of common proteases and multicatalytic protease systems, as shown in bacterial and animal cells (Levine et al., 1990; Pacifici and Davies, 1990; Stadtman, 1993). A similar protein degradation mechanism was also observed in plants when chloroplasts were exposed to photo-oxidative conditions (Virgin et al., 1991; Casano, Lascano, and Trippi, 1994) and cell wall proteins in the extracellular fluid were exposed to oxidative stress (Gomez, Casano, and Trippi, 1995). However, under stress conditions, with proteases inhibited by inactivation or decreased production, the oxidized proteins can accumulate to toxic levels that can induce oxidative stress, thus impairing cellular metabolism (Stadtman and Oliver, 1991). The oxidative stress that resulted from the accumulation of inactivated or unfolded proteins in the endoplasmic reticulum was suggested to result in the translocation of the transcription factor NF-kB to the nucleus, and the subsequent induction of specific target genes (Pahl and Baeuerle, 1995). Several reports indicated the inhibition of various enzymes in cells exposed to chilling or oxidative stress conditions (Fucci et al., 1983; Kenis, Morlans, and Trippi, 1989), implying that ROS can promote the oxidation of proteins in the cell.

It was shown in maize seedlings that the protein carbonyl content, a reflection of protein oxidation, was induced by more than twofold in non-acclimated seedlings under severe chilling stress (Prasad, 1996) (see Figure 1.4). However, there was reduced protein oxidation in acclimated seedlings, similar to controls, because of the induced antioxidant system that scavenged the accumulating ROS. Since seedlings treated with oxidants also have increased protein oxidation, it is suggested that in vivo-generated ROS could be responsible for the twofold induction in carbonyl content in nonacclimated chilled seedlings (see Figure 1.5). The changes in the levels of inactivated proteins in nonacclimated seedlings negatively correlate with protease activities. Nonacclimated seedlings had 35 to 40 percent less aminopeptidase and endopeptidase activities during chilling stress than the controls and acclimated seedlings (Prasad, 1996) (see Figure 1.4). These data suggest that the accumulation of oxidized proteins in nonacclimated seedlings was due to the loss of protease activity that otherwise would promote protein degradation. This loss of protease activity in nonacclimated seedlings was likely due to the inhibitory effect of in vivo-generated oxidative stress. This conclusion was supported by the fact that maize seedlings treated with H_2O_2 or menadione (at 0.5 and 1 mM concentrations) also exhibited a 35 to 40 percent decrease in protease activity (Prasad, 1996) (see Figure 1.5). Similarly, Gomez and colleagues (1995) reported the inhibition of extracellular endopeptidase activity by peroxide treatment in bean hypocotyl cells.

Conflicting evidence suggests that chilling promotes protein degradation in some plant species (Levitt, 1980). Schaffer and Fisher (1988) reported that one of the mRNA (messenger ribonucleic acid) species that accumulates in tomato during low temperature encodes a thiol protease related to papain. They suggested that the protein product might play a part in the degradation of polypeptides that were inactivated or disassembled as a result of low-temperature stress. It is possible that while some of the proteases are down-regulated, others are activated by the chilling-induced oxidative stress. However, cells have molecular chaperones, such as CPN60, HSP70, and HSC70, that are also induced by low temperatures presumably to protect the proteins from chilling-induced protein denaturation (Neven, Haskel, and Guy, 1990; Prasad, unpublished results). Although the involvement of molecular chaperones in protein protection from high temperature stress was previously demonstrated (Martin, Horwich, and Hartl, 1992), their role in protein protection during low-temperature stress is still unknown. Nevertheless, this induction of molecular chaperones in response to low temperatures suggests that protein protection from low-temperature damage is likely a part of the tolerance mechanism in plant cells.

ANTIOXIDANT GENE EXPRESSION RESPONSIVE TO CHILLING STRESS

Environmental stresses are important determinants of crop productivity. This observation emphasizes the importance of understanding the molecular and biochemical mechanisms involved in the regulation of gene expression in response to stress signals. Several reports indicate that the effects of some stresses, such as herbicides, atmospheric pollutants, temperature, light, waterlogging, drought, salt, pathogens, and plant hormones, are mediated by oxidative stress in plant tissues (McKersie, 1991; Bowler, Van Montagu, and Inzé, 1992; Scandalios, 1992, 1993; Doke et al., 1994). Plants induce various antioxidants in response to chilling, and these antioxidants induce chilling tolerance (Bowler, Van Montagu, and Inzé, 1992).

Several antioxidant enzymes and proteins of unknown function have been reported to increase during low temperatures (Guy, 1990; Welin et al., 1996). Several cold-regulated *(cor)* genes whose expression was changed during freezing acclimation (Thomashow, 1990; Welin et al., 1996) have been isolated and characterized. Similarly, analogous genes expressed differentially during chilling acclimation, designated as chilling acclimation-responsive genes, have been isolated from various plants including maize (Anderson et al., 1994; Bowler, Van Montagu, and Inzé, 1992; Prasad et al., 1994). Some of the induced genes are known to be regulated by transcription and posttranscription processes (Prasad et al., 1994; Hughes and Dunn, 1996). However, it is not known whether this induction is controlled by enzyme stability.

Functional cooperation between antioxidant enzymes is important for effective scavenging of harmful ROS. ROS are suggested to be responsible for cold injury involving abrupt increases in H_2O_2 in winter wheat leaves (Okuda et al., 1991) and other plant species (Bowler, Van Montagu, and Inzé, 1992). A detailed molecular analysis of the regulation of SODs was reported by Tsang and colleagues (1991) and Bowler and colleagues (1992). cDNA (complementary DNA) clones encoding mitochondrial MnSOD, cytosolic Cu(copper)/Zn(zinc)SOD, and chloroplastic Fe(iron)SOD have been isolated and used to study gene expression in response to various treatments that are suggested to increase ROS in plant tissues. These studies indicate increases in the level of mRNAs encoding these various isoforms of SOD (Bowler, Van Montagu, and Inzé, 1992; Tsang et al., 1991) and demonstrate some degree of specificity in the responses among different SOD isoforms. For example, treatments increasing mitochondrial activity increase transcripts of MnSOD (Bowler et al., 1989). Chilling, heat shock, and paraquat treatments increased transcripts of cyt Cu/ZnSOD and Fe-SOD in high light. In low light, a selective increase in transcript levels of Cu/ZnSOD occurred. These results suggest that genes of different SOD isoforms are regulated according to the subcellular compartment in which the treatments are expected to increase the accumulation of ROS (Tsang et al., 1991). SOD activities were reported to be induced when rice plants were transferred from 5 to 25°C, but CAT activity decreased in both sensitive and resistant rice cultivars at 5°C stress (Saruyama and Tanida, 1995), similar to pea, cucumber, maize, rye, and wheat (Mishra, Mishra, and Singhal, 1993; Feierabend, Schaan, and Hertwig, 1992; Omran, 1980). The cold lability of CAT in both rice cultivars and a corresponding increase in APX in the resistant cultivar suggested the significance of the cooperative function of APX in compensating for the loss of CAT activity in those cultivars during chilling stress (Saruyama and Tanida, 1995). The importance of APX was also reported for many other plants (Feierabend, Schaan, and Hertwig, 1992; Sen Gupta, Webb, et al., 1993). In mustard cotyledons, APX was regulated by phytochrome (Thomsen, Drumm-Herel, and Mohr, 1992) and by high light intensity (Gillham and Dodge, 1987). The increase in APX isozymes could be blunted by feeding the seedling with the singlet oxygen/peroxidation quenchers p-benzoquinone and α-tocopherol (Thomsen, Drumm-Herel, and Mohr, 1992). This suggests that the induction of APX depends on the generation of ROS as well as light itself. Ethylene-induced APX in mung bean conferred protection against H_2O_2, O_3 (ozone), and paraquat (Mehlhorn, 1990). Covello and colleagues (1988) showed the accumulation of a 31 kilodalton (kDa) polypeptide, a component of light-harvesting complex II (Hayden, Cavello, and Baker, 1988), in maize leaf thylakoids exposed to 5°C at high light intensity, conditions that cause photoinhibition as a result of oxidative stress in maize leaves.

Increased activities of SOD, APX, monodehydroascorbate reductase (MDHAR), and higher zeaxanthin levels occur in cold-hardened spinach leaves (Schöner and Krause, 1990), which would suggest that low temperatures enhance radiationless dissipation of excess excitation energy and protection against oxidative damage. Similar effects are seen in *Zea mays* and *Zea diploperennis*, where chilling causes increases in the activities of GR, APX, and MDHAR in either or both species (Jhanke, Hull, and Long, (1991). Cold acclimation increases tolerance to ROS in cereals (Bridger et al., 1994) and correlates with an increase in antioxidant enzymes. In chilling-sensitive plants, the ability to defend against oxidative damage has been shown to be inhibited by a reduction in antioxidants such as AsA, GSH, and α-tocopherol (Wise and Naylor, 1987), CAT (Omran, 1980; Fadzillah et al., 1996), and SOD (Michalski and Kaniuga, 1982). Chilling tolerance improved when GSH, peroxidase, and CAT levels were enhanced (Upadhyaya et al., 1989). Induced SOD and APX activities were correlated with increased chilling tolerance in paclobutrazol-induced chilling tolerance in chilling-sensitive maize inbreds (Pinhero et al., 1997).

GR is necessary for the regeneration of ascorbate, which is required for the activity of APX. GR has also been shown to be important in protection against oxidative stress in plants (Guy and Carter, 1984; Aono et al., 1993; Foyer et al., 1991). GR levels increased by two- to threefold in response to oxidative stress caused by paraquat and O_3 in peas (Edwards et al., 1994). Induction of GR by H_2O_2 and paraquat in maize leaves appears to be controlled at the level of transcription (Pasturi and Trippi, 1992). On the other hand, GR activity was decreased at 4°C in rice cultivars (Fadzillah et al., 1996). Similar to chilling tolerance, paraquat resistance is correlated in many instances with increased activity of antioxidant enzymes, including SOD, CAT, APX, and GR (Gilham and Dodge, 1987; Shaaltiel et al., 1988; Malan, Greyling, and Gressel, 1990). Treating plant tissues with O_3, a stress that also induces ROS, induced various antioxidants that are correlated with O_3 tolerance (Bender et al., 1994; Conklin and Last, 1995).

Differential expression of *cat* genes was studied in great detail in tobacco (Willekens et al., 1994), maize (Scandalios, 1992), and cotton (Ni and Trelease, 1991). CAT activities were affected by various environmental signals, such as light (Skadsen and Scandalios, 1987), temperature (Matters and Scandalios, 1986; Prasad et al., 1994), and other factors (Scandalios et al., 1984). Willekens and colleagues (1994) demonstrated that while *cat2* and *cat3* mRNAs were not significantly affected by cold treatment (6°C for 24 h), they were slightly induced during the recovery period in tobacco. In contrast, *cat1* mRNA levels were induced during chilling. In maize, chilling acclimation (at 14°C) induced *cat3* mRNA and protein in mesocotyls of dark-grown seedlings (Prasad et al., 1994). Besides, acclimation-induced tolerance has been reported to be developmentally regulated in maize seedlings. This developmentally regulated tolerance was correlated with the induced CAT activity, and in that, younger seedlings were shown to have higher CAT activity than the older seedlings. These differences in CAT levels were correlated with a higher percentage of the younger seedlings surviving chilling stress. A similar senario of developmental regulation exists in pea leaves, for which resistance to paraquat was correlated with leaf age, GR, and Cu/ZnSOD transcript levels (Donahue et al., 1997). Elevated CAT activity was shown to play a role in chilling tolerance in maize (Prasad, 1997). Kendall and colleagues (1983) reported that CAT-deficient barley mutants failed to survive severe oxidative stress caused by photorespiratory conditions.

Although the ascorbate-glutathione cycle is known to be responsible primarily for H_2O_2 scavenging in chloroplasts, its importance in the cytosol of photosynthetic cells and nonphotosynthetic tissues has also been suggested (Law, Charles, and Halliwell, 1983; Alscher, 1989; Foyer, Descourvières, and Kunert, 1994; Alscher, Donahue, and Cramer, 1997). The capacity of this cycle depends on the maintenance of reduced AsA and

GSH pools by the action of dehydroascorbate reductase (DHAR) and GR. AsA is a powerful reducing agent, and an AsA-deficient mutant, *Sox1*, was shown to be susceptible to oxidative stress (Conklin, Williams, and Last, 1996). Jhanke and colleagues (1991) reported that the activities of both DHAR and GR were severely inhibited at 5°C; thus, a depletion of AsA and GSH and a corresponding increase in dehydroascorbate (DHA) and oxidized glutathione (GSSG) are expected in chilled maize seedlings. However, in preemergent maize seedlings, reduced levels of AsA or DHA were observed in mesocotyls or coleoptiles of both acclimated and non-acclimated seedlings during chilling (Anderson, Prasad, and Stewart 1995). On the other hand, there was a net increase in the total glutathione pool in acclimated but not in nonacclimated seedlings. This increase was more significant in coleoptile than in mesocotyl regions of acclimated seedlings. Since chilling increases the accumulation of H_2O_2 in the absence of induced antioxidants in nonacclimated tissues, the net increase in the glutathione pool (in the absence of induced antioxidant enzymes) was correlated with reduced H_2O_2 levels and increased chilling tolerance in coleoptile regions of acclimated seedlings (Anderson, Prasad, and Stewart, 1995). Similarly, Walker and McKersie (1993) observed an increased synthesis of total glutathione and a maintenance of the reduced form in chilling-resistant tomato, while the susceptible tomato species exhibited only oxidation of GSH with no increase in the glutathione pool.

POSSIBLE ROLES OF HYDROGEN PEROXIDE IN SIGNALING MECHANISMS

The first stable by-product of oxygen metabolism, H_2O_2, has been shown to play a regulatory role in bacterial (Storz, Tartaglia, and Ames, 1990), mammalian (Schreck et al., 1991), and plant systems (Allen et al., 1994; Prasad et al., 1994; Mehdy et al., 1995). ROS are implicated as second messengers in signal transduction mechanisms in those systems. However, it may seem unlikely that highly reactive compounds such as H_2O_2 or other ROS might function as secondary messengers. The effects of ROS could be either cytotoxic or beneficial to the cell, depending on the levels and duration of exposure. If the cells were subjected to severe oxidative stress, the result could be metabolic dysfunctions, such as protein denaturation, lipid peroxidation, and DNA damage (Halliwell and Gutteridge, 1986). However, the emerging view is that, although a severe oxidative stress induced by high levels of ROS is toxic to the cell, a mild oxidative stress with optimal levels could be beneficial (Prasad et al., 1994; Karpinski et al., 1997).

Optimal concentrations of peroxide in the cell can induce antioxidant gene expression and oxidative stress tolerance in tissue exposed to temperature stress. In chilling-sensitive maize seedlings pretreated with low concentrations of H_2O_2 or menadione, tolerance against subsequent chilling-induced oxidative stress damage was induced. This induced chilling tolerance was correlated with the induction of *cat3* transcripts and CAT and peroxidase (POX) activities. It was also demonstrated that H_2O_2, generated in vivo either by pretreating the tissues with pro-oxidants or by treating tissues with the CAT-specific inhibitor aminotriazole, acts as a signal molecule by inducing *cat3* gene expression (Prasad et al., 1994). H_2O_2 was also shown to induce CAT activity by at least twofold in the scutellum of maize seedlings, and this was shown to be the result of de novo synthesis of CAT protein (Scandalios et al., 1984). The treatment of winter wheat with low concentrations of H_2O_2 and a CAT inhibitor induced the synthesis of proteins similar to those found in plants exposed to low temperatures (Matsuda, Okuda, and Sagisaka, 1994). Potato nodal cuttings incubated with H_2O_2 for 1 h developed growth inhibition in the clonal-microplant-population derived from these nodal cuttings. However, nodal explants from these H_2O_2-treated microplants developed resistance to 15 h heat shock at 42°C even after four weeks of cultures (Foyer et al., 1997). Sen Gupta, Webb, and colleagues (1993) reported a threefold increase in APX activities in Cu/ZnSOD overexpressing transgenic plants, implying that H_2O_2 generated from the dismutation of superoxide by overexpressing SOD might be responsible for inducing APX. More interesting, a similar induction in APX activity could also be demonstrated by treating wild-type tobacco leaves with exogenous peroxide (Allen et al., 1994). A gene encoding glutathione-S-transferase *(GST6)*, another antioxidant enzyme, was shown to be induced by H_2O_2 in *Arabidopsis* (Chen, Chao, and Singh, 1996).

Production of H_2O_2 in plants also appears to be an important defense mechanism against pathogens (Mehdy et al., 1995; Wu et al., 1995). Leaf tissues and cell cultures produce an oxidative burst within minutes of exposure to bacteria (Devlin and Gustine, 1992) and fungal cell wall components (Apostol, Heinstein, and Low, 1989; Svalheim and Robertson, 1993). Exposure of cells to biotic stress produces superoxides through activation of the plasma membrane-associated NADPH(nicotinamide adenine dinucleotide phosphate, reduced)-dependent superoxide synthase (Doke et al., 1994; Alvarez and Lamb, 1997), or a cell wall peroxidase (Bolwell et al., 1995). NADPH oxidase has been shown to be involved in superoxide generation that is dependent on Ca^{2+} (calcium) influx in tobacco cells (Pugin et al., 1997). This superoxide in turn dismutased to H_2O_2 either enzymatically (SOD) or nonenzymatically in the apoplast. Such an oxidative burst may serve as a first line of defense in resistant plants by directly attacking the pathogen as it enters the cytosol during the earliest stages of infection. Subsequently, massive changes occur in the pattern of plant gene expression, leading to the transcriptional activation of genes involved in defense mechanisms (Lamb et al., 1989; Jabs et al., 1997). Chen and colleagues (1993) showed that salicylic acid (SA)-induced systemic acquired resistance (SAR) in tobacco was modulated by H_2O_2 through the induction of pathogenesis-related (PR) proteins associated with SAR. However, more recent studies have disputed this observation and presented evidence that H_2O_2 is not involved in the down stream of SA-induced SAR (Bi et al., 1995; Neuenschwander et al., 1995). Nevertheless, endogenously generated ROS or externally supplied H_2O_2 has been shown to induce phytoalexin accumulation in the absence of a fungal elicitor (Apostol, Heinstein, and Low, 1989). Jabs and colleagues (1997) suggested that elicitor-stimulated ROS production was necessary for the phytoalexin production in parsley cells. Exogenous H_2O_2 also induced the transcription of phenylalanine ammonia-lyase (PAL), chalcone synthase, chalcone isomerase, and a basic endochitinase in bean suspension cultures (Mehdy et al., 1995) and may be involved in the stability of defense-related transcripts (Zhang et al., 1993).

Some of the antioxidant and disease resistance genes are known to be induced by various biotic and abiotic stresses. For example, MnSOD was induced by SA, pathogens, and chilling (Bowler et al., 1989; Bowler, Van Montagu, and Inzé, 1992). Likewise, homologous PR proteins were induced by H_2O_2, fungal pathogen, and paraquat in soybean (Crowell et al., 1992). The *GST6* gene was shown to be induced by auxin, SA, and H_2O_2 in *Arabidopsis* (Chen, Chao, and Singh, 1996). These studies suggest that ROS-, and specifically H_2O_2, induced gene expression is not just specific to disease resistance but plays a role in the general plant defense mechanisms that confer protection against oxidative stress. Given that several genes are induced by diverse stress conditions, it is conceivable that a common signal transduction mechanism is involved in the induction of a general plant defense mechanism.

Although H_2O_2 is known to induce gene expression, its mechanism of gene regulation is largely unknown in plants. However, evidence recently obtained in prokaryotic and eukaryotic systems suggests at least some of the molecules are likely involved in such signal transduction mechanisms (see Figure 1.6).

The model mechanism depicted in Figure 1.6, for example, was developed after taking several pieces of evidence from bacterial, plant, and

FIGURE 1.6. Schematic Representation of Signal Transduction Mechanisms Responsive to Oxidative Stress in Plant Cells

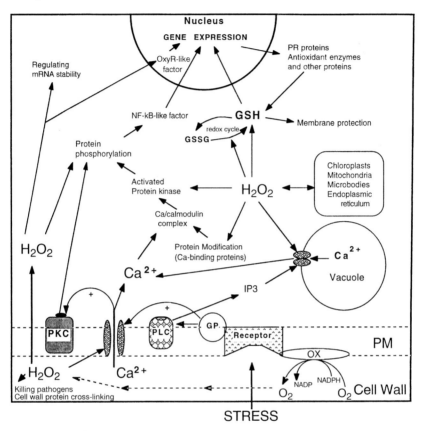

mammalian systems into consideration. When a cell perceives stress, it activates the receptor-coupled NADPH-dependent superoxide synthase or oxidase on the plasma membrane (PM) or a cell wall peroxidase that reduces oxygen to produce superoxide. This superoxide is quickly dismutased, either enzymatically by apoplastic SOD or nonenzymatically to H_2O_2. H_2O_2 produced in apoplast would serve as a first line of defense that involves killing pathogens and participating in cell wall protein cross-linking processes. H_2O_2 diffuses into the cytosol and also modulates Ca^{2+} channels on the plasma membrane to release Ca^{2+} from the cell wall. This Ca^{2+} influx from the cell wall could also be achieved by activating Ca^{2+} channels through the G-protein (GP)-coupled receptor on the membrane.

Ca^{2+} efflux from the vacuole is activated by inositol-3-phosphate (IP3), the product of the hydrolysis of the membrane phosphoinositides by the activated phospholipase C (PLC), and also through the modulation of the Ca^{2+} channel on the tonoplast by the cytosolic H_2O_2 that is diffused from the cell wall and organelles. Thus, the elevated cytosolic Ca_{2+} in combination with calmodulin activates protein kinases on the membrane or in the cytosol that can lead to the events of specific protein phosphorylation. On the other hand, elevated levels of H_2O_2 in the cytosol can either directly activate transcription factors in the cytosol that eventually are translocated into the nucleus or diffuse into the nucleus to activate transcription factors that drive antioxidant and other defense-related gene expression. H_2O_2 can also induce GSH levels in the cytosol. However, continuous recycling of GSSG to GSH is also required to maintain high levels of antioxidant function in the cell. Similar to H_2O_2, GSH diffuses freely through biological membranes and induces defense gene expression as well as protects the membrane from lipid peroxidation, along with various other small antioxidant molecules.

OxyR/SoxRS-Like Mechanisms

Bacterial cells have an adaptive response to high concentrations of H_2O_2 and menadione (Christman et al., 1985; Greenberg et al., 1990) that appears to be specific for each compound. One of the thirty-three proteins induced by H_2O_2 is CAT HP1, which is regulated by the OxyR regulon. OxyR protein is directly activated by H_2O_2 to become the transcriptional activator, making it both the sensor of oxidative stress and the mediator of enhanced transcription of genes involved in the protective response. H_2O_2 interacts with the OxyR protein to change its oxidation state, causing a conformational change, so the protein can now interact with its target promoters (Storz, Tartaglia, and Ames, 1990). Although *cat* is one of the genes induced by peroxide in plants, the molecular mechanism of its induction is unknown. One of the nine proteins induced by menadione is MnSOD, which is regulated by the SoxRS regulon (Hidalgo and Demple, 1994), which was encoded by *SoxR* and *SoxS* genes. In the SoxRS regulon, SoxR is hypothesized to be acting as both a sensor for superoxide stress and a transcriptional activator of the *SoxS* gene, which induces the genes involved in the protection mechanism (Hidalgo and Demple, 1994). Although adaptive mechanisms for both H_2O_2 and superoxide-generating compounds exist in plants (Bowler, Van Montagu, and Inzé, 1992; Prasad et al., 1994; Scandalios, 1993; Pinhero et al., 1997), it is not known whether they have analogous OxyR and SoxRS regulons. Given the evolutionarily conserved nature of antioxidant mechanisms in prokaryotes and eukaryotes, possibly analogous regulatory proteins exist that directly or indirectly activate gene expression in plants as well.

NF-kB-Like Mechanism

In mammalian cells, the DNA-binding activity of transcription factors such as Fos, Jun (Abate et al., 1990), and NF-kB (Baeuerle, 1991), is regulated in vitro by posttranslational modification and influenced by the redox process. NF-kB is one of the best characterized transcription activators in mammalian systems that participates in the response to oxidative stress (Baeuerle, 1991; Baeuerle and Baltimore, 1996; Baldwin, 1996). NF-kB is a heterotrimer protein complex present in the cytosol in an inactive form (Baeuerle, 1991; Baeuerle and Baltimore, 1996). In response to H_2O_2, directly or indirectly, one of the subunits of the complex, the inhibitory IkB, is dissociated (Baeuerle, 1991; Ghosh and Baltimore, 1990), and the resulting activated heterodimer then enters the nucleus and activates gene expression (Baeuerle, 1991). This transcription activator appears to function in the organism to rapidly regulate the synthesis of defense and signaling proteins upon exposure to various stresses. NF-kB was shown to be activated by H_2O_2, Ca^{2+} ionophore, ultraviolet (UV) light, and viral and bacterial proteins (Schreck, Rieber, and Baeuerle, 1991; Baeuerle and Baltimore, 1996), and its activation was blocked by antioxidants (Meyer, Schreck, and Baeuerle, 1993), which suggests that diverse agents activating NF-kB by distinct intracellular pathways might all act through a common mechanism involving ROS.

In a surprising discovery, Ryals and colleagues (1997) isolated a gene, *NIM1,* from *Arabidopsis* that shows a sequence homology to the IkB-α protein of NF-kB/IkB complex in mammals and *Drosophila*. The product of *NIM1* is involved in the signal transduction cascade leading to SAR. However, unlike in mammals and flies, in which the activated NF-kB translocates into the nucleus to stimulate gene expression, the mutations in the *NIM1* gene product suppressed the SAR transduction pathway. The authors suggested that the transcription factor targeted by *NIM1* serves as a suppressor of SAR gene expression and thus affects disease resistance directly or indirectly. Similarly, other plant-resistant genes, such as *N* (Witham et al., 1994), *L6* (Lawrence et al., 1995), and *Rpp5* (Bent, 1996), were reported to have significant homology with Toll and the IL1 receptor, which are also involved in NF-kB/IkB signaling pathways in *Drosophila* and humans, respectively. Although the presence of NF-kB-like factor in plant cells is yet to be demonstrated, these reports suggest that analogous NF-kB/IkB signaling pathways do exist in plant defense mechanisms.

GSH-Mediated Mechanism

A number of enzyme activities are now known to be modified by the formation of mixed disulfides with GSH, a tripeptide, and other cellular disulfides. GSH appears to play an important role as an antioxidant and, simultaneously, as the inducer of gene expression. It has been implicated in the adaptation of plants to stresses such as drought, air pollutants, and extreme temperatures (Alscher, 1989; Alscher, Donahue, and Cramer, 1997). GSH is synthesized in two steps: catalysis by γ-glutamylcysteine synthetase forms γ-glutamylcysteine, which is further utilized by GSH synthetase to produce GSH. GSH synthesis has been shown to be induced in response to H_2O_2, either directly or indirectly (Smith, 1985; May and Leaver, 1993). Extracellular GSH was shown to activate the transcription of genes encoding cell wall hydroxyproline–rich glycoproteins (HRGPs) and phenylpropanoid biosynthetic enzymes in suspension-cultured cells or protoplasts of bean, soybean, and alfalfa (Wingate, Lawton, and Lamb, 1988; Choudhary, Lamb, and Dixon, 1990), and PAL and phytoalexin synthesis in legume cell cultures (Edwards, Blount, and Dixon, 1991). Because GSH was induced by H_2O_2 (Smith, 1985; Guo et al., 1993; May and Leaver, 1993), it is possible to speculate that, in the previous studies, GSH and H_2O_2 interact in a cellular redox balance that regulates gene expression. However, Herouart and colleagues (1993) reported that the expression of cytosolic Cu/ZnSOD gene in the protoplasts of transgenic tobacco leaves was induced by GSH, but not by H_2O_2. On the other hand, GSH was shown to down-regulate GR and cytosolic/chloroplastic Cu/ZnSOD in pine needles (Wingsle and Karpinski, 1996), and APX under excess light in *Arabidopsis* (Karpinski et al., 1997). As H_2O_2-dependent and -independent effects on gene expression mediated by GSH do exist, it is reasonable to suggest that the expression of some of the H_2O_2-responsive genes could also be mediated by GSH in plant systems. Recently, the transcription activator AP1 was shown to be induced at a transcriptional level by both H_2O_2 and antioxidants, via the H_2O_2-dependent pathway and antioxidant-responsive pathway (Meyer, Schreck, and Baeuerle, 1993). This suggests that some of the plant genes could also be regulated by both oxidant- and antioxidant-responsive mechanisms.

Ca^{2+}-Mediated Mechanism

Ca^{2+} has been implicated as a second messenger in response to stress in the signal transduction mechanisms in a variety of organisms, including plants (Trewavas and Gilroy, 1991; Bush, 1993). Rapid changes in cytosolic Ca^{2+} levels occur in plants in response to endogenous stimuli, such as plant hormones (Gehring, Irving, and Parish, 1990; Gilroy, Reed, and Trewavas, 1990; McAinsh et al., 1995), as well as to exogenous stimuli, such as light, heat, cold, elicitors, and touch (Knight et al., 1991; Knight, Smith, and Trewavas, 1992; Bush, 1993), and involve protein phosphorylation (Daminov et al., 1992; Sarokin and Chua, 1992; Monroy, Sarhan, and Dhindsa, 1993). Since all of these biotic and abiotic stresses are suggested to be mediated by oxidative stress, a possible connection exists between the increased ROS and Ca^{2+} levels and induction of a common signal transduction mechanism.

Evidence suggests that some of the effects of ROS, H_2O_2 in particular, are mediated by Ca^{2+} ions. Ca^{2+} ions seem to be necessary to induce a temperature-sensing mechanism that enables the plant to better withstand cold stress. Knight and colleagues (1996) reported that pretreatment of *Arabidopsis* plants with H_2O_2 produced an elevated Ca^{2+} (cytosolic) signature similar to the one associated with cold acclimation. Ca^{2+} signaling has also been implicated in cold acclimation in chilling-resistant plants (Monroy, Sarhan, and Dhindsa, 1993; Monroy and Dhindsa, 1995; Knight, Trewavas, and Knight, 1996). Acclimation at $14°C$ or pretreated with H_2O_2 improved chilling tolerance of maize seedlings by inducing CAT activity (Prasad et al., 1994). These results suggest that cold acclimation and H_2O_2 act through a common pathway (Ca^{2+}-mediated) in chilling tolerance mechanisms. Similarly, both Ca^{2+} and H_2O_2 were reported to activate a mammalian transcription factor, NF-kB, which also suggests that H_2O_2 and Ca^{2+} signal transduction mechanisms are similar.

The source of H_2O_2 within the cell may depend on the type of stress imposed. Once generated, H_2O_2 can diffuse freely between organelles (Halliwell and Gutteridge, 1986) and can mobilize intracellular Ca^{2+} into the cytosol (Richter and Frei, 1985), possibly by modifying a channel protein. Ca^{2+} binds to calmodulin and other Ca^{2+}-binding proteins, and the resulting Ca^{2+}-protein complexes can participate in cellular processes. These events result in the phosphorylation or dephosphorylation of specific proteins that regulate gene expression. Involvement of protein phosphorylation and dephosphorylation in gene expression has been reported in eukaryotic cells (Hunter, 1995). ROS may be involved in the activation of redox-sensitive protein kinases or phosphatases that have been implicated in signaling in mammalian cells (Kass, Duddy, and Orrenius, 1989; Keyse and Emslie, 1992). Schieven and colleagues (1993) reported that the p72[syk] tyrosine kinase was activated by H_2O_2, and this kinase was believed to be contributing to cellular tyrosine phosphorylation and Ca^{2+} signaling in lymphocytes. Transiently elevated cytosolic Ca^{2+} ions coupled to protein kinases, such as protein kinase C (Kikkawa and Nishizuka, 1986) and

Ca^{2+}/calmodulin-dependent protein kinases (Hanson and Schulman, 1992), have also been reported in animal cells. It was also reported that H_2O_2 specifically activated tyrosine kinases and enhanced in vivo phosphorylation of tyrosine residues. These events potentiated the effects of insulin on intact Fao cells (Heffetz and Zick, 1989). Using protein kinase and phosphatase inhibitors, Conrath and colleagues (1997) demonstrated that *PR-1* gene induction can be mediated by dephosphorylation of serine/ threonine residue(s) of two or more unidentified phosphoproteins in tobacco. It was shown that binding of *trans*-acting factors to a *cis*-acting element (H box), involved in chalcone synthase gene regulation, was modulated by phosphorylation that could be in response to ROS, as elicitors and other stimuli generate ROS in plant cells (Jabs et al., 1997). Genes encoding tyrosine phosphatase were cloned from *Chlamydomonas eugametos* (Haring et al., 1995), and the serine/threonine kinase, *pto,* that confers resistance to *Pseudomonas syringae* was cloned from tomato (Martin et al., 1993).

The plant cell wall and vacuoles act as the major Ca^{2+} storage sites. Both plasma membrane and tonoplast possess Ca^{2+} channels and are activated by various stresses. The source of cytosolic Ca^{2+} has been shown to be dependent on the type of stimulus: extracellular (cell wall) Ca^{2+} in response to cold, pathogens, and O_3, and intracellular Ca^{2+} in response to cold, wind, and mechanical pressure (Castillo and Heath, 1990; Knight, Smith, and Trewavas, 1992; Knight, Trewavas, and Knight, 1996). The plant transmembrane signaling mechanism via phophatidylinositol 4,5-bis-phosphate hydrolysis, and the involvement of intracellular Ca^{2+} in such a signal transduction mechanism in response to extracellular signals, has been proposed (Einspahr and Thompson, 1990; Trevawas and Gilroy, 1991). Using inhibitors (neomycin and lithium), Knight and colleagues (1996) suggested a role for inositol 3-phosphate-mediated vacuolar Ca^{2+} release in *Arabidopsis*. Because H_2O_2 can interact with Ca^{2+} channels and pumps in plasma and organelle membranes to mobilize Ca^{2+} in the cell, it is likely that H_2O_2 responses are mediated by the Ca^{2+}-induced events. GTP(guanosine 5'-triphosphate)-binding proteins and protein kinase receptors are believed to be involved in transmembrane signal transduction pathways in plant systems (Devary et al., 1993).

Lipid Peroxides As Signal Molecules

Both biotic and abiotic stresses are known to induce oxidative stress, which increases lipid peroxidation in plants (Kondo, Miyazawa, and Mitzutani, 1992; Prasad, 1996). Methyl jasmonate and other derivatives of lipid peroxides are known to act as intracellular signals that induce plant defense mechanisms (Doke et al., 1994; Doares et al., 1995; Rickauer et al., 1997). Methyl jasmonate was shown to induce chilling tolerance in rice seedlings (Lee et al., 1996). Treatment of tobacco suspension cells with high doses of SA resulted in rapid lipid peroxidation that, in turn, was responsible for *PR-1* gene expression (Conrath, Silva, and Klessig, 1997). These studies suggest a potential role for lipid peroxides in mediating gene expression in relevance to plant defense mechanisms.

SUMMARY OF CHILLING STRESS AND TOLERANCE MECHANISMS IN MAIZE

As described previously, chilling induces oxidative stress in tissues. At least some of the effects of chilling stress and tolerance mechanisms are believed to be mediated by chilling-induced ROS. Results from the maize model system suggest that ROS play a dual role in response to chilling (see Figure 1.7), and that these responses are dependent on the severity of the stress. Signaling mechanisms responsive to H_2O_2 described earlier are relevant only when tissues are exposed to acclimation ($14°C$) or pretreated with pro-oxidants at optimal temperatures ($14°C$ or above) that provide ideal conditions to induce antioxidant gene expression. However, H_2O_2 accumulated during below-threshold temperatures, such as $4°C$, fails to induce antioxidant gene expression in chilling-sensitive plants, including maize. This lack of induction of antioxidant enzymes by ROS is likely due to involvement of a temperature-sensitive ($4°C$ stress) signal transduction mechanism that is responsive to ROS. In nonacclimated maize seedlings, ROS accumulate to toxic levels due to the lack of induction of antioxidant enzymes during $4°C$ stress (see Figure 1.2). This accumulated ROS potentially increases lipid peroxidation and thus causes membrane instability. Furthermore, it inactivates various proteins and enzymes through oxidation. These oxidized proteins may accumulate and exert cytotoxic effects (see Figure 1.4). As a consequence of these destructive events, cells of nonacclimated tissues become more susceptible to diseases upon recovery from chilling stress (see Figure 1.1). On the other hand, in acclimated maize seedlings, the accumulating ROS during the early stages of acclimation induces antioxidant gene expression, CAT3 in particular (see Figure 1.3), thus elevating intracellular levels of antioxidant enzymes and small-molecule antioxidants. Induced antioxidants scavenge the ROS generated during acclimation followed by chilling at $4°C$ stress (see Figures 1.2 and 1.3). Since ROS accumulation was prevented, destructive cellular processes, such as protein oxidation, lipid peroxidation, and the inhibition of protease activities, as seen in nonacclimated tissues, are reduced to control

FIGURE 1.7. A Proposed Sequence of Events That Leads to Chilling Damage or Tolerance in Preemergent Maize Seedlings Subjected to Chilling Stress or Acclimation

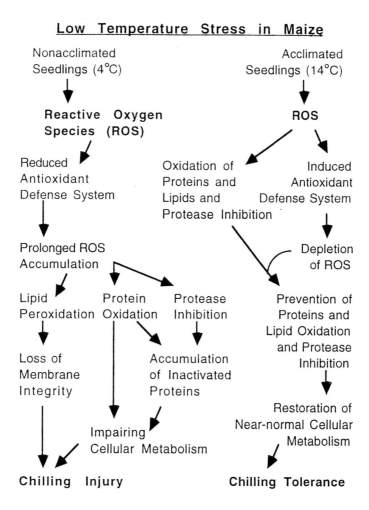

levels (see Figure 1.4), resulting in survival of seedlings (see Figure 1.1). Similar to maize, the acclimation phenomenon has been demonstrated in many crop plants. Therefore, it is reasonable to suggest that similar ROS-induced tolerance mechanisms exist in other plants. In this regard, many studies indicate the induction of antioxidant enzymes, or lack of it, in response to cold stress and correlate it with chilling tolerance or suceptibility in several crop plants.

GENETIC ENGINEERING
OF CHILLING (COLD) TOLERANCE

Increasing evidence suggests that environmental stress, and oxidative stress in particular, is a major limiting factor in crop productivity. Oxidative stress is generated in various cell organelles. The oxidative stress generated in one compartment of the cell can affect the function of the other organelles and the entire cell. To prevent such oxidative damage by ROS, plant cells have evolved with an antioxidant defense mechanism. Antioxidant enzymes and small-molecule antioxidants are critical components in preventing oxidative stress in the cell. Using gene transfer technology, we are beginning to understand the functions of individual enzymes in various organelles under diverse stress conditions. Evidence thus far suggests that manipulating the ROS scavenging systems in various cell compartments can lead to significant changes in oxidative stress tolerance (Foyer, Descourvières, and Kunert, 1994; Allen, 1995). Although studies with single gene transfers provide some evidence to indicate that molecular approaches can be used to improve the general performance of crop plants against stresses such as chilling, altering only one component of the antioxidant system may not always guarantee success. Nevertheless, this gene transfer approach is clearly promising and would be useful in examining the response of transgenic plants to environmental stress.

SODs

SOD isozymes are present in various organelles. Whereas chloroplasts contain Cu/ZnSOD and FeSOD, mitochondria and cytosol contain MnSOD and Cu/ZnSOD, respectively. cDNAs and genes encoding various SOD isozymes have been cloned from plants (Bowler, Van Montagu, and Inzé, 1992; Sakamoto et al., 1992; Van Camp et al., 1994, 1996), and attempts have been made to manipulate the levels of all of these isozymes. Overproduction of SOD in chloroplasts results in improved tolerance against oxidative stress generated by paraquat in association with light in transgenic tobacco (Bowler, Van Montagu, and Inzé, 1992; Sen Gupta, Heinen, et al., 1993; Foyer, Descourvières, and Kunert, 1994; Van Camp et al., 1994, 1996; Slooten et al., 1995), alfalfa (McKersie et al., 1993), potato (Perl et al., 1993), and cotton (Allen, 1995). SOD overexpressed in mitochondria and cytosol provided similar protection against oxidative stress for alfalfa (McKersie et al., 1993) and potato (Perl et al., 1993), respectively. Overexpression of SOD also conferred protection against freezing in alfalfa (McKersie et al., 1993) and against chilling stress in tobacco, in darkness (Foyer, Descourvières, and Kunert, 1994) and in light (Sen Gupta, Heinen, et al., 1993).

Different isozymes of SOD have different properties that may be related to their subcellular location. Two isozymes from different organelles may not have the same degree of efficiency when overexpressed independently in a particular organelle. While overproduction of Cu/ZnSOD in chloroplast provides increased protection from oxidative stress (Sen Gupta, Heinen, et al., 1993), overproduction of MnSOD in chloroplasts gives even higher protection against oxidative stress caused by herbicide treatment. However, this enzyme is less effective than Cu/ZnSOD in providing protection from photooxidation under photoinhibitory conditions (Allen, 1995). Although overproduction of FeSOD in chloroplasts of transgenic tobacco protected both plasmalemma and photosystem II against methyl-viologen-induced oxidative stress (Van Camp et al., 1996), overproduction of MnSOD in chloroplasts protected only plasmalemma (Slooten et al., 1995). These authors suggest that the variability in effectiveness is linked to difference in membrane affinities of transgenic FeSOD and MnSOD.

Conflicting results indicate that overproduction of any particular SOD isozyme does not always result into induced tolerance against oxidative stress. Pitcher and colleagues (1991) found no protection against O_3-induced foliar injury in W38 transgenic tobacco, which overproduced Cu/ZnSOD in chloroplasts at fifteenfold higher than the level of endogenous SOD. However, production of SOD in the cytosol at two- to sixfold resulted in significant protection against O_3 (Pitcher and Zilinskas, 1996). Tepperman and Dunsmuir (1990) reported no protection in Cu/ZnSOD transgenic tobacco against methylviologen-induced oxidative stress, even though SOD activity was fiftyfold higher than in wild-type controls. However, only a threefold increase in Cu/ZnSOD protected chloroplasts against paraquat associated with light in transgenic tobacco (Sen Gupta, Heinen, et al., 1993). These differences suggest that simply elevating SOD activity in an organelle may not be sufficient to confer protection. The type of SOD increased and the balance in the amounts of different isoforms are also important factors in achieving the intended tolerance in plants.

APXs

APX is an efficient scavenger of H_2O_2 in cytosol and chloroplasts (Asada, 1992). Isozymes of APX are present in chloroplasts, cytosol, and glyoxysomes of plant cells (Asada, 1992; Bunkelmann and Trelease,

1996). cDNAs and genes encoding most of the isozymes have been identified from different plant species (Mittler and Zilinskas, 1992; Kubo et al., 1995; Ishikawa et al., 1996). Overexpression of cytosolic APX conferred protection against methylviologen-induced oxidative stress in transgenic tobacco plants (Pitcher, Repetti, and Zilinskas, 1994; Webb and Allen, 1996). Moreover, the down-regulation of cytosolic APX increased susceptibility to O_3 stress in transgenic tobacco (Orvar and Ellis, 1997). Sen Gupta, Webb, and colleagues (1993) reported a threefold increase in APX activity and mRNA in transgenic plants that overexpressed chloroplastic Cu/ZnSOD, and this was suggested to contribute to tolerance against chilling-induced oxidative stress. Overexpression of cytosolic APX in either cytosol or chloroplasts of transgenic tobacco plants confers similar levels of resistance against oxidative stress (Webb and Allen, 1996). Similar to SOD isozymes, overexpression of APX did not always induce stress tolerance. Torsethaugen and colleagues (1997) reported no protection against O_3 in transgenic tobacco plants overexpressing pea cytosolic APX targeted to the chloroplast at a level tenfold higher than in the wild type. In many of these studies, the extent of tolerance against oxidative stress in transgenic plants is moderate. This suggests that more aggressive H_2O_2 scavenging may be required, or that the increased involvement of another component of the ROS scavenging system is required (Allen, 1995).

GRs

GR is an important enzyme involved in recycling glutathione in the cell. GR isozymes are localized in chloroplasts, mitochondria, and cytosol (Edwards, Rawsthorne, and Mullineaux, 1990; Anderson, Prasad, and Stewart, 1995). cDNAs for most of the isozymes have been isolated from pea and tobacco (Aono et al., 1993; Creissen, Edwards, and Mullineaux, 1994). Transgenic tobacco plants overexpressing bacterial GR have been produced (Aono et al., 1993). These transgenic plants contained about threefold higher GR activity in the cytosol than in the wild-type controls and showed increased tolerance to methylviologen and SO_2 (sulfur dioxide), but not to O_3. On the other hand, Foyer and colleagues (1994) did not observe an increased tolerance to paraquat, nor were they able to detect any changes in the glutathione pool in transgenic tobacco that overexpressed GR in the cytosol. Transgenic tobacco overexpressing pea GR was produced that had a range of elevated levels of GR activity in cytosol, in chloroplasts, or in both chloroplasts and mitochondria (Broadbent et al., 1995). These authors demonstrated that enhancing GR levels in transgenic plants may be beneficial for reducing sensitivity to some stresses. However, not all lines that had higher GR activity displayed a positive response to oxidative stress. Even for those GR lines which did, no correlation could be seen between the degree of response and the elevated levels of GR. Once again, conflicting results were reported concerning GR over-expression with relevance to its role in oxidative stress tolerance. Levels of GR activity that were increased in *Escherichia coli* by increasing the copy number of the GR gene did not result in tolerance against oxidative stress (Kunert et al., 1990). Similarly, no chilling tolerance was detected in transgenic tobacco plants overexpressing GR (Broadbent et al., 1995).

CATs

CAT is an important enzyme in scavenging peroxide in the cell. CAT isozymes are present in the mitochondria, peroxisomes, and glyoxysomes (Scandalios et al., 1984; Prasad, Anderson, and Stewart, 1995). cDNAs and genes encoding various CAT isozymes have been isolated (Scandalios, 1992; Guan and Scandalios, 1993; Abler and Scandalios, 1994; Prasad et al., 1994; Guan, Polidoros, and Scandalios, 1996). Using a CAT-specific inhibitor, aminotriazole, CAT was shown to play a major role in chilling tolerance in maize seedlings (Prasad, 1997). Kendall and colleagues (1983) reported that CAT-deficient mutant barley did not survive oxidative stress produced under high photorespiratory conditions. Although transgenic tobacco plants deficient in CAT levels were tested in relation to PR gene expression and disease resistance (Chamnongpol et al., 1996; Takahashi et al., 1997), the importance of CAT in chilling tolerance was not tested in these transgenic plants. CAT's function in vivo in alleviating chilling tolerance in any of the plant species has yet to be demonstrated, although this could be achieved by using transgenic plants that over-produce CAT activity.

Lipid Composition

It has been well documented that membranes may undergo changes in both lipid composition and degree of unsaturation in response to low-temperature changes (Wada, Gombos, and Murata, 1990; Shewfelt, 1992; Riken, Dillwith, and Bergman, 1993; Uemura, Joseph, and Steponkus, 1995). These changes are expected to impose membrane phase transitions. Genetic manipulation of fatty acid unsaturation was known to alter the chilling sensitivity of prokaryotes (Wada, Gombos, and Murata, 1990). Using transgenic tobacco that contains the chloroplast enzyme glycerol-3-phosphate acyltransferase, Murata and colleagues (1992) showed that chilling tolerance can also be induced by altering unsaturated lipid contents. However, the source of the introduced gene differed in inducing chilling tolerance in these transgenic plants. When the gene cloned from chilling-sensitive squash was introduced into tobacco, it increased the chilling sensitivity of the transgenic plants. On the other hand, when the gene cloned from *Arabidopsis* (chilling tolerant) was introduced, it increased the chilling tolerance in those transgenic tobacco plants. These results suggest that chilling tolerance can be improved in sensitive plants, if the correct genes are selected and introduced. Using the same transgenic plants, Moon and colleagues (1995) demonstrated that chilling tolerance was affected by the levels of unsaturated membrane lipids. These authors showed that an increased level of unsaturated fatty acids in phophatidyl-glycerol in thylakoid membranes stabilizes the photosynthetic machinery against low temperatures in those transgenic tobacco leaves. On the other hand, transgenic tobacco that contains the ω-3 fatty acid desaturase gene *(fad7)* from *Arabidopsis* expresses increased levels of polyunsaturated lipids and shows an increased tolerance against chilling stress (Kodama et al., 1994).

CONCLUDING REMARKS

Increasing evidence suggests that some of the responses of many biotic and abiotic stresses are mediated by oxidative stress. However, the responses to oxidative stress are diverse and concentration dependent. We are beginning to understand that whereas a severe oxidative stress is injurious, a mild oxidative stress is indeed beneficial and positions the cell to cope with subsequent severe oxidative stress.

Although ROS are implicated as causative agents in all of the stresses, no substantial progress has been made thus far to understand the mechanisms by which plant cells sense environmental (including biotic and abiotic) adversity and then transduce the stress signals into a change in gene expression. Much progress has been made in understanding the oxidative stress signal transduction mechanism in bacteria and mammalian cells. Oxidative stress-responsive DNA-binding transcription activators, such as OxyR and SoxR in bacterial systems and NF-kB and AP1 in mammalian systems, have been extensively characterized. We are beginning to realize that some of the defense mechanisms against oxidative stress that are conserved among bacterial and mammalian systems also exist in plant systems. Based on evidence from bacterial, mammalian, and plant systems, it is conceivable that H_2O_2 can participate directly or indirectly as a second messenger in plant signal transduction mechanisms. In particular, evidence from plant and mammalian systems suggests that some of the effects of H_2O_2 are mediated by, or coordinated with, Ca^{2+} ions, leading to protein phosphorylation events.

The ability to manipulate the levels of specific enzymes involved in tolerance to oxidative stress using gene transfer technology would provide meaningful insights into the specific functions of individual enzymes. Overexpression of antioxidant and other low-temperature-induced genes has been explored as a solution to generate stress-tolerant plants (Bowler, Van Montagu, and Inzé, 1992; Scandalios, 1993; Allen, 1995; Artus et al., 1996; Welin et al., 1996). The degree of resistance to oxidative stress was shown to be moderate in many of the reported studies with single gene transfers. However, it should be kept in mind that crops in the field are vulnerable to various types of stresses and that different antioxidant enzymes respond differently to a particular type of stress. Hence, introducing a cascade of genes encoding various antioxidant enzymes into plants might be one of the alternative approaches to improve oxidative stress tolerance (Shaaltiel and Gressel, 1986). This approach, however, might eventually prove to be rather difficult to put into practice because of the complexities in gene expression.

In view of the recent discoveries that ROS, in particular H_2O_2, could serve as the second messenger in plant systems under various stress conditions, the tolerance mechanisms could best be enhanced by manipulating regulatory processes that are responsive to oxidative stress. Since several genes encoding antioxidant enzymes have been cloned from various plant species, the identification and the characterization of *cis*-acting elements and *trans*-acting factors regulating their expression should advance understanding of signal transduction mechanisms in plants responsive to oxidative stress. A homologous 11-bp antioxidant-responsive element (ARE: 5'-puGTGACNNNGC-3'), a *cis*-acting element responsive to oxidative stress in the 5' flanking regions of several plant and animal genes, has been identified (Guan and Scandalios, 1993; Meyer, Schreck, and Baeuerle, 1993; Rushmore, Morton, and Pickett, 1991) and is required for inducible gene expression by H_2O_2 and antioxidants (Meyer, Schreck, and Baeuerle, 1993; Rushmore, Morton, and Pickett, 1991). However, it was recently shown that this 11-bp sequence alone is insufficient to mediate induction (Wasserman and Fahl, 1997). By working with the mutational analysis of the murine glutathione-S-transferase Ya ARE, these authors demonstrated that additional elements outside of ARE are required for functional ARE. In the final analysis, a better understanding of a signal transduction mechanism(s) that is responsive to oxidative stress would provide clues for designing strategies to genetically manipulate crop plants for better growth and productivity during environmental challenges.

Cross-tolerance has been shown in various plant varieties and seems to be an important feature for generating plants that are resistant to multiple stresses. Because several genes were shown to be induced by distinct stress conditions, further studies on gene regulation might suggest a common signal transduction mechanism that is induced under various stress conditions and is also responsive to ROS. If these molecular mechanisms of gene regulation are known, gene expression can be manipulated by gene transfer technology to develop stress-tolerant plants. In addition, selecting or breeding programs for tolerance against oxidative stress, although cumbersome and possibly leading to a compromise between high growth and yield productivity, may also be a rewarding approach (Bowler, Van Montagu, and Inzé, 1992) for obtaining plants that are resistant to a wide spectrum of environmental adversities.

REFERENCES

Abate, C., L. Patel, F.J. Rauscher III, and T. Curran (1990). Redox regulation of Fos and Jun DNA-binding activity in vitro. *Science* 249: 1157-1161.

Abler, M.A. and J.G. Scandalios (1994). Isolation and characterization of a genomic sequence encoding the maize *Cat3* catalase gene. *Plant Molecular Biology* 22: 1031-1038.

Allen, R.D. (1995). Dissection of oxidative stress tolerance using transgenic plants. *Plant Physiology* 107: 1049-1054.

Allen, R.D., A. Sen Gupta, R.P. Webb, and A.S. Holaday (1994). Protection of plants from oxidative stress using SOD transgenes: Interactions with endogenous enzymes. In *Frontiers of Reactive Oxygen Species in Biology and Medicine,* eds. K. Asada and T. Yoshikawa. Amsterdam: Excerpta Medica, pp. 321-322.

Alonso, A., C.S. Queiroz, and A.C. Magalhaes (1997). Chilling stress leads to increased cell-membrane rigidity in roots of coffee (*Coffea arabica* L.) seedlings. *Biochimica et Biophysica Acta—Biomembranes* 1323: 75-84.

Alscher, R.G. (1989). Biosynthesis and antioxidant function of glutathione in plants. *Physiologia Plantarum* 77: 457-464.

Alscher, R.G., J.L. Donahue, and C.L. Cramer (1997). Reactive oxygen species and antioxidants—relationships in green cells. *Physiologia Plantarum* 100: 224-233.

Alvarez, M.E. and C. Lamb (1997). Oxidative burst-mediated defense response in plant disease resistance. In *Oxidative Stress and the Molecular Biology of Antioxidant Defenses,* ed. J. Scandalios. New York: Cold Spring Harbor Laboratory Press, pp. 815-839.

Anderson, M.D., T.K. Prasad, B.A. Martin, and C.R. Stewart (1994). Differential gene expression in chilling acclimated maize seedlings and evidence for the involvement of abscisic acid in chilling tolerance. *Plant Physiology* 105: 331-339.

Anderson, M.D., T.K. Prasad, and C.R. Stewart (1995). Changes in isozyme profiles of catalase, peroxidase, and glutathione reductase during acclimation to chilling in mesocotyls of maize seedlings. *Plant Physiology* 109: 1247-1257.

Aono, M., A. Kubo, H. Saji, K. Tanaka, and N. Kondo (1993). Enhanced tolerance to photoxidative stress of transgenic *Nicotiana tabacum* with high chloroplastic glutathione reductase activity. *Plant Cell Physiology* 34: 129-135.

Apostol, I., P.F. Heinstein, and P.S. Low (1989). Rapid stimulation of an oxidative burst during elicitation of cultured plant cells. *Plant Physiology* 90: 109-116.

Armond, P.A. and L.A. Staehelin (1979). Lateral and vertical displacement of integral membrane proteins during lipid phase transition in *Anacystis nidulans*. *Proceedings of the National Academy of Sciences, USA* 76: 1901-1905.

Artus, N.N., M. Uemura, P.L. Steponkus, S.J. Gilmour, C. Lin, and M.F. Thomashow (1996). Constitutive expression of the cold-regulated *Arabidopsis thaliana COR15a* gene affects both chloroplast and protoplast freezing tolerance. *Proceedings of the National Academy of Sciences, USA* 93: 13404-13409.

Asada, K. (1992). Ascorbate peroxidased—A hydrogen peroxide-scavenging enzyme in plants. *Physiologia Plantarum* 85: 235-241.

Baeuerle, P.A. (1991). The inducible transcription activator NF-*k*B: regulation by distinct protein subunits. *Biochimica et Biophysica Acta* 1072: 63-80.

Baeuerle, P. and D. Baltimore (1996). NF-*k*B: Ten years after. *Cell* 87: 13-20.

Baldwin, A. (1996). The NF-*k*B and I*k*B proteins: New discoveries and insights. *Annual Review of Immunology* 14: 649-681.

Bedi, S. and A.S. Basra. (1993). Chilling injury in germinating seeds: Basic mechanisms and agricultural implications. *Seed Science Research* 3: 219-229.

Bender, J., H.J. Weigel, U. Wegner, and H.J. Jager (1994). Response of cellular antioxidants to ozone in wheat flag leaves at different stages of plant development. *Environmental Pollution* 84: 15-21.

Bent, A. (1996). Plant disease resistance genes: Function meets structure. *Plant Cell* 8: 1757-1771.

Bi, Y.M., P. Kenton, L. Mur, R. Darby, and J. Draper (1995). Hydrogen peroxide does not function downstream of salicylic acid in the induction of PR protein expression. *Plant Journal* 8: 235-245.

Bolwell, G.P., V.S. Buti, D.R. Davies, and A. Zimmerlin (1995). The origin of the oxidative burst in plants. *Free Radicals Research* 23: 517-532.

Bowler, C., T. Alliotte, M. De Loose, M. Van Montagu, and D. Inzé (1989). The induction of manganese superoxide dismutase in response to stress in *Nicotiana plumbaginifolia*. *EMBO Journal* 8: 31-38.

Bowler, C., M. Van Montagu, and D. Inzé (1992). Superoxide dismutase and stress tolerance. *Annual Review of Plant Physiology and Plant Molecular Biology* 43: 83-116.

Bridger, G.M., W. Yang, D.E. Falk, and B.D. McKersie (1994). Cold acclimation increases tolerance of activated oxygen in winter cereals. *Journal of Plant Physiology* 144: 235-240.

Broadbent, P., G.P. Creissen, B. Kular, A.R. Wellburn, and P.M. Mullineaux (1995). Oxidative stress responses in transgenic tobacco containing altered levels of glutathione reductase activity. *Plant Journal* 8: 247-255.

Bunkelmann, J.R. and R.N. Trelease (1996). Ascorbate peroxidase. A prominent membrane protein in oilseed glyoxysomes. *Plant Physiology* 110: 589-598.

Bush, D.S. (1993). Regulation of cytosolic calcium in plants. *Plant Physiology* 103: 7-13.

Cadenas, E. (1989). Biochemistry of oxygen toxicity. *Annual Review of Biochemistry* 58: 79-110.

Casano, L.M., H.R. Lascano, and V.S. Trippi (1994). Hydroxyl radicals and a thylakoid-bound endopeptidase are involved in light-oxygen-induced proteolysis in oat chloroplasts. *Plant Cell Physiology* 35: 145-152.

Castillo, F.J. and R.L. Heath (1990). Ca^{2+} transport in membrane vesicles from pinto bean leaves and its alteration after ozone exposure. *Plant Physiology* 94: 788-795.

Chamnongpol, S., H. Willekens, C. Langebartels, M. Van Montagu, D. Inzé, and W. Van Camp (1996). Transgenic tobacco with a reduced catalase activity develops necrotic lesions and induces pathogenesis-related expression under high light. *Plant Journal* 10: 491-503.

Chen, W., G. Chao, and K.B. Singh (1996). The promoter of a H_2O_2-inducible, *Arabidopsis* glutathione *S*-transferase gene contains closely linked OBF- and OBP1-binding sites. *Plant Journal* 10: 955-966.

Chen, Z., H. Silva, and D.F. Klessig (1993). Active oxygen species in the induction of plant systemic acquired resistance by salicylic acid. *Science* 262: 1883-1886.

Choudhary, A.D., C.J. Lamb, and R.A. Dixon (1990). Stress responses in alfalfa (*Medicago sativa* L.). VI. Differential responsiveness of chalcone synthase induction to fungal elicitor or glutathione in electroporated protoplasts. *Plant Physiology* 94: 1802-1807.

Christman, M.F., R.W. Morgan, F.S. Jacobson, and B.N. Ames (1985). Positive control of a regulon for defense against oxidative stress and some heat shock proteins in *Salmonella typhimurium*. *Cell* 41: 753-762.

Collins, G.G., X.L. Nie, and M.E. Saltveit (1995). Heat shock proteins and chilling sensitivity of mung bean hypocotyls. *Journal of Experimental Botany* 46: 795-802.

Conklin, P.L. and R.L. Last (1995). Differential accumulation of antioxidant mRNAs in *Arabidopsis thaliana* exposed to ozone. *Plant Physiology* 109: 203-212.

Conklin, P.L., E.H. Williams, and R.L. Last (1996). Environmental stress sensitivity of an ascorbic acid-deficient *Arabidopsis* mutant. *Proceedings of the National Academy of Sciences, USA* 93: 9970-9974.

Conrath, V., H. Silva, and D.F. Klessig (1997). Protein dephosphorylation mediates salicylic acid-induced expression of PR-1 genes in tobacco. *Plant Journal* 11: 747-757.

Covello, P.S., D.B. Haydon, and N.R. Baker (1988). The roles of low temperature and light in accumulation of 31-kDda polypeptide in the light-harvesting apparatus of maize leaves. *Plant Cell and Environment* 11: 481-486.

Creencia, R.P. and W.J. Bramlage (1971). Reversibility of chilling injury to corn seedlings. *Plant Physiology* 47: 389-392.

Creissen, G.P., A. Edwards, and P.M. Mullineaux (1994). Glutathione reductase and ascorbate peroxidase. In *Causes of Photooxidative Stress and Amelioration of Defense Systems in Plants,* eds. C.H. Foyer and P.M. Mullineaux. Boca Raton, FL: CRC Press, pp. 343-364.

Crowell, D.N., M.E. John, D. Russell, and R.M. Amasino (1992). Characterization of a stress-induced, developmentally regulated gene family from soybean. *Plant Molecular Biology* 18: 459-466.

Daminov, J.A., L. Stenzler, S. Lee, J.J. Schwartz, S. Leisner, and S.H. Howell (1992). Cytokinins and auxins control the expression of a gene in *Nicotiana plumbaginifolia* cells by feedback regulation. *Plant Cell* 4: 451-461.

Devary, Y., C. Rosette, J.A. Di Donato, and M. Karin (1993). NF-*k*B activation by ultraviolet light not dependent on a nuclear signal. *Science* 261: 1442-1445.

Devlin, W.S. and D.L. Gustine (1992). Involvement of the oxidative burst in phytoalexin accumulation and the hypersensitive reaction. *Plant Physiology* 100: 1189-1195.

Dhindsa, R.S. and W. Mattowe (1981). Drought tolerance in two mosses: Correlation with enzymatic defence against lipid peroxidation. *Journal of Experimental Botany* 32: 79-93.

Doares, S.H., T. Syrovets, E.W. Weiler, and C.A. Ryan (1995). Oligogalacturonides and chitosan activate plant defense genes through the octadecanoid pathway. *Proceedings of the National Academy of Sciences, USA* 92: 4095-4098.

Doke, N., Y. Miura, L.M. Sanchez, and K. Kawakita (1994). Involvement of superoxide in signal transduction: Responses to attack by pathogens, physical and chemical shocks and UV irradiation. In *Causes of Photooxidative Stress and Amelioration of Defense Systems in Plants,* eds., C.H. Foyer and P. Mullineaux. Boca Raton, FL: CRC Press, pp. 178-197.

Donahue, J.L., C.M. Okpodu, C.L. Cramer, E.A. Grabau, and R.G. Alscher (1997). Responses of antioxidants to paraquat in pea leaves. *Plant Physiology* 113: 249-257.

Duke, S.H., L.E. Shrader, and M.G. Miller (1977). Low temperature effects on soybean [*Glycine max* (L.) Merr. cv. Wells] mitochondrial respiration and several dehydrogenases during imbibition and germination. *Plant Physiology* 60: 716-722.

Edwards, E.A., C. Enard, G.P. Creissen, and P.M. Mullineaux (1994). Synthesis and properties of glutathione reductase in stressed peas. *Planta* 192: 137-143.

Edwards, E.A., S. Rawsthorne, and P.M. Mullineaux (1990). Subcellular distribution of multiple forms of glutathione reductase in leaves of pea (*Pisum sativum* L.). *Planta* 180: 278-284.

Edwards, R., J.W. Blount, and R.A. Dixon (1991). Glutathione and elicitation of the phytoalexin response in legume cell cultures. *Planta* 184: 403-409.

Einspahr, K. and G.A. Thompson Jr. (1990). Transmembrane signaling via phosphatidylinositol 4,5-bisphosphate hydrolysis in plants. *Plant Physiology* 93: 361-366.

Elstner, E.F (1991). Mechanisms of oxygen activation in different compartments of plant cells. In *Active Oxygen/Oxidative Stress and Plant Metabolism,* eds.

E.J. Pell and K.L. Steffen. Rockville, MD: American Society of Plant Physiologists, pp. 13-25.

Elthon, T.E., C.R. Stewart, C.A. McCoy, and W.D. Bonner Jr. (1986). Alternative respiratory path capacity in plant mitochondria: Effect of growth temperature, the electrochemical gradient, and assay pH. *Plant Physiology* 80: 378-383.

Fadzillah, N.M., V. Gill, R.P. Finch, and R.H. Burdon (1996). Chilling, oxidative stress and antioxidant responses in shoot cultures of rice. *Planta* 199: 552-556.

Feierabend, J., C. Schaan, and B. Hertwig (1992). Photoinactivation of catalase occurs under both high- and low-temperature stress conditions and accompanies photoinhibition of photosystem II. *Plant Physiology* 100: 1554-1561.

Foyer, C.H., P. Descourvières, and K.J. Kunert (1994). Protection against oxygen radicals an important defense mechanism studied in transgenic plants. *Plant Cell and Environment* 17: 507-523.

Foyer, C.H., M. Lelandais, C. Galap, and K.J. Kunert (1991). Effects of elevated cytosolic glutathione reductase activity on the cellular glutathione pool and photosynthesis in leaves under normal and stress conditions. *Plant Physiology* 97: 863-872.

Foyer, C.H., H. Lopezdelgado, J.F. Dat, and I.M. Scott (1997). Hydrogen peroxide- and glutathione-associated mechanisms of acclimatory stress tolerance and signaling. *Physiologia Plantarum* 100: 241-254.

Fridovich, I (1978). The biology of oxygen radicals. *Science* 201: 875-880.

Fucci, L., C. Oliver, M. Coon, and E. Stadtman (1983). Inactivation of key metabolic enzymes by mixed-function oxidation reactions: Possible implication in protein turnover and aging. *Proceedings of the National Academy of Sciences, USA* 80: 1521-1525.

Gehring, C.A., H.R. Irving, and R.W. Parish (1990). Effects of auxin and abscisic acid on cytosolic calcium and pH in plant cells. *Proceedings of the National Academy of Sciences, USA* 87: 9645-9649.

Ghosh, S. and D. Baltimore (1990). Activation *in vitro* of NF-kB by phosphorylation of its inhibitor IkB. *Nature* 344: 678-682.

Gillham, D.J. and A.D. Dodge (1987). Chloroplast superoxide and hydrogen peroxide scavenging systems from pea leaves: Seasonal variations. *Plant Science* 50: 105-109.

Gilmour, S.J., R.K. Hajela, and M.F. Thomashow (1988). Cold acclimation in *Arabidopsis thaliana*. *Plant Physiology* 87: 745-750.

Gilroy, S., K.D. Reed, and A.J. Trewavas (1990). Elevation of cytoplasmic calcium by caged calcium or caged inositol triphosphate initiates stomata closure. *Nature* 346: 766-771.

Gomez, L.D., L.M. Casano, and V.S. Trippi (1995). Effect of hydrogen peroxide on degradation of cell wall-associated proteins in growing bean hypocotyls. *Plant Cell Physiology* 36: 1259-1264.

Greaves, J.A (1996). Improving suboptimal temperature tolerance in maize—The search for variation. *Journal of Experimental Botany* 47: 307-323.

Greenberg, J.T., P. Monach, J.H. Chou, P.D. Josephy, and B. Demple (1990). Positive control of a global antioxidant defense regulon activated by superoxide-generating agents in *Escherichia coli. Proceedings of the National Academy of Sciences, USA* 87: 6181-6185.

Guan, L., A.N. Polidoros, and J.G. Scandalios (1996). Isolation, characterization and expression of the maize *Cat2* catalase gene. *Plant Molecular Biology* 30: 913-924.

Guan, L. and J.G. Scandalios (1993). Characterization of catalase antioxidant defense gene *Cat1* of maize, and its developmentally regulated expression in transgenic tobacco. *Plant Journal* 3: 527-536.

Guo, Z.-J., S. Nakagawara, K. Sumitani, and Y. Ohta (1993). Effect of intracellular glutathione level on the production of 6-methoxymellein in cultured carrot *(Daucus carota)* cells. *Plant Physiology* 102: 45-51.

Guy, C.L (1990). Cold acclimation and freezing stress tolerance: Role of protein metabolism. *Annual Review of Plant Physiology and Plant Molecular Biology* 41: 187-223.

Guy, C.L. and J.V. Carter (1984). Characterization of partially purified glutathione reductase from cold-hardened and nonhardened spinach leaf tissue. *Cryobiology* 21: 454-464.

Guy, C.L., K.J. Niemi, and R. Brambl (1998). Altered gene expression during cold acclimation in spinach. *Proceedings of the National Academy of Sciences, USA* 82: 3673-3677.

Guye, M.G., L. Vigh, and J.M. Wilson (1987). Recovery after chilling: An assessment of chilling tolerance in *Phaseolus* species. *Journal of Experimental Botany* 38: 691-701.

Halliwell, B. and J.M.C. Gutteridge (1986). Oxygen free radicals and iron in relation to biology and medicine: Some problems and concepts. *Archives of Biochemistry and Biophysics* 246: 501-514.

Hanson, P.I. and H. Schulman (1992). Neuronal Ca^{2+}/calmodulin-dependent protein kinases. *Annual Review of Biochemistry* 61: 559-601.

Haring, M.A., M. Siderius, C. Jonak, H. Hirt, K.M. Walton, and A. Musgrave (1995). Tyrosine phosphatase signaling in a lower plant: Cell-cycle and oxidative stress-regulated expression of the *Chlamydomonas eugametos* VH-PTP13 gene. *Plant Journal* 7: 981-988.

Hayden, D.B., P.S. Cavello, and N.R. Baker (1988). Characterization of 31-kDa polypeptide that accumulates in the light-harvesting apparatus of maize leaves during chilling. *Photosynthesis Research* 15: 257-270.

Heffetz, D. and Y. Zick (1989). H_2O_2 potentiates phosphorylation of novel putative substrates for the insulin receptor kinase in intact Fao cells. *Journal of Biological Chemistry* 264: 10126-10132.

Herouart, D., M. Van Montagu, and D. Inzé (1993). Redox-activated expression of the cytosolic copper/zinc superoxide dismutase gene in *Nicotiana. Proceedings of the National Academy of Sciences, USA* 90: 3108-3112.

Hidalgo, E. and B. Demple (1994). An iron-sulfur center essential for transcriptional activation by the redox-sensing SoxR protein. *EMBO Journal* 13: 138-146.

Howarth, C.J. and H.J. Ougham (1993). Gene expression under temperature stress. *New Phytologist* 125: 1-26.

Hughes, M.A. and M.A. Dunn (1996). The molecular biology of plant acclimation to low temperature. *Journal of Experimental Botany* 47: 291-305.

Hughes, M.A., M.A. Dunn, R.S. Pearce, A.J. White, and L. Zhang (1992). An abscisic acid-responsive, low temperature barley gene has homology with a maize phospholipid transfer protein. *Plant Cell and Environment* 15: 861-865.

Hugly, S. and C. Somerville (1992). A role for membrane lipid polyunsaturation in chloroplast biogenesis at low temperature. *Plant Physiology* 99: 197-202.

Hunter, T. (1995). Protein kinases and phosphatases: The yin and yang of protein phosphorylation and signaling. *Cell* 80: 225-236.

Ishikawa, T., K. Sakai, K. Yoshimura, T. Takeda, and S. Shigeoka (1996). cDNAs encoding spinach stromal and thylakoid-bound ascorbate peroxidase, differing in the presence or absence of their 3'-coding regions. *FEBS Letters* 384: 289-293.

Jabs, T., M. Tschope, C. Colling, K. Hahlbrock, and D. Scheel (1997). Elicitor-stimulated ion fluxes and O_2^- from the oxidative burst are essential components in triggering defense gene activation and phytoalexin synthesis in parsley. *Proceedings of the National Academy of Sciences, USA* 94: 4800-4805.

Jhanke, L.S., M.R. Hull, and S.P. Long (1991). Chilling stress and oxygen metabolizing enzymes in *Zea mays* and *Zea diploperennis*. *Plant Cell and Environment* 14: 97-104.

Karpinski, S., C. Escobar, B. Karpinska, G. Creissen, and P.M. Mullineaux (1997). Photosynthetic electron transport regulates the expression of cytosolic ascorbate peroxidase genes in *Arabidopsis* during excess light stress. *Plant Cell* 9: 627-640.

Kass, E.N., S.K. Duddy, and S. Orrenius (1989). Activation of hepatocyte protein kinase C by redox-cycling quinones. *Biochemical Journal* 260: 499-507.

Kendall, A.C., A.J. Keys, J.C. Turner, P.J., Lea, and B.J. Miflin (1983).The isolation and characterization of a catalase-deficient mutant of barley (*Hordeum vulgare* L.). *Planta* 159: 505-511.

Kenis, J.D., J.D. Morlans, and V.S. Trippi (1989). Effect of hydrogen peroxide on nitrate reductase activity in detached oat leaves in darkness. *Physiologia Plantarum* 76: 216-220.

Kenrick, J.R. and D.G. Bishop (1986). The fatty acid composition of phosphatidylglycerol and sulfoquinovosyldiacylglycerol of higher plants in relation to chilling sensitivity. *Plant Physiology* 81: 946-949.

Keyse, S.M. and E.A. Emslie (1992). Oxidative stress and heat shock induce a human gene encoding a protein-tyrosine phosphatase. *Nature* 359: 644-647.

Kiener, C.M. and W.J. Bramlage (1981). Temperature effects on the activity of the alternative respiratory pathway in chill-sensitive *Cucumis sativus*. *Plant Physiology* 68: 1474-1478.

Kikkawa, U. and Y. Nishizuka (1986). The role of protein kinase C in transmembrane signalling. *Annual Review of Cell Biology* 2: 149-178.

Knight, H., A.J. Trewavas, and M.R. Knight (1996). Cold calcium signaling in *Arabidopsis* improves two cellular pools and a change in calcium signature after acclimation. *Plant Cell* 8: 489-503.

Knight, M.R., A.K. Campbell, S.M. Smith, and A.J. Trewavas (1991). Transgenic plant aequorin reports the effects of touch and cold shock and elicitors on cytoplasmic calcium. *Nature* 352: 524-526.

Knight, M.R., S.M. Smith, and A.J. Trewavas (1992). Wind-induced plant motion immediately increases cytosolic calcium. *Proceedings of the National Academy of Sciences, USA* 89: 4967-4971.

Kodama, H., T. Hamada, G. Horiguchi, M. Nishimura, and K. Iba (1994). Genetic enhancement of cold tolerance by expression of a gene for chloroplast ω-fatty acid desaturase in transgenic tobacco. *Plant Physiology* 105: 601-605.

Kondo, Y., T. Miyazawa, and J. Mitzutani (1992). Detection and time-course analysis of phospholipid hydroperoxide in soybean seedlings after treatment with fungal elicitor, by chemiluminescence-HPLC assay. *Biochimica et Biophysica Acta* 1127: 227-232.

Kubo, A., H. Saji, K. Tanaka, and N. Kondo (1995). Expression of *Arabidopsis* cytosolic ascorbate peroxidase gene in response to ozone or sulfur dioxide. *Plant Molecular Biology* 29: 479-486.

Kunert, K.J., C.F. Cresswell, A. Schmidt, P.M. Mullineaux, and C.H. Foyer (1990). Variations in the activity of glutathione reductase and the cellular glutathione content in relation to sensitivity to methylviologen in *E. coli*. *Archives Biochemistry and Biophysics* 282: 233-238.

Lamb, C.J., M.A. Lawton, M. Dron, and R.A. Dixon (1989). Signals and transduction mechanisms for activation of plant defenses against microbial attack. *Cell* 56: 215-224.

Law, M.Y., S.A. Charles, and B. Halliwell (1983). Glutathione and ascorbic acid in spinach (*Spinacia oleracea*) chloroplasts: The effect of hydrogen peroxide and of paraquat. *Biochemical Journal* 210: 899-903.

Lawrence, G.J., E.J. Finnegan, M.A. Ayliffe, and J.G. Ellis (1995). The *L6* gene for flax rust resistance is related to the *Arabidopsis* bacterial resistance gene *RPS2* and the tobacco viral resistance gene *N*. *Plant Cell* 7: 1195-1206.

Lee, T.M., H.S. Lur, Y.H. Lin, and C. Chu (1996). Physiological and biochemical changes related to methyl jasmonate-induced chilling tolerance of rice (*Oryza sativa* L.) seedlings. *Plant Cell and Environment* 19: 65-74.

Leheny, E.A. and S.M. Theg (1994). Apparent inhibition of chloroplast protein import by cold temperatures is due to energetic considerations not membrane fluidity. *Plant Cell* 6: 427-437.

Leprince, O., R. Deltour, P.C. Thorpe, N.M. Atherton, and G.A.F. Hendry (1990). The role of free radicals and radical processing systems in loss of desiccation tolerance in germinating maize (*Zea mays* L.). *New Phytologist* 116: 573-580.

Levine, R.I., D. Garland, C. Oliver, A. Amici, I. Climent, A. Lenz, B. Ahn, S. Shaltiel, and E.R. Stadtman (1990). Determination of carbonyl content in oxidatively modified proteins. *Methods in Enzymology* 186: 464-478.

Levitt, J (1980). *Responses of Plants to Environmental Stress*, Volume 1: *Chilling, Freezing and High Temperature Stress*. New York: Academic Press.

Lin, G.C., W.W. Guo, E. Everson, and M.F. Thomashow (1990). Cold acclimation in *Arabidopsis thaliana*. *Plant Physiology* 94: 1078-1083.

Lyons, J.M (1973). Chilling injury in plants. *Annual Review of Plant Physiology* 24: 445-466.

Lyons, J.M. and J.K. Raison (1970). Oxidative activity of mitochondria isolated from plant tissues sensitive and resistant to chilling injury. *Plant Physiology* 45: 386-389.

Lyons, J.M., T.A. Wheaton, and H.K. Pratt (1964). Relationship between the physical nature of mitochondrial membranes and chilling sensitivity in plants. *Plant Physiology* 39: 262-268.

Malan, C., M.M. Greyling, and J. Gressel (1990). Correlation between Cu,Zn superoxide dismutase and glutathione reductase and environmental and xenobiotic stress tolerance in maize inbreds. *Plant Science* 69: 157-166.

Markhart, A.H. III (1986). Chilling injury: A review of possible causes. *HortScience* 21: 1329-1333.

Martin, G.B., S.H. Brommonschenkel, J. Chunwongse, A. Frary, M.W. Ganal, R. Spivy, T. Wu, E.D. Earle, and S.D. Tanksley (1993). Map-based cloning of a protein kinase gene conferring disease resistance in tomato. *Science* 262: 1432-1436.

Martin, J., A.L. Horwich, and F.U. Hartl (1992). Prevention of protein denaturation under heat stress by the chaperonin HSP60. *Science* 258: 995-998.

Matsuda, Y., T. Okuda, and S. Sagisaka (1994). Regulation of protein synthesis by hydrogen peroxide in crowns of winter wheat. *Bioscience, Biotechnology and Biochemistry* 58: 906-909.

Matters, G.L. and J.G. Scandalios (1986). Effect of elevated temperature on catalase and superoxide dismutase during maize development. *Differentiation* 30: 190-196.

May, M.J., and C.J. Leaver (1993). Oxidative stimulation of glutathione synthesis in *Arabidopsis thaliana* suspension cultures. *Plant Physiology* 103: 621-627.

McAinsh, M.R., A.A.R. Webb, J.E. Tylor, and A.M. Hetherington (1995). Stimulus-induced oscillations in guard cell cytosolic calcium. *Plant Cell* 7: 1207-1219.

McKersie, B.D (1991). The role of oxygen free radicals in mediating freezing and desiccation stress in plants. In *Active Oxygen/Oxidative Stress and Plant Metabolism,* eds. E.J. Pell and K.L. Steffen. Rockville, MD: American Society of Plant Physiologists, pp. 107-118.

McKersie, B.D., Y. Chen, M. De Beus, S.R. Bowley, C. Bowler, D. Inzé, K. D'Halluin, and J. Botterman (1993). Superoxide dismutase enhances tolerance of freezing stress in transgenic alfalfa (*Medicago sativa* L.). *Plant Physiology* 103: 1155-1163.

Mehdy, M.C., Y.K. Sharma, K. Sathasivan, and N.W. Bays (1995). The role of activated oxygen species in plant disease resistance. *Physiologia Plantarum* 98: 365-374.

Mehlhorn, H. (1990). Ethylene-promoted ascorbate peroxidase activity protects plants against hydrogen peroxide, ozone and paraquat. *Plant Cell Environment* 13: 971-976.

Meyer, M., R. Schreck, and P.A. Baeuerle (1993). H_2O_2 and antioxidants have opposite effects on activation of NF-*k*B and AP-1 in intact cells: AP-1 as a secondary antioxidant-responsive factor. *EMBO Journal* 12: 2005-2015.

Michalski, W.P. and Z. Kaniuga (1982). Photosynthetic apparatus of chilling-sensitive plants. XI. Reversibility by light of cold- and dark-induced inactivation of cyanide-sensitive superoxide dismutase activity in tomato leaf chloroplasts. *Biochimica et Biophysica Acta* 680: 250-257.

Mishra, N.P., R.K. Mishra, and G.S. Singhal (1993). Changes in the activities of antioxidant enzymes during exposure of intact wheat leaves to strong visible light at different temperatures in the presence of protein synthesis inhibitors. *Plant Physiology* 102: 903-910.

Mittler, R. and B. Zilinskas (1992). Molecular cloning and characterization of a gene encoding pea cytosolic ascorbate peroxidase. *Journal of Biological Chemistry* 267: 21802-21807.

Mohapatra, S.S., R.J. Poole, and R.S. Dhindsa (1987). Cold acclimation, freezing resistance and protein synthesis in alfalfa (*Medicago sativa* L. cv. Saranac). *Journal of Experimental Botany* 38: 1697-1703.

Monroy, A.F. and R.S. Dhindsa (1995). Low temperature signal transduction induction of cold acclimation-specific genes of alfalfa by calcium at 25°C. *Plant Cell* 7: 321-331.

Monroy, A.F., F. Sarhan, and R.S. Dhindsa (1993). Cold-induced changes in freezing tolerance, protein phosphorylation, and gene expression: Evidence for a role of calcium. *Plant Physiology* 102: 1227-1235.

Moon, B.Y., S.I. Higashi, Z. Gombos, and N. Murata (1995). Unsaturation of the membrane lipids of chloroplasts stabilizes the photosynthetic machinery against low-temperature photoinhibition in transgenic tobacco plants. *Proceedings of the National Academy of Sciences, USA* 92: 6219-6223.

Murata, N., O. Ishizaki-Nishizawa, S. Higashi, H. Hayashi, Y. Tasaka, and I. Nishida (1992). Genetically engineered alteration in the chilling sensitivity of plants. *Nature* 356: 710-713.

Neuenschwander, U., B. Vernooji, L. Friedrich, S. Uknes, H. Kessmann, and J. Ryals (1995). Is hydrogen peroxide a second messenger of salicylic acid in systemic acquired resistance? *Plant Journal* 8: 227-233.

Neven, L.G., D. Haskel, and C.L. Guy (1990). A heat shock cognate comes out in the cold. *Cryobiology* 27: 661-668.

Ni, W. and R.N. Trelease (1991). Post-transcriptional regulation of catalase isozyme expression in cotton seeds. *Plant Cell* 3: 737-744.

Nicchitta, C.V. and G. Blobel (1989). Nascent secretory chain binding and translocation are distinct processes: Differentiation by chemical alkylation. *Journal of Cell Biology* 108: 789-795.

Nykiforuk, C.L. and A.M. Johnson-Flanagan (1998). Low temperature emergence in crop plants: Biochemical and molecular aspects of germination and early seedling growth. *Journal of Crop Production* 1: 249-289.

O'Kane, D., V. Gill, P. Boyd, and R.H. Burdon (1996). Chilling, oxidative stress and antioxidant responses in *Arabidopsis thaliana* callus. *Planta* 198: 366-370.

Okuda, T., Y. Matsuda, A. Yamamaka, and S. Sagisaka (1991). Abrupt increase in the level of hydrogen peroxide in leaves of winter wheat is caused by cold treatment. *Plant Physiology* 97: 1265-1267.

Omran, R.J (1980). Peroxide levels and the activities of catalase, peroxidase, and indole acetic acid oxidase during and after chilling cucumber seedlings. *Plant Physiology* 65: 407-408.

Orvar, L. and B.E. Ellis (1997). Transgenic tobacco expressing antisense RNA for cytosolic ascorbate peroxidase show increased susceptibility to ozone injury. *Plant Journal* 11: 1297-1305.

Pacifici, R.E. and K.J.A. Davies (1990). Protein degradation as an index of oxidative stress. *Methods in Enzymology* 186: 485-502.

Pahl, H.L. and P.A. Baeuerle (1995). A novel signal transduction pathway from the endoplasmic reticulum to the nucleus is mediated by transcription factor NF-*k*B. *EMBO Journal* 14: 2580-2588.

Pasturi, G.M. and V.S. Trippi (1992). Oxidative stress induces a high rate of glutathione reductase synthesis in a drought-resistant maize strain. *Plant Cell Physiology* 33: 957-961.

Perl, A., R. Perl-Treves, S. Galili, D. Aviv, E. Shalgi, S. Malkin, and E. Galun (1993). Enhanced oxidative-stress defense in transgenic potato expressing tomato Cu,Zn superoxide dismutases. *Theoretical and Applied Genetics* 85: 568-576.

Pinhero, R.G., M.V. Rao, G. Paliyath, D.P. Murr, and R.A. Fletcher (1997). Changes in activities of antioxidant enzymes and their relationship to genetic and paclobutrazol-induced chilling tolerance of maize seedlings. *Plant Physiology* 114: 695-704.

Pitcher, L.H., E. Brennan, A. Hurley, P. Dunmuir, J.M. Tepperman, and B.A. Zilinskas (1991). Overproduction of petunia chloroplastic copper/zinc superoxide dismutase does not confer ozone tolerance in transgenic tobacco. *Plant Physiology* 97: 452-455.

Pitcher, L.H., P. Repetti, and B.A. Zilinskas (1994). Overexpression of ascorbate peroxidase protects plants from oxidative stress. *Plant Physiology* (suppl.) 105: 116.

Pitcher, L.H. and B.A. Zilinskas (1996). Overexpression of copper/zinc superoxide dismutase in the cytosol of transgenic tobacco confers partial resistance to ozone-induced foliar necrosis. *Plant Physiology* 110: 583-588.

Prasad, T.K (1996). Mechanisms of chilling-induced oxidative stress injury and tolerance: Changes in antioxidant system, oxidation of proteins and lipids and protease activities. *Plant Journal* 10: 1017-1026.

Prasad,T.K (1997). Role of catalase in inducing chilling tolerance in preemergent maize seedlings. *Plant Physiology* 114: 1369-1376.

Prasad, T.K., M.D. Anderson, B.A. Martin, and C.R. Stewart (1994). Evidence for chilling-induced oxidative stress in maize seedlings and a regulatory role for hydrogen peroxide. *Plant Cell* 6: 65-74.

Prasad, T.K., M.D. Anderson, and C.R. Stewart (1994). Acclimation, hydrogen peroxide and abscisic acid protect mitochondria against irreversible chilling injury in maize seedlings. *Plant Physiology* 105: 619-627.

Prasad, T.K., M.D. Anderson, and C.R. Stewart (1995). Localization and characterization of peroxidases in the mitochondria of chilling-acclimated maize seedlings. *Plant Physiology* 108: 1597-1605.

Pugin, A., J.M. Frachisse, E. Tavernier, R. Bligny, E. Gout, R. Douce, and J. Guern (1997). Early events induced by the elicitor cryptogein in tobacco cells: Involvement of a plasma membrane NADPH oxidase and activation of glycolysis and the pentose phosphate pathway. *Plant Cell* 9: 2077-2091.

Puntarulo, S., M. Galleano, R.A. Sanchez, and A. Boveris (1991). Superoxide anion and hydrogen peroxide metabolism in soybean embryonic axes during germination. *Biochimica et Biophysica Acta* 1074: 277-283.

Purvis, A.C (1997). Role of alternative oxidase in limiting superoxide production by plant mitochondria. *Physiologia Plantarum* 100: 165-170.

Purvis, A.C. and R.L. Shewfelt (1993). Does alternative pathway ameliorate chilling injury in sensitive plant tissues? *Physiologia Plantarum* 88: 712-718.

Purvis, A.C., R.L. Shewfelt, and J.W. Gegogeine (1995). Superoxide production by mitochondria isolated from green bell pepper fruit. *Physiologia Plantarum* 94: 743-749.

Raison, J.K. and G.R. Orr (1986). Phase transitions in thylakoid polar lipids of chilling-sensitive plants. *Plant Physiology* 80: 638-645.

Richter, C. and B. Frei (1985). Ca^{2+} movements induced by hydroperoxides in mitochondria. In *Oxidative Stress*, ed. H. Sies. London: Academic Press, pp. 221-241.

Rickauer, M., W. Brodschelm, A. Bottin, C. Veronesi, H. Grimal, and M.T. Esqerre-Tugaye (1997). The jasmonate pathway is involved differentially in the regulation of different defense responses in tobacco cells. *Planta* 202: 155-162.

Riken, A., J.W. Dillwith, and D.K. Bergman (1993). Correlation between the circadian rhythm of resistance to extreme temperatures and changes in fatty acid composition in cotton seedlings. *Plant Physiology* 101: 31-36.

Roughan, P.G (1985). Phosphatidylglycerol and chilling sensitivity in plants. *Plant Physiology* 77: 740-746.

Rushmore, T.H., M.R. Morton, and C.B. Pickett (1991). The antioxidant response element. *Journal of Biological Chemistry* 266: 11632-11639.

Ryals, J., K. Weymann, K. Lawton, L. Friedrich, D. Ellis, H.-Y., Steiner, J. Johnson, T.P. Delaney, T. Jesse, P. Vos, and S. Uknes (1997). The *Arabidopsis NIM1* protein shows homology to the mammalian transcription factor inhibitor IkB. *Plant Cell* 9: 425-439.

Rybka, Z. (1989). Changes in the cyanide-sensitive and cyanide-resistant oxygen uptake in crowns of winter wheat seedlings during hardening to frost. *Journal of Plant Physiology* 134: 17-19.

Sakamoto, A., T. Okumura, H. Ohsuga, and K. Tanaka (1992). Genomic structure of the gene for copper/zinc-superoxide dismutase in rice. *FEBS Letters* 301: 185-189.

Sarokin, L.P. and N.H. Chua (1992). Binding sites for two novel phosphoproteins, 3AF5 and 3AF3, are required for *rbcS-3A* expression. *Plant Cell* 4: 473-483.

Saruyama, H. and M. Tanida (1995). Effect of chilling on activated oxygen-scavenging enzymes in low temperature-sensitive and -tolerant cultivars of rice (*Oryza sativa* L.). *Plant Science* 109: 105-113.

Scandalios, J.G. (1992). Regulation of the antioxidant defense genes *cat* and *sod* of maize. In *Molecular Biology of Free Radical Scavenging Systems,* ed. J.G. Scandalios. New York: Cold Spring Harbor Laboratory Press, pp. 117-152.

Scandalios, J.G. (1993). Oxygen stress and superoxide dismutases. *Plant Physiology* 101: 7-12.

Scandalios, J.G. (1994). Regulation and properties of plant catalases. In *Causes of Photooxidative Stress and Amelioration of Defense Systems in Plants,* eds. C.H. Foyer and P.M. Mullineaux. Boca Raton, FL: CRC Press, pp. 275-316.

Scandalios, J.G., A.S. Tsaftaris, J.M. Chandlee, and R.W. Skadsen (1984). Expression of the developmentally regulated catalase (*cat*) genes in maize. *Developmental Genetics* 4: 281-293.

Schaffer, M.A. and R.L. Fischer (1988). Analysis of mRNAs that accumulated in response to low temperature identifies a thiol protease in tomato. *Plant Physiology* 87: 431-436.

Schieven, G.L., J.M. Kirihara, D.L. Burg, R.L. Geahlen, and J.A. Ledbetter (1993). p72[syk] tyrosine kinase is activated by oxidizing conditions that induce lymphocyte tyrosine phosphorylation and Ca^{2+} signals. *Journal of Biological Chemistry* 22: 16688-16692.

Schöner, S. and G.H. Krause (1990). Protective system against active oxygen species in spinach: Response to cold acclimation in excess light. *Planta* 180: 383-389.

Schreck, R., P. Rieber, and P.A. Baeuerle (1991). Reactive oxygen intermediates as apparently widely used messengers in the activation of the NF-*k*B transcription factor and HIV-1. *EMBO Journal* 10: 2247-2258.

Sen Gupta, A, J.L. Heinen, A.S. Holaday, J.J. Burke, and R.D. Allen (1993). Increased resistance to oxidative stress in transgenic plants that overexpress chloroplastic Cu/Zn superoxide dismutase. *Proceedings of the National Academy of Sciences, USA* 90: 1629-1633.

Sen Gupta, A., R.P. Webb, A.S. Holaday, and R.D. Allen (1993). Overexpression of superoxide dismutase protects plants from oxidative stress: Induction of ascorbate peroxidase in superoxide dismutase-overexpressing plants. *Plant Physiology* 103: 1067-1073.

Senaratna, T., B.D. McKersie, and A. Borochov (1987). Desiccation and free radical mediated changes in plant membranes. *Journal of Experimental Botany* 38: 2005-2014.

Shaaltiel, T., N.H. Chua, S. Gepstein, and J. Gressel (1988). Dominant pleiotrophy controls enzymes co-segregating with paraquat resistance in *Conyza bonariensis*. *Theoretical and Applied Genetics* 75: 850-856.

Shaaltiel, Y. and J. Gressel (1986). Multienzyme oxygen radical detoxifying system correlated with paraquat resistance in *Conyza bonariensis*. *Pesticide Biochemistry and Physiology* 26: 22-28.

Sharon, M., C. Willemot, and J.E. Thompson (1994). Chilling injury induces lipid phase changes in membranes of tomato fruit. *Plant Physiology* 105: 305-308.

Shewfelt, R.L (1992). Response of plant membranes to chilling and freezing. In *Plant Membranes*, ed. Y. Y. Leshem. Dordrecht: Kluwer, pp. 192-219.

Shewfelt, R.L. and M.C. Erickson (1991). Role of lipid peroxidation in the mechanism of membrane-associated disorders in edible plant tissue. *Trends in Food Science Technology* 2: 152-154.

Skadsen, R.W. and J.G. Scandalios (1987). Translational control of photo-induced expression of the *Cat2* catalase gene during leaf development in maize. *Proceedings of National Academy of Sciences, USA* 84: 2785-2789.

Slooten, L., K. Capiau, W. Van Camp, M. Van Montagu, C. Sybesma, and D. Inzé (1995). Factors affecting the enhancement of oxidative stress tolerance in transgenic tobacco overexpressing manganese superoxide dismutase in the chloroplasts. *Plant Physiology* 107: 737-750.

Smirnoff, N. (1993). The role of active oxygen in the response of plants to water deficit and desiccation. *New Phytologist* 125: 27-58.

Smith, I.K. (1985). Stimulation of glutathione synthesis in photorespiring plants by catalase inhibitors. *Plant Physiology* 79: 1044-1047.

Stadtman, E.R. (1993). Oxidation of free amino acids and amino acid residues in proteins by radiolysis and by metal-catalyzed reactions. *Annual Review of Biochemistry* 62: 797-821.

Stadtman, E.R. and C.N. Oliver (1991). Metal-catalyzed oxidation of proteins. Physiological consequences. *Journal of Biological Chemistry* 266: 2005-2008.

Steponkus, P.L. (1984). Role of plasma membrane in freezing injury and cold acclimation. *Annual Review of Plant Physiology* 35: 543-584.

Stewart, C.R., B.A. Martin, L. Reding, and S. Cerwick (1990a). Respiration and alternative oxidase in corn seedling tissues during germination at different temperatures. *Plant Physiology* 92: 755-760.

Stewart, C.R., B.A. Martin, L. Reding, and S. Cerwick (1990b). Seedling growth, mitochondrial characteristics, and alternative respiratory capacity of corn genotypes differing in cold tolerance. *Plant Physiology* 92: 761-766.

Storz, G., L.A. Tartaglia, and B.N. Ames (1990). Transcriptional regulator of oxidative stress-inducible genes: Direct activation by oxidation. *Science* 248: 189-194.

Svalheim, O. and B. Robertson (1993). Elicitation of H_2O_2 production in cucumber hypocotyl segments by oligo-1,4-α-D-galacturonides and an oligo-β-glucan preparation from cell walls of *Phytophthera megasperma* sp. *glycinea*. *Physiologia Plantarum* 88: 675-681.

Takahashi, H., Z. Chen, H. Du, Y. Liu, and D.F. Klessig (1997). Development of necrosis and activation of disease resistance in transgenic tobacco plants with severely reduced catalase levels. *Plant Journal* 11: 993-1005.

Tepperman, J.M. and P. Dunsmuir (1990). Transformed plants with elevated levels of chloroplastic SOD are not more resistant to superoxide toxicity. *Plant Molecular Biology* 14: 501-511.

Thieringer, R., H. Shio, Y. Han, G. Cohen, and P.B. Lazarow (1991). Peroxisomes in *Saccharomyces cerevisie*: Immunofluorescence analysis and import of catalase A into isolated peroxisomes. *Molecular and Cellular Biology* 11: 510-522.

Thomashow, M.F (1990). Molecular genetics of cold acclimation in higher plants. *Advanced Genetics* 28: 99-131.

Thomsen, B., H. Drumm-Herrel, and H. Mohr (1992). Control of the appearance of ascorbate peroxidase (EC 1.11.1.11) in mustard seedling cotyledons by phytochrome and photooxidative treatments. *Planta* 186: 600-608.

Torsethaugen, G., L.H. Pitcher, B.A. Zilinskas, and E.J. Pell (1997). Overproduction of ascorbate peroxidase in the tobacco chloroplast does not provide protection against ozone. *Plant Physiology* 114: 529-537.

Trewavas, A.J. and S. Gilroy (1991). Signal transduction in plant cells. *Trends in Genetics* 7: 356-361.

Tsang, E.W.T., C. Bowler, D. Herouart, W. Van Camp, R. Villarroel, C. Genetello, M. Van Montagu, and D. Inzé (1991). Differential regulation of superoxide dismutases in plants exposed to environmental stress. *Plant Cell* 3: 783-792.

Uemura, M., R.A. Joseph, and P.L. Steponkus (1995). Cold acclimation of *Arabidopsis thaliana*. Effect on plasma membrane lipid composition and freeze-induced lesions. *Plant Physiology* 109: 15-30.

Upadhyaya, A., T.D. Davis, R.H. Walser, A.B. Galbraith, and N. Sankhla (1989). Uniconazole-induced alleviation of low-temperature damage in relation to antioxidant activity. *HortScience* 24: 955-957.

Van Camp, W., K. Capiau, M. Van Montagu, D. Inzé, and L. Slooten (1996). Enhancement of oxidative stress tolerance in transgenic tobacco plants overproducing Fe-superoxide dismutase in chloroplasts. *Plant Physiology* 112: 1703-1714.

Van Camp, W., H. Willekens, C. Bowler, M. Van Montagu, and D. Inzé (1994). Elevated levels of superoxide dismutase protect transgenic plants against ozone damage. *Bio/Technology* 12: 165-168.

Van de Venter, H.A. (1985). Cyanide-resistant respiration and cold resistance in seedlings of maize (*Zea mays* L.). *Annals of Botany* 56: 561-563.

Vanlerberghe, G.C. and L. McIntosh (1992). Lower growth temperature increases alternative pathway capacity and alternative oxidase protein in tobacco. *Plant Physiology* 100: 115-119.

Vanlerberghe, G.C. and L. McIntosh (1996). Signals regulating the expression of the nuclear gene encoding alternative oxidase of plant mitochondria. *Plant Physiology* 111: 589-595.

Vanlerberghe, G.C., A.E. Vanlerberghe, and L. McIntosh (1994). Molecular genetic alteration of plant respiration. Silencing and overexpression of alternative oxidase in transgenic tobacco. *Plant Physiology* 106: 1503-1510.

Virgin, I., H. Slater, D.F. Ghanotakis, and B. Andersson (1991). Light-induced D1 protein degradation is catalyzed by a serine-type protease. *FEBS Letters* 287: 125-128.

Wada, H., Z. Gombos, and N. Murata (1990). Enhancement of chilling tolerance of a cyanobacterium by genetic manipulation of fatty acid desaturation. *Nature* 347: 200-203.

Walker, M.A. and B.D. McKersie (1993). Role of ascorbate-glutathione antioxidant system in chilling resistance of tomato. *Journal of Plant Physiology* 141: 234-239.

Wang, C.Y. (1982). Physiological and biochemical responses of plants to chilling stress. *HortScience* 17: 173-186.

Wasserman, W.W. and W.E. Fahl (1997). Functional antioxidant responsive elements. *Proceedings of the National Academy of Sciences, USA* 94: 5361-5366.

Webb, M.S., D.V. Lynch, and B.R. Green (1992). Effects of temperature on the phase behavior and permeability of thylakoid lipid vesicles. Relevance to chilling stress. *Plant Physiology* 99: 912-918.

Webb, R.P. and R.D. Allen (1996). Overexpression of pea cytosolic ascorbate peroxidase confers protection against oxidative stress in transgenic *Nicotiana tabacum*. *Plant Physiology* (suppl.) 111: 64.

Welin, B., P. Heino, K.N. Henriksson, and E.T. Palva (1996). Plant cold acclimation: Possibilities for engineering freezing tolerance. *AgBiotech News and Information* 8: 15-22.

Willekens, H., C. Langebartels, C. Tire, M. Van Montagu, D. Inzé, and W. Van Camp (1994). Differential expression of catalase genes in *Nicotiana plumbaginifolia* (L.). *Proceedings of the National Academy of Sciences, USA* 91: 10450-10454.

Wingate, V.P.M., M.A. Lawton, and C.J. Lamb (1988). Glutathione causes a massive and selective induction of plant defense genes. *Plant Physiology* 87: 206-210.

Wingsle, G. and S. Karpinski (1996). Differential redox regulation by glutathione of glutathione reductase and Cu/Zn superoxide dismutase genes expression in *Pinus sylvestris* (L.) needles. *Planta* 198: 151-157.

Wise, R.R. and A.W. Naylor (1987). Chilling-enhanced photooxidation. The peroxidative destruction of lipids during chilling injury to photosynthesis and ultrastructure. *Plant Physiology* 83: 272-277.

Witham, S., S.P. Dinesh-Kumar, D. Choi, R. Hehl, C. Corr, and B. Baker (1994). The product of the tobacco mosaic virus resistance gene N: Similarity to Toll and the interleukin-1 receptor. *Cell* 78: 1101-1115.

Wu, G., B.J. Short, E.B. Lawrence, E.B. Levine, K.C. Fitzsimmons, and D.M. Shaw (1995). Disease resistance conferred by expression of a gene encoding H_2O_2-generating glucose oxidase in transgenic potato plants. *Plant Cell* 7: 1357-1368.

Wu, J. and J. Browse (1995). Elevated levels of high-melting-point phosphati-dylglycerols do not induce chilling sensitivity in an *Arabidopsis* mutant. *Plant Cell* 7: 17-27.

Zhang, J. and M.B. Kirkham (1994). Drought stress-induced changes in activities of superoxide dismutase, catalase, and peroxidase in wheat species. *Plant Cell Physiology* 35: 785-791.

Zhang, S., J. Sheng, Y. Liu, and M.C. Mehdy (1993). Fungal elicitor-induced bean proline-rich protein mRNA down-regulation is due to destabilization that is transcription and translation dependent. *Plant Cell* 5: 1089-1099.

Zhu, D. and J.G. Scandalios (1992). Expression of the maize *MnSOD* (*Sod3*) gene in MnSOD-deficient yeast rescues the mutant yeast under oxidative stress. *Genetics* 131: 803-809.

Chapter 2

Chilling Effects
on Active Oxygen Species
and Their Scavenging Systems
in Plants

D. Mark Hodges

AN OXYGENATED ENVIRONMENT

Oxygen is an essential component in the life of all aerobic organisms. However, life with oxygen carries with it a potential danger. Although molecular oxygen (O_2) itself is not toxic, it can produce by-products which are highly reactive and which can pose a potential for severe cellular damage or lethality. Although these by-products may be intrinsically involved in cell signaling processes, if their production exceeds their scavenging potential, they may play significant roles in the peroxidation of essential phospholipids (Senaratna, McKersie, and Borochov, 1987; Matsuo, Kashiwaki, and Itoo, 1990; Zheng and Yang, 1991), and in both nucleic acid (Elstner, 1982; Imlay and Lin, 1988; Monk, Fagerstedt, and Crawford, 1989) and protein (Lesser and Shick, 1989; Casano and Trippi, 1992) denaturation. For plants, typical symptoms of the presence of excess amounts of these active oxygen species are inhibition of chloroplast development (Poskuta et al., 1974), early leaf and cotyledon senescence (Dhindsa, Plumb-Dhindsa, and Thorpe, 1981; Li and Mei, 1989), chloroplast membrane damage (Knox and Dodge, 1985), damage to cellular lipids and DNA (deoxyribonucleic acid) (Prinsze, Bubbleman, and Van StSteveninck, 1990), and the inactivation of many types of enzymes result-

The author gratefully acknowledges the valuable advice of Dr. Willy Kalt, Atlantic Food and Horticulture Research Centre, Agriculture and Agri-Food Canada, Kentville, Nova Scotia, Canada, in preparing this manuscript.

53

ing from the alteration of proteins (Halliwell, 1981) and subsequent attacks by proteases (Landry and Pell, 1993).

Plants produce active oxygen species in greater quantities when exposed to environmental stresses such as drought, high light intensities, high ambient O_2 pressures, ozone or sulfur dioxide, some pathogens (Bowler, Van Montagu, and Inzé, 1992; Pastori and Trippi, 1992), salt (Gossett, Millhollon, and Lucas, 1994), herbicides, metals (Foyer, Lelandais, and Kunert, 1994), and low temperatures (Wise and Naylor, 1987; Prasad et al., 1994; Sonoike and Terashima, 1994; Terashima, Funayama, and Sonoike, 1994).

Active oxygen species arise from O_2 as products of its reduction or its excitation to the singlet state. A complete reduction of oxygen requires four electrons, with water being the end product. When the reduction of oxygen proceeds in univalent steps, reactive intermediates are produced. Among these are the superoxide anion ($O_2{}^-$), hydrogen peroxide (H_2O_2), and the hydroxyl radical ($\cdot OH$). Excitation of O_2 can lead to the production of singlet oxygen ($^1O_2$).

ACTIVE OXYGEN SPECIES

Superoxide

As a result of oxygen's spin restriction, its tendency to react with radical species and unpaired electrons is much greater than with substrates that donate pairs of electrons. Therefore, the simplest route of O_2 reduction is via the univalent pathway, leading to the production of the superoxide anion ($O_2{}^-$). Molecules of $O_2{}^-$ can also be produced from the univalent oxidation of H_2O_2 (Salin, 1988). The generation of $O_2{}^-$ has been observed in mitochondria, chloroplasts, glyoxysomes, microsomes, peroxisomes, and nuclei of many types of living organisms (del Rio et al., 1989). These $O_2{}^-$ molecules can arise as catalytic by-products from normal aerobic metabolism involving electron transport chains (Foyer, Lelandais, and Kunert, 1994; Foyer, Descourvières, and Kunert, 1994) and enzymes such as nitropropane dioxygenase, aldehyde oxidase, lipoxygenase, galactose oxidase, and xanthine oxidase (Halliwell, 1981; Palm et al., 1991; Thompson et al., 1991; Ushimaru, Shibaska, and Tsuji, 1992; Scandalios, 1993). Superoxide may also arise from nonenzymatic processes involving the autooxidation of such substrates as ferredoxins, hydroquinones, thiols, and reduced hemoproteins (Fridovich, 1976).

Superoxide itself is not as highly toxic as are some of the other active oxygen species (e.g., $\cdot OH$, $^1O_2$), although within biological membranes it can act as a powerful nucleophile and base (Alscher and Amthor, 1988). More important, however, O_2^- will react nonenzymatically with H_2O_2 in the Haber-Weiss reaction to form the hydroxyl radical, a highly destructive species. This reaction can only occur in the presence of certain transition metal ions or metal chelates (Alscher and Amthor, 1988). Superoxide also appears to convert the manganese ion Mn^{2+} into a more reactive species (Halliwell, 1981). As well, O_2^- is recognized as being associated with the elicitation of plant stress responses (for a review, see Foyer, Descourvières, and Kunert, 1994).

The dismutative elimination of O_2^- can occur spontaneously or under the catalytic influence of the enzyme superoxide dismutase (SOD; EC 1.15.1.1), which catalyzes a reaction leading to the production of H_2O_2 and O_2 (see Figure 2.1). Compounds such as ascorbate and α-tocopherol also remove O_2^- directly through noncatalyzed reactions.

FIGURE 2.1. Some of the Antioxidants Involved in Scavenging Active Oxygen Species in Plants

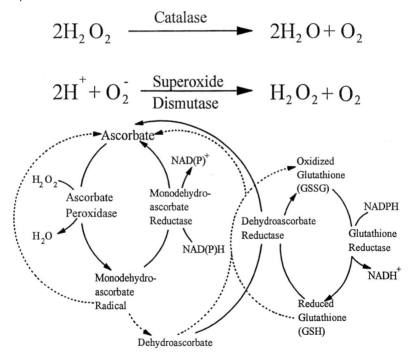

Note: Solid lines and dashed lines indicate enzymatically catalyzed and non-enzymatic reactions, respectively.

Hydrogen Peroxide

Hydrogen peroxide (H_2O_2) is the most stable of the active oxygen species because its two outer orbitals are completely filled. Similar to O_2^-, H_2O_2 acts as both a mild reductant and an oxidant, although it is a much stronger nucleophilic oxidizing agent. In addition, as with O_2^-, H_2O_2 is relatively unreactive; although it does tend to form complexes with transition metals, it does not react efficiently with organic substrates. However, H_2O_2 can play a significant role in the oxidation of thiol groups. As well, H_2O_2 can inhibit key enzymes of the Calvin cycle and can interact with O_2^- in the Haber-Weiss reaction to form the highly reactive and toxic $\cdot OH$ radical. As H_2O_2 can readily diffuse across membranes (Bowler, Van Montagu, and Inzé, 1992), the presence of this compound can lead to indirect damage. In general, however, any observed toxic effects directly attributable to H_2O_2 seem to occur only at nonphysiological concentrations, although the toxicity is greatly enhanced in the presence of metal catalysts (Fridovich, 1976). Some pathways require H_2O_2 as an oxidizer for coupling of such aromatic compounds in plant cell walls as cinnamic acids esterified to polysaccharides, cross-linking tyrosine residues in structural proteins, and the monomeric precursors of lignin (Liu, Eriksson, and Dean, 1995). The presence of H_2O_2, which can be mediated by salicylic acid (Sánchez-Casa and Klessig, 1994), has been postulated to be a secondary messenger in its own right (for a review, see Foyer, Descourvières, and Kunert, 1994) and has been demonstrated to provoke an increase in concentrations of free cytosolic calcium, a ubiquitous secondary messenger for cellular responses (Price et al., 1994).

Molecules of H_2O_2 can be produced in chloroplasts (Cakmak and Marschner, 1992), peroxisomes (Badger, 1985), and mitochondria (Halliwell, 1981; Prasad et al., 1994). Production of H_2O_2 can also occur through ß-oxidation of fatty acids and peroxisomal photorespiration reactions involving glyoxylate oxidation (Scandalios, 1993). Catalase (CAT; EC 1.11.1.6) is the enzyme directly responsible for the removal of H_2O_2 with water (H_2O) being the primary reaction product (see Figure 2.1). Ascorbate can also directly remove H_2O_2 in a reaction requiring the enzyme ascorbate peroxidase (APX; EC 1.11.1.11) (Nakano and Asada, 1987). Glutathione and the enzymes guaiacol peroxidase, monodehydroascorbate reductase (MDHAR; EC 1.6.5.4), dehydroascorbate reductase (DHAR; EC 1.8.5.1), and glutathione reductase (GR; EC 1.6.4.2) are all indirectly involved in H_2O_2 removal.

Hydroxyl Radical

A free radical contains an unpaired electron and will always beget another radical when it reacts with a nonradical. A chain reaction will thus be produced that will only come to an end when stable products are formed: hydroxylated products and hydroperoxides in the case of oxygen-dependent reactions, and fragmentation or dismutation products when no oxygen is involved (Saran, Michel, and Bors, 1988).

The hydroxyl radical ($\cdot$OH) is the product of the univalent reduction of H_2O_2 and is one of the strongest oxidizing agents known. This reduction, known as the Haber-Weiss reaction,

$$O_2^- + H_2O_2 \rightarrow \cdot OH + OH^- + O_2$$

involves a reaction between O_2^- and H_2O_2 which is catalyzed by certain transition metal ions or metal chelates (Alscher and Amthor, 1988). Copper and iron ions present in the plant systems can serve this function (Alscher and Amthor, 1988).

The hydroxyl radical is an extremely unspecific, highly reactive species that tends to react with the first available substrate. These substrates include enzymes, nucleic acids, and lipids, making the potential for damage due to $\cdot$OH relatively high. In particular, $\cdot$OH is very reactive toward proteins and may cause modification of almost all amino acid residues, covalent cross-linking, and fragmentation, resulting in loss of function and susceptibility to degradation by proteolytic enzymes (Prinsze, Bubbleman, and Van Steveninck, 1990).

Molecules of $\cdot$OH are very reactive and will not undergo many collisions before reacting. Thus, specific scavengers for $\cdot$OH are not very feasible; the best strategy is to prevent the production of, to minimize the damage caused by, and/or repair any damage done by this active oxygen species (Rabinowitch and Fridovich, 1983). Ascorbate, glutathione, and α-tocopherol are all compounds that can react directly with and neutralize $\cdot$OH.

Singlet Oxygen

When electronically excited species of oxygen are formed due to one of the outer shell electrons being elevated to a higher orbital and inverting its spin, the antiparallel spin that results is referred to as the singlet state. Singlet oxygen ($^1O_2$) has two excited states, the second being extremely short-lived and rapidly inactivated by collisional quenching to form the first singlet state, which has a long enough lifetime to allow for chemical reactions with other compounds.

Molecules of $^1O_2$ can be produced from various sources, such as by-products of lipoxygenase activity (Thompson et al., 1991). However, the conditions that most favor singlet oxygen production are found within the actively photosynthesizing chloroplast (Knox and Dodge, 1985). The major mechanism of formation is by energy transfer from photoexcited compounds such as chlorophyll. The photosynthetic transfer of energy or electrons occurs when photoexcited chlorophyll is in the singlet state. However, singlet-state chlorophyll may also convert to the triplet chlorophyll state that can then interact with ground-state O_2 to form $^1O_2$. Alternatively, chlorophyll can undergo radiative decay (fluorescence) (Larson, 1988).

Singlet oxygen can take part in addition reactions to enes and dienes to form hydroperoxides and endoperoxides, which propagate free radicals in chain reactions (Larson, 1988), and can result in peroxidized lipids, leading to weakened and altered membranes. Specific amino acids, such as histidine, methionine, and tryptophan, are susceptible to oxidation by singlet oxygen, possibly resulting in protein damage and enzyme deactivation (Knox and Dodge, 1985).

Carotenoid pigments are the primary means of quenching singlet oxygen within the chloroplast. In terms of quantity, lutein and ß-carotene are the most important and, to a lesser extent, violaxanthin and α-carotene. Compounds such as ascorbate, glutathione, and α-tocopherol can also react directly with and quench $^1O_2$.

ACTIVE OXYGEN SCAVENGING SYSTEMS

During the course of evolution, plants have developed complex antioxidant systems to protect themselves from active oxygen species. Essentially, the antioxidant system includes three classes: (1) the antioxidant enzymes, (2) the water-soluble antioxidant compounds, and (3) the lipid-soluble, membrane-associated antioxidant compounds.

The antioxidant enzymes include superoxide dismutase (SOD), catalase (CAT), ascorbate peroxidase (APX), monodehydroascorbate reductase (MDHAR), dehydroascorbate reductase (DHAR), and glutathione reductase (GR) (see Figure 2.1). SOD serves to dismutate O_2^- to H_2O_2 and O_2. Isozymes of SOD have been found in chloroplasts, mitochondria, and the cytosol (Acevedo and Scandalios, 1991). The enzyme CAT converts two molecules of H_2O_2 to two of H_2O and one of O_2. Activities of this enzyme have been detected primarily in glyoxysomes and peroxisomes (Scandalios, 1993), though isozymes of this enzyme have been identified in the mitochondria of maize (Prasad et al., 1994). APX catalyzes the conversion of H_2O_2 to H_2O through the concomitant oxidation of ascorbate; activities of this enzyme are located mainly in chloroplasts and in the cytosol (Foyer, Descourvières, and Kunert, 1994). The enzymes MDHAR and DHAR are concerned with the reduction of oxidized ascorbate, and GR serves to reduce oxidized glutathione. Activities of MDHAR have been reported in the cytosol, mitochondria, microsomes, and chloroplasts of plant cells (Alscher and Amthor, 1988). DHAR has been detected in chloroplasts (Nakano and Asada, 1981; Schöner and Krause, 1990) and in the cytosol (Foyer and Halliwell, 1977). The majority of GR activity is found within chloroplasts, though smaller amounts are present in the mitochondria and cytosol (Alscher and Amthor, 1988; Edwards et al., 1994).

Ascorbate, glutathione, and some of the flavonoids are examples of water-soluble antioxidants. Ascorbate has been found in relatively high concentrations in chloroplasts, cytosol, vacuoles, and apoplastic spaces of leaf cells (Foyer, Descourvières, and Kunert, 1994). In addition to redox cycling of ascorbate within the chloroplast, the chloroplast can also take up ascorbate by virtue of a translocator located on the chloroplast envelope. Ascorbate is an extremely important antioxidant in plants. It can react directly with and detoxify O_2^-, $\cdot OH$, and $^1O_2$ and can remove H_2O_2 through the processes of the glutathione-ascorbate cycle (see Figure 2.1). Reduced ascorbate can be regenerated by the enzymes MDHAR and DHAR and by the oxidation of glutathione.

Glutathione is present in high concentrations in both chloroplasts and cytosol (Alscher and Amthor, 1988). Gillham and Dodge (1986) demonstrated in pea (*Pisum sativum* L.) that 90 percent of the total cellular glutathione can exist outside the chloroplasts. Glutathione can be oxidized by reacting directly with $^1O_2$ and $\cdot OH$ species, as well as by regenerating reduced ascorbate (see Figure 2.1). The resulting oxidized glutathione may then be converted back to the reduced form by GR.

Some of the phenolic components, such as flavonoids and hydroxycinnamic acid derivatives, have recently been regarded as biologically active substances capable of assuming antioxidant properties (Tamura and Yamagami, 1994; Yamasaki, Uefuji, and Sakihama, 1996; Wang, Cao, and Prior, 1997), including a protective function against iron-induced free-radical reactions (Cao, Sofic, and Prior, 1997). By virtue of their relatively higher content, the antioxidant activities of flavonoids have been suggested to be of greater value in protecting against oxidative stress in fruits and vegetables than those of ascorbic acid, ß-carotene, and α-tocopherol (Wang, Cao, and Prior, 1996). This, however, remains speculative. Flavonoids, such as anthocyanins, are primarily located within the vacuoles and/or anthocyanoplasts (Hodges and Nozzolillo, 1996). The vast majority of active oxygen species production occurs outside the vacuole, and these species would tend to react with nearby cellular components, as evinced by their extremely short life span. It is highly unlikely that any of these active oxygen species would travel to and enter the vacuole without undergoing a reaction. The exception may be H_2O_2, as this compound is relatively unreactive and can readily diffuse across membranes (Bowler, Van Montagu, and Inzé, 1992). Further work to elucidate the effectiveness of flavonoids in protecting plants against oxidative stress is required.

The lipid-soluble, membrane-associated antioxidants include carotenoids and α-tocopherol, as well as certain members of the flavonoids. The membrane-associated antioxidant α-tocopherol can scavenge O_2^-, ·OH, and $^1O_2$. It also acts as a lipid oxidation chain-breaker and can trap fatty acyl peroxy radicals formed during lipid peroxidation. Although present in smaller concentrations than ascorbate, it is considerably more lipophilic and has been found to be the more potent antioxidant, particularly with respect to lipid peroxidation (Packer, Slater, and Willson, 1979). Ascorbate can reduce tocopherolquinone radicals produced through the oxidation of α-tocopherol, thus allowing for its regeneration (Packer, Slater, and Willson, 1979).

Another lipid-soluble antioxidant compound is ß-carotene, which not only acts as a photoreceptive antenna pigment but can also quench both $^1O_2$ and the excess chlorophyll excitation energy not readily passed on through the photosystem. The rate constant for ß-carotene's quenching of $^1O_2$ exceeds that for the reaction of this active oxygen species with most unsaturated fatty acids by four to five orders of magnitude, thus allowing a relatively low concentration of ß-carotene to effectively protect membrane lipids from $^1O_2$-induced lipid peroxidation (Larson, 1988).

The components of the xanthophyll cycle are also important in the photoprotective processes involving the dissipation of photo-energy (Demmig-Adams, 1990; Adams and Demmig-Adams, 1995). Under excessive light, violaxanthin is de-epoxidized primarily by the pH-dependent enzyme violaxanthin de-epoxidase to zeaxanthin by way of the intermediate antheraxanthin. Zeaxanthin, primarily only produced under stress conditions, may then provide an alternative pathway to dissipate excess photosystem energy.

Photosynthesis is the major source of potentially toxic oxygen species in plants. Although antioxidant enzymes and compounds can be found in various organelles, those located in the chloroplasts thus hold special significance.

CHILLING EFFECTS ON PLANTS

Temperature is an important environmental factor that determines both latitudinal and longitudinal distributions of plants (Öquist, 1983). Those plants seriously injured or killed by temperatures above the freezing point of the tissue are considered chilling sensitive, while those able to continue growth near $0°C$ are deemed chilling resistant (Graham and Patterson, 1982).

Many important crops of tropical or subtropical origin, such as maize (*Zea mays* L.) and tomato (*Lycopersicum esculentum* L.), are sensitive to low temperatures (Koscielnak, 1993; Walker and McKersie, 1993; Koroleva, Brüggemann, and Krause, 1994). The physiological dysfunctions that may occur upon exposure of chilling-sensitive species to low temperatures lead to a variety of visible symptoms, the extent of which is a function of temperature extremes (Lyons, Raison, and Steponkus, 1979). Chilling can affect the germination capacity of seeds of such sensitive species as cotton, corn, and tomatoes (Lyons, Raison, and Steponkus, 1979; Eagles and Brooking, 1981; Hodges, Hamilton, and Charest, 1994; Hodges et al., 1997a). Once the plant has become established, chilling can result in water loss and wilting (Wilson, 1976; Koscielniak, 1993); the closure of leaf stomata at chilling temperatures, along with increased water viscosity and reduced root permeability (Miedema, 1983), impairs the ability to transport water (Lyons, Raison, and Steponkus, 1979; Stamp, 1984). Reduced accumulation of dry mass in roots and shoots (Stamp, 1984; Hodges, Hamilton, and Charest, 1995; Hodges et al., 1997a) and bleaching (Stamp, 1984) are also common visible symptoms of chilling stress.

At the cellular level, apparent loss of cell turgor, vacuolization, reduction in the volume of both the cytoplasm and the vacuole, deposition of new material in the cell walls, disorganization of organelles, and a general loss of cytoplasmic structure have all been reported (Iker et al., 1976). Possible mechanisms of chilling injury include lipid membrane phase separations (Parkin et al., 1989; Hariyadi and Parkin, 1993), weakened hydrophobic bonding affecting protein-protein and protein-lipid interactions (Patterson and Graham, 1987; Parkin et al., 1989), and effects on the compartmentalization of secondary messengers of chilling stress (Price et al., 1994; Sánchez-Casa and Klessig, 1994).

CHILLING EFFECTS ON PHOTOSYNTHESIS AND ACTIVE OXYGEN SPECIES PRODUCTION

Examples of photosynthetic processes affected by low temperatures include enzymatic steps of the electron transport chain in chloroplast thylakoids (Brüggemann and Dauborn, 1993), coupling to photophosphorylation, enzymes in the carbon reduction cycle (Humbeck, Melis, and

Krupinska, 1994; Byrd, Ort, and Ogren, 1995; Kingston-Smith et al., 1997), and transport mechanisms of the photosynthetic products from the chloroplasts (Long, 1983; Öquist, 1983).

It has been demonstrated that chilling of sensitive plants in light is much more damaging to the photosynthetic apparatus than chilling in darkness (Wise and Naylor, 1987; Krause, 1988; Mishra, Mishra, and Singhal, 1993). Chilling under light leads to an increase in the amount of active oxygen species present in plant systems (MacRae and Ferguson, 1985; Hodgson and Raison, 1991; Tsang et al., 1991; Havaux and Davaud, 1994; Sonoike and Terashima, 1994; Terashima, Funayama, and Sonoike, 1994). That chilling under even moderate light intensities is more damaging to plants than chilling in darkness is primarily attributed to the process of chilling-induced photoinhibition.

Under nonchilling conditions, photoinhibition occurs with overreduction of the primary electron acceptor of photosystem II (PSII) (Q_A) by strong illumination. This inhibition of the normal electron transfer through Q_A leads to the recombination of the primary charge pair and, hence, to triplet formation (Sonoike and Terashima, 1994). The triplet reacts with oxygen to form $^1O_2$, which can cause the destruction of P-680, the reaction-center chlorophylls of PSII. Degradation of the D1 protein, one of the reaction-center subunits, may also be induced. Under chilling conditions, internal CO_2 concentrations may be low due to stomatal closure (Öquist, 1983), and both energy-consuming carbon metabolism (Cakmak and Marschner, 1992; Elstner and Osswald, 1994; Wise, 1995) and repair processes in the chloroplast can become restricted (Öquist, 1983). This results in an overenergization of the photosystem reaction-centers, caused by an inadequate supply of the natural electron acceptor $NADP^+$ (nicotinamide adenine dinucleotide phosphate, oxidized) (Wise, 1995). This chilling-induced photoinhibition, at least initially, is thought to occur at PSI (photosystem I), not PSII (Havaux and Davaud, 1994; Sonoike and Terashima, 1994; Terashima, Funayama, and Sonoike, 1994). Molecular oxygen may act as an electron acceptor both for the PSI electron transport chain at the iron-sulphur centers and for reduced ferredoxin, producing superoxide (Long, 1983) that then destroys the iron-sulphur centers of PSI, causing a cascade production of toxic oxygen compounds (Sonoike and Terashima, 1994). Alternatively, the destruction of the iron-sulphur centers may occur first, leading to the recombination of an early electron acceptor and P-700 and resulting in the formation of triplet chlorophyll and, hence, $^1O_2$, which then may inactivate P-700 itself (Sonoike and Terashima, 1994).

EFFECTS OF CHILLING ON ANTIOXIDANT SYSTEMS

As mentioned previously, chilling in the light leads to an increase in the amount of active oxygen species present in plant systems. This is primarily due to chilling-induced photoinhibition. However, membrane phase separations allowing for electron leakage from electron transport chain and other pathways and low-temperature inhibition of other metabolic and repair processes, including a potential decrease in scavenger system capacities, can all contribute to active oxygen proliferation in chilled plant systems. The species and/or cultivar under study, along with the duration and severity of the chilling treatment, the developmental stage of the stressed plant, and the presence of an acclimatory period influence observed antioxidant responses to chilling-induced increases in active oxygen.

Activities of catalase (CAT) have been observed to decrease in response to low temperatures in various crop plants (MacRae and Ferguson, 1985; Volk and Feierabend, 1989; Schöner and Krause, 1990; Fadzillah et al., 1996). However, activities of chilled material have also been shown to remain similar to those of controls (O'Kane et al., 1996), or even to increase (Gianinetti, Lorenzoni, and Marocco, 1993; Anderson, Prasad, and Stewart, 1995). It has been shown that H_2O_2 levels are elevated in chilled tissue (Prasad et al., 1994; Anderson, Prasad, and Stewart, 1995; Fadzillah et al., 1996), and that up-regulation of CAT is an important means of maintaining low H_2O_2 levels in dark-grown, chilling-acclimated maize tissue (Prasad et al., 1994). These differences in observed responses of CAT are probably dependent upon the species and its particular developmental stage, the duration (i.e., short or long term), and/or severity of the applied chilling treatment. Activity of CAT relative to controls was shown to be highly dependent upon developmental stage in both short-term (Hodges et al., 1997b) and long-term (Prasad, 1996; Hodges et al., 1997b) chilled maize. Demonstrated decreases in the activity of CAT under both short-term (Feierabend, Schann, and Hertwig, 1992; Mishra, Mishra, and Singhal, 1993) and long-term chilling treatments (Schöner and Krause, 1990; Prasad, 1996; Hodges et al., 1997b) have been suggested to occur through photo-inactivation of this enzyme when it is exposed to both light and cold (Volk and Feierabend, 1989; Feierabend, Schaan, and Hertwig, 1992). This photo-inactivation is mediated through light absorption by the enzyme's heme group and chloroplast pigments (Volk and Feierabend, 1989). Concomitant resynthesis of CAT can compensate for the loss of CAT and can maintain constant levels of this enzyme under light (Hertwig, Streb, and Feierabend, 1992; Mishra, Mishra, and Singhal, 1993). However, low-temperature inhibition of CAT biosynthesis may well slow down the mechanisms that, under normal control conditions, would compensate for the loss of the active enzyme.

Activities of monodehydroascorbate reductase (MDHAR) have been reported to increase under both short-term (*Triticum aestivum* L.; Mishra, Mishra, and Singhal, 1993) and long-term (*Spinacia oleracea* L.; Schöner and Krause, 1990) chilling stress, or not to change at all (*Lycopersicon* spp.; Walker and McKersie, 1993). Activities of MDHAR in differentially chilling-sensitive maize lines subjected to short- or long-term chilling either increased or decreased relative to controls, depending upon the developmental stage (Hodges et al., 1997b) and the relative inherent sensitivity of the line to chilling stress (Hodges et al., 1997b, c). In comparing a chilling-sensitive maize cultivar with the more tolerant *Zea diploperennis* L., Jhanke and colleagues (1991) found only a transient increase in MDHAR activity compared to control in the maize chill stressed for one to five days. As MDHAR activity is known to be regulated by its metabolites (Hossain, Nakano, and Asada, 1984), changes in activity may parallel similar changes in production of MDHAR's substrate, the mondehydroascorbate radical. The production of this radical is dependent upon the rate of scavenging of H_2O_2 by ascorbate and ascorbate peroxidase (APX).

Relatively short-term stresses leading to production of active oxygen species resulted in a decrease in activities of superoxide dismutase (SOD) (Del Longo et al., 1993; Li, Wu, and He, 1995), and both increases (Pastori and Trippi, 1992; Pinhero et al., 1997) and decreases (Jhanke, Hull, and Long, 1991) in activities of APX and glutathione reductase (GR). As with MDHAR, Hodges and colleagues (1997b, c) found a growth stage-dependent and an inherent chilling susceptibility-dependent increase or decrease in activities of SOD and APX measured in differentially chilling-sensitive maize lines. Activities of GR were observed to increase under short-term chilling. Prolonged oxidative stress (i.e., long-term chilling) has been demonstrated to generally induce or enhance activities of SOD (Schöner and Krause, 1990; Bowler, Van Montagu, and Inzé, 1992; Burdon et al., 1994; Foyer, Lelandais, and Kunert, 1994), ASX (Melhorn et al., 1987; Cakmak and Marschner, 1992; Mishra, Mishra, and Singhal, 1993; O'Kane et al., 1996), and GR (Schmidt and Kunert, 1986; Pastori and Trippi, 1992; Foyer, Lelandais, and Kunert, 1994; Gossett, Milhollon, and Lucas, 1994; O'Kane et al., 1996). This was not the case, however, with SOD and APX in work by Hodges and colleagues (1997b, c) with maize.

Activities of dehydroascorbate reductase (DHAR) in tissue chilled for either short- or long-term periods have not been demonstrated to vary significantly from those of controls (Schöner and Krause, 1990; Mishra, Mishra, and Singhal, 1993; Walker and McKersie, 1993). Although the production of dehydroascorbate is nonenzymatic, its precursors are under the control of antioxidant enzymes such as APX and MDHAR, which are affected by chilling. Thus, stressed systems that have, for example, increased levels of ascorbate and activities of APX and MDHAR (e.g., Mishra, Mishra, and Singhal, 1993) may very well have their scavenging potential limited by DHAR.

Levels of ascorbate (Schöner and Krause, 1990; Foyer, Lelandais, and Kunert, 1994), glutathione (Smith, 1985; Schmidt and Kunert, 1986; May and Leaver, 1993; Hodges et al., 1996; Wildi and Lütz, 1996), and carotenoids (Mishra, Mishra, and Singhal, 1993; Walker and McKersie, 1993) have all been observed to increase conditions that impose oxidative stress. Although this was also generally the case for glutathione concentrations of maize grown under both short-term (Brunner et al., 1995; Hodges et al., 1996; Kocsy et al., 1996) and long-term (Hodges et al., 1996) chilling regimes, ascorbate levels were found to increase only at the first-leaf stage of development, after which they began to decrease, in work by Hodges and colleagues (1996). A similar decrease in ascorbate levels of nonacclimated chilled coleoptiles and leaves of maize seedlings was observed by Anderson and colleagues (1995). The increase of glutathione over controls in virtually all chilled material studied to date is highly significant, as this compound, with ascorbate, is the key component in the scavenging of H_2O_2 and other toxic oxygen compounds. A large pool of reduced glutathione would presumably also allow for sufficient reduction of ascorbate in the absence of high activites of GR or high concentrations of NADPH (nicotinamide adenine denucleotide phosphate, reduced) (Anderson, Prasad, and Stewart, 1995). A large store of immediately available reduced ascorbate may then be unnecessary. However, increased glutathione levels alone are probably not adequate in reducing chilling-enhanced photooxidative stress.

Decreases in carotenoid content have been observed under short-term chilling in the relatively chilling-sensitive species tomato (Walker and McKersie, 1993) and chilling-tolerant wheat (Mishra, Mishra, and Singhal, 1993) and, as well, under long-term chilling in the chilling-tolerant species spinach (Schöner and Krause, 1990). Decreases in ß-carotene content of long-term-chilled maize have also been shown by Haldimann and colleagues (1995) and Haldimann (1996). Work by Hodges and colleagues (1996) demonstrated that, though ß-carotene levels of differentially chilling-sensitive maize lines grown under both short- and long-term chilling treatments increased or decreased depending upon developmental stage, levels were significantly lowest in the most chilling-sensitive line. As ß-carotene is vital in directly scavenging $^1O_2$ and inhibiting its formation through the quenching of excess photosystem excitation energy, a low content of this lipophilic antioxidant is strongly suggested to be an important contributor to chilling sensitivity.

Wise and Naylor (1987), in experiments with chill-stressed cucumber (*Cucumis sativus* L.) and pea (*Pisum sativum* L.), demonstrated that α-tocopherol appeared to be implicated as the first line of defense in protection of photosynthetic pigments. Levels of α-tocopherol in chilling-sensitive cucumber dropped drastically soon after the chilling treatment began, whereas those of chilling-tolerant pea were hardly affected throughout the duration of the experiment. However, though α-tocopherol levels generally decreased during long-term chilling in maize, no significant differences in these levels were found between chilling-sensitive and -tolerant lines (Hodges et al., 1996). This suggests that, apparently, at least in maize, α-tocopherol is not one of the key factors that limits chilling tolerance. The relative importance of α-tocopherol as an active oxygen species scavenger in chilled tissue compared with other antioxidants requires further exploration.

Anthocyanin accumulation in vegetative tissues is a well-recognized indicator of environmental stress (Christie, Alfenito, and Walbot, 1994; Hodges and Nozzolillo, 1996). The abundance of anthocyanin and transcripts of genes whose products function in the phenylpropanoid and anthocyanin pathways was found to increase in sheaths of maize in relation to the severity and duration of cold treatments (Christie, Alfenito, and Walbot, 1994). The recent regard for anthocyanins as biologically active antioxidants, and more than just simple plant secondary pigments, suggests that the accumulation of anthocyanins under chilling and other stresses contributes to the overall defense against oxidative stress. What portion of this contribution can be attributed to anthocyanin scavenging of active oxygen species and/or other free radicals remains to be elucidated.

Xanthophyll cycle carotenoids were demonstrated to increase over controls in third-leaf-stage maize subjected to long-term chilling under high light (Haldimann, Fracheboud, and Stamp, 1995; Haldimann, 1996). Zeaxanthin, in particular, accumulated in large amounts. As zeaxanthin functions to mediate de-excitation of the singlet state of chlorophyll before active oxygen species can be produced, increases of this compound under chilling would presumably facilitate the curtailment of oxidative stress.

Often an alteration in antioxidant system biosynthesis is brought about by the presence of active oxygen species (Schöner and Krause, 1990; Perl-Treves and Galun, 1991; Foyer, Descourvières, and Kunert, 1994). The production of active oxygen in plants is also speculated to be part of a general alarm signal that serves to alert metabolism and gene expression for possible modifications (Schöner and Krause, 1990; Foyer, Descourvières, and Kunert, 1994). Usually, antioxidant capacities will increase when active oxygen levels are induced by a relatively low application of stress, or by an acclimatory period before subjection to a relatively higher degree of stress. However, if the production of active oxygen exceeds scavenging potential, such as that generated by exposure to extremely high degrees of stress without prior acclimation, severe damage can result. General increases observed in activities of such enzymes as SOD, APX, and GR may not, however, be solely due to increases in gene expression or steady state mRNA (messenger ribonucleic acid) (Tsang et al., 1991; Foyer, Descourvières, and Kunert, 1994) or protein levels, but may be the result of subtle changes in the intracellular distribution and differential sensitivity to photo-oxidation between different isozymic forms with different substrate affinities (Edwards et al., 1994; Foyer, Lelandais, and Kunert, 1994). The varied antioxidant response of a species to a particular severity and/or duration of chilling treatment is undoubtedly attributable not only to its inherent genetic complement but also to its capacity to express these genes.

The question of whether plants having higher antioxidant capacities can withstand chilling to a greater degree than plants with lower antioxidant capacities is of great interest. However, the vast majority of past research attempting to determine different responses of antioxidant systems between relatively chilling-sensitive and -tolerant plants has compared two or more different species (MacRae and Ferguson, 1985; Wise and Naylor, 1987; Hodgson and Raison, 1991; Jhanke, Hull, and Long, 1991; Walker and McKersie, 1993). When comparing responses in different species, it must be taken into account that the genetic and metabolic mechanisms behind one species' response may be quite different from another's. The advantage of working with different lines or cultivars of one species exhibiting differential responses is that the mechanisms involved in both relative sensitivity and resistance can be studied in a similar genotype, thus reducing the complexity of genetic differences and variability. In the last few years, work comparing antioxidant capacities between differentially drought-sensitive maize lines (Del Longo et al., 1993) and differentially salt-sensitive cotton (*Gossypium hirsutum* L.) cultivars (Gossett, Millhollon, and Lucas, 1994) has been reported. Research comparing antioxidant capacities between two populations of *Echinochloa crus-galli* from contrasting climates (Hakam and Simon, 1996) and low-temperature-sensitive and -tolerant cultivars of rice (*Oryza sativa* L.) (Saruyama and Tanida, 1995) has also appeared. A recent spate of papers comparing antioxidant profiles of differentially chilling-sensitive inbred and/or hybrid maize (Massacci et al., 1995; Zhang et al., 1995; Hodges et al., 1996; Kocsy et al., 1996; Hodges et al., 1997b, c; Pinhero et al., 1997) has provided valuable information about antioxidants and chilling tolerance in regard to this species. However, compilation and scrutiny of the data from these and other publications strongly suggests that no one universal factor confers chilling tolerance, although most of these endeavors, a few being similar in experimental design, have shown that the most chilling sensitive of the species tested exhibited the lowest activities of CAT and APX and contents of glutathione and ß-carotene. The optimal configuration of different antioxidants to eliminate the threat of oxidative stress is apparently dependent on the species and developmental stage, and this optimal profile may change with severity and duration of the applied chilling stress.

CONCLUSIONS

Molecular oxygen plays critical roles in the normal metabolic processes of plants. Active oxygen species, however, have the potential to cause severe cellular damage if their production exceeds the scavenging potential of the cell, regardless of whether they also play important roles in such metabolic events as eliciting stress responses. Antioxidant capacities of plants must be able to maintain the balance of active oxygen both under normal metabolism and under stressful conditions.

Although studies of transgenic plants with increased levels of particular antioxidant enzymes, such as SOD, have met with mixed success in terms of stress tolerance (Tepperman and Dunsmuir, 1990; Sen Gupta, Webb, et al., 1993; Sen Gupta, Heinen, et al., 1993; Slooten et al., 1995; Van Camp et al., 1996), the increased expression of more than one antioxidant enzyme or compound may lead to a more enhanced tolerance of photooxidative stress. Many interacting factors are involved in the overall defense against oxidative stress, and it may not always be possible to enhance stress tolerance through manipulation of single components, as they may not be the only factor limiting the ability to tolerate the stress. Research has repeatedly shown that the various antioxidants work in concert in an attempt to prevent damage due to active oxygen. Inherent contents and activities, capacities to up-regulate and regenerate, turnover, and the level of codependence of the antioxidant enzymes and compounds to create a complete system are all components of a concerted active oxygen scavenging action. The optimal configuration of different antioxidants to eliminate the threat of oxidative stress induced by chilling is apparently species and developmental stage dependent, and this optimal profile may change with severity and duration of the applied stress. Elucidation of the variable components of chilling tolerance, and subsequent breeding or genetic manipulation of germplasm, could eventually lead to the development of species and/or cultivars exhibiting increased chilling resistance. This would represent a significant achievement for growers and producers of economically important crops in temperate countries such as Canada and those of northern Europe. Increased chilling tolerance would allow for sowing in more northerly areas and would lower existing mortality rates of crops sown into cool, spring soils, both of which would lead to greater overall crop and economic yields.

REFERENCES

Acevedo, A. and J.G. Scandalios (1991). Catalase and superoxide dismutase gene expression and distribution during stem development in maize. *Developmental Genetics* 12: 423-430.

Adams, W.W., and B. Demmig-Adams (1995). The xanthophyll cycle and sustained thermal energy dissipation activity in *Vinca minor* and *Euonymus kiautschoicus* in winter. *Plant, Cell and Environment* 18: 117-127.

Alscher, R.G. and J.S. Amthor (1988). The physiology of free-radical scavenging: Maintenance and repair processes. In *Air Pollution and Plant Metabolism*, eds. S. Schulte Hostede, N.M. Darrall, L.W. Blank, and A.R. Wellburn. London, England: Elsevier Applied Science, pp. 94-115.

Anderson, M.D., T. Prasad, and C.R. Stewart (1995). Changes in isozyme profiles of catalase, peroxidase, and glutathione reductase during acclimation to chilling in mesocotyls of maize seedlings. *Plant Physiology* 109: 1247-1257.

Badger, M.R (1985). Photosynthetic oxygen exchange. *Annual Review of Plant Physiology* 36: 27-53.

Bowler, C., M. Van Montagu, and D. Inzé (1992). Superoxide dismutase and stress tolerance. *Annual Reviews of Plant Physiology and Plant Molecular Biology* 43: 83-116.

Brüggemann, W. and B. Dauborn (1993). Long-term chilling of young tomato plants under low light. III. Leaf development as reflected by photosynthesis parameters. *Plant and Cell Physiology* 34: 1251-1257.

Brunner, M., G. Kocsy, A. Rüegsegger, D. Shmutz, and C. Brunold (1995). Effect of chilling on assimilatory sulfate reduction and glutathione synthesis in maize. *Journal of Plant Physiology* 146: 743-747.

Burdon, R.H., V. Gill, P.A. Boyd, and D. O'Kane (1994). Chilling, oxidative stress and antioxidant enzyme responses in *Arabidopsis thaliana*. In *Oxygen and Environmental Stress in Plants*, eds. R.M.N. Crawford, G.A.F. Hendry, and B.A. Goodman. Edinburgh, UK: Royal Society of Edinburgh, Proceedings of the Royal Society of Edinburgh, Vol. 102, pp. 177-185.

Byrd, G.T., D.R. Ort, and W.L. Ogren (1995). The effects of chilling in the light on ribulose-1,5-biphosphate carboxylase/oxygenase activation of tomato (*Lycopersicum esculentum* Mill.). *Plant Physiology* 107: 585-591.

Cakmak, I. and H. Marschner (1992). Magnesium deficiency and high light intensity enhance activities of superoxide dismutase, ascorbate peroxidase, and glutathione reductase in bean leaves. *Plant Physiology* 98: 1222-1227.

Cao, G., E. Sofic, and R.L. Prior (1997). Antioxidant and proxidant behaviour of flavonoids: Structure-activity relationships. *Free Radical Biology and Medicine* 22: 749-760.

Casano, L.M. and V.S. Trippi (1992). The effect of oxygen radicals on proteolysis in isolated oat chloroplasts. *Plant and Cell Physiology* 33: 329-332.

Christie, P., M.R. Alfenito, and V. Walbot (1994). Impact of low-temperature stress on general phenylpropanoid and anthocyanin pathways: Enhancement of transcript abundance and anthocyanin pigmentation in maize seedlings. *Planta* 194: 541-549.

Del Longo, O.T., C.A. González, G.M. Pastori, and V.S. Trippi (1993). Antioxidant defences under hyperoxygenic and hyperosmotic conditions in leaves of two lines of maize with differential sensitivity to drought. *Plant and Cell Physiology* 34: 1023-1028.

del Rio, L.A., V.M. Fernandez, F.L. Ruperez, L.M. Sandalio, and J.M. Palma (1989). NADH induces the generation of superoxide radicals in leaf peroxisomes. *Plant Physiology* 89: 728-731.

Demmig-Adams, B. (1990). Carotenoids and photoprotection in plants: A role for the xanthophyll zeaxanthin. *Biochemica et Biophysica Acta* 1020: 1-24.

Dhindsa, R.S., P. Plumb-Dhindsa, and T.A. Thorpe (1981). Leaf senescence: Correlated with increased levels of membrane permeability and lipid peroxidation, and decreased levels of superoxide dismutase and catalase. *Journal of Experimental Botany* 32: 93-101.

Eagles, H.A. and I.R. Brooking (1981). Populations of maize with more rapid and reliable seedling emergence than cornbelt dents at low temperature. *Euphytica* 30: 755-763.

Edwards, E.A., C. Enard, G.P. Creissen, and P.M. Mullineaux (1994). Synthesis and properties of glutathione reductase in stressed peas. *Planta* 192: 137-143.

Elstner, E.F. (1982). Oxygen activation and oxygen toxicity. *Annual Review of Plant Physiology* 33: 73-96.

Elstner, E.F. and W. Osswald (1994). Mechanisms of oxygen activation during plant stress. In *Oxygen and Environmental Stress in Plants*, eds. R.M.N. Crawford, G.A.F. Hendry, and B.A. Goodman. Edinburgh, UK: Royal Society of Edinburgh, Proceedings of the Royal Society of Edinburgh, Vol. 102, pp. 131-154.

Fadzillah, N.M., V. Gill, R. Finch, and R.H. Burdon (1996). Chilling, oxidative stress and antioxidant responses in shoot cultures of rice. *Planta* 199: 552-556.

Feierabend, J., C. Schann, and B. Hertwig (1992). Photoinactivation of catalase occurs under both high- and low-temperature stress conditions and accompanies photoinhibition of photosystem II. *Plant Physiology* 100: 1534-1561.

Foyer, C.H., P. Descourvières, and K.J. Kunert (1994). Protection against oxygen radicals: An important defence mechanism studied in transgenic plants. *Plant, Cell and Environment* 17: 507-523.

Foyer, C.H. and B. Halliwell (1977). Purification and properties of dehydroascorbate reductase from spinach leaves. *Phytochemistry* 16: 1347:1350.

Foyer, C.H., M. Lelandais, and K.J. Kunert (1994). Photooxidative stress in plants. *Physiologia Plantarum* 92: 696-717.

Fridovich, I. (1976). Superoxide dismutases. *Annual Review of Biochemistry* 44: 146-159.

Gianinetti, A, C. Lorenzoni, and A. Marocco (1993). Changes in superoxide dismutase and catalase activities in response to low temperature in tomato mutants. *Journal of Genetics and Breeding* 47: 353-356.

Gillham, D.J. and A.D. Dodge (1986). Hydrogen-peroxide scavenging systems within pea chloroplasts. *Planta* 167: 246-251.

Gossett, D.R., E.P. Millhollon, and M.C. Lucas (1994). Antioxidant response to NaCl stress in salt-tolerant and salt-sensitive cultivars of cotton. *Crop Science* 34: 706-714.

Graham, D. and B.D. Patterson (1982). Responses of plants to low, non-freezing temperatures: Proteins, metabolism, and acclimation. *Annual Review of Plant Physiology* 33: 347-372.

Hakam, N. and J.P. Simon (1996). Effect of low temperature on the activity of oxygen-scavenging enzymes in two populations of the C_4 grass *Echinochloa crus-galli*. *Physiologia Plantarum* 97: 209-216.

Haldimann, P. (1996). Effects of changes in growth temperature on photosynthesis and carotenoid composition in *Zea mays* leaves. *Physiologia Plantarum* 97: 554-562.

Haldimann, P., Y. Fracheboud, and P. Stamp (1995). Carotenoid composition in *Zea mays* developed at sub-optimal temperature and different light intensities. *Physiologia Plantarum* 95: 409-414.

Halliwell, B. (1981). *Chloroplast Metabolism: The Structure and Function of Chloroplasts in Green Leaf Cells.* Oxford, England: Clarendon Press.

Hariyadi, P. and K.L. Parkin (1993). Chilling-induced oxidative stress in cucumber (*Cucumis sativus* L. cv. Calypso) seedlings. *Journal of Plant Physiology* 141: 733-738.

Havaux, M. and A. Davaud (1994). Photoinhibition of photosynthesis in chilled potato leaves is not correlated with a loss of photosystem-II activity. *Photosynthesis Research* 40: 75-92.

Hertwig, B., P. Streb, and J. Feierabend (1992). Light dependence of catalase synthesis and degradation in leaves and the influence of interfering stress condition. *Plant Physiology* 100: 1547-1553.

Hodges, D.M., C.J. Andrews, D.A. Johnson, and R.I. Hamilton (1996). Antioxidant compound responses to chilling stress in differentially sensitive inbred maize lines. *Physiologia Plantarum* 98: 685-692.

Hodges, D.M., C.J. Andrews, D.A. Johnson, and R.I. Hamilton (1997a). Sensitivity of maize hybrids to chilling and their combining abilities at two developmental stages. *Crop Science* 37: 850-856.

Hodges, D.M., C.J. Andrews, D.A. Johnson, and R.I. Hamilton (1997b). Antioxidant enzyme responses to chilling stress in differentially sensitive inbred maize lines. *Journal of Experimental Botany* 48: 1105-1113.

Hodges, D.M., C.J. Andrews, D.A. Johnson, and R.I. Hamilton (1997c). Antioxidant enzyme and compound responses to chilling stress and their combining abilities in differentially sensitive maize hybrids. *Crop Science* 37: 857-863.

Hodges, D.M., R.I. Hamilton, and C. Charest (1994). A chilling resistance test for inbred maize lines. *Canadian Journal of Plant Science* 74: 687-691.

Hodges, D.M., R.I. Hamilton, and C. Charest (1995). A chilling response test for early growth phase maize. *Agronomy Journal* 87: 970-974.

Hodges, D.M. and C. Nozzolillo (1996). Anthocyanin and anthocyanoplast content of cruciferous seedlings subjected to mineral nutrient deficiencies. *Journal of Plant Physiology* 147: 749-754.

Hodgson, R.A.J. and J.K. Raison (1991). Superoxide production by thylakoids during chilling and its implication in the susceptibility of plants to chilling-induced photoinhibition. *Planta* 183: 222-228.

Hossain, M.A., Y. Nakano, and K. Asada (1984). Monodehydroascorbate reductase in spinach chloroplasts and its participation in regeneration of ascorbate for scavenging hydrogen peroxide. *Plant and Cell Physiology* 25: 385-395.

Humbeck, K., A. Melis, and K. Krupinska (1994). Effects of chilling on chloroplast development in barley primary foliage leaves. *Journal of Plant Physiology* 143: 744-799.

Iker, R., A.J. Waring, J.M. Lyons, and R.W. Breidenbach (1976). The effects of chilling temperatures on chloroplasts. *Protoplasma* 90: 229-252.

Imlay, J.A. and S. Lin (1988). DNA damage and oxygen radical toxicity. *Science* 240: 1302-1309.

Jhanke, L.S., M.R. Hull, and S.P. Long (1991). Chilling stress and oxygen metabolizing enzymes in *Zea mays* and *Zea diploperennis*. *Plant, Cell and Environment* 14: 97-104.

Kingston-Smith, A.H., J. Harbinson, J. Williams, and C.H. Foyer (1997). Effects of chilling on carbon assimilation, enzyme activation, and photosynthetic electron transport in the absence of photoinhibition in maize leaves. *Plant Physiology* 114: 1039-1046.

Knox, J.P. and A.D. Dodge (1985). Singlet oxygen and plants. *Phytochemistry* 24: 889-896.

Kocsy, G., M. Brunner, A. Rüegsegger, P. Stamp, and C. Brunold (1996). Glutathione synthesis in maize genotypes with different sensitivities to chilling. *Planta* 198: 365-370.

Koroleva, O.Y., W. Brüggemann, and G.H. Krause (1994). Photoinhibition, xanthophyll cycle and *in vivo* chlorophyll fluorescence quenching of chilling-tolerant *oxyria digyna* and chilling-sensitive *Zea mays*. *Physiologia Plantarum* 92: 577-584.

Koscielnak, J. (1993). Effects of low night temperatures on photosynthetic activity of maize seedlings (*Zea mays* L.). *Journal of Agronomy and Crop Science* 171: 73-81.

Krause, G.H. (1988). Photoinhibition of photosynthesis: An evaluation of damaging and protective mechanisms. *Physiologia Plantarum* 74: 566-574.

Landry, L.G. and E.J. Pell (1993). Modification of rubisco and altered proteolytic activity in O_3-stressed hybrid poplar (*Populus maximowizii × trichocarpa*). *Plant Physiology* 101: 1355-1362.

Larson, R.A. 1988. The antioxidants of higher plants. *Phytochemistry* 27: 969-978.

Lesser, M.P. and J.M. Shick (1989). Effects of irradiance and ultraviolet radiation on photoadaption in the zooxanthellae of *Aiptasia pallida:* Primary production, photoinhibition, and enzymatic defenses against oxygen toxicity. *Experimental Marine Biology and Ecology* 102: 243-255.

Li, B.L. and H.S. Mei (1989). Relationship between oat leaf senescence and activated oxygen metabolism. *Acta Phytophysiologia Sinica* 15: 6-12.

Li, X., Z. Wu, and G. He (1995). Effects of low temperature and physiological age on superoxide dismutase in water hyacinth (*Eichhornia crassipes* Solms). *Aquatic Botany* 50: 193-200.

Liu, L., K.L. Eriksson, and J.F.D. Dean (1995). Localization of hydrogen peroxide production in *Pisum sativum* L. using epi-polarization microscopy to follow cerium perhydroxide deposition. *Plant Physiology* 107: 501-506.

Long, S.P. (1983). C4 photosynthesis at low temperatures. *Plant, Cell and Environment* 6: 345-363.

Lyons, J.M., J.K. Raison, and P.L. Steponkus (1979). The plant membrane in response to low temperature: An overview. In *Low Temperature Stress in Crop Plants: The Role of the Membrane,* eds. J.M. Lyons, D. Graham, and J.K. Raison. New York: Academic Press, pp. 1-24.

MacRae, E.A. and I.B. Ferguson (1985). Changes in catalase activity and hydrogen peroxide concentration in plants in response to low temperature. *Physiologia Plantarum* 65: 51-56.

Massacci, A., M. Iannelli, F. Pietrini, and F. Loreto (1995). The effect of growth at low temperature on photosynthetic characteristics and mechanisms of photoprotection of maize leaves. *Journal of Experimental Botany* 46: 119-127.

Matsuo, T., Y. Kashiwaki, and S. Itoo (1990). A mechanism of mitochondrial damage induced by tert-butyl hydroperoxide and microsomes in vitro. *Physiologia Plantarum* 80: 226-232.

May, M.J. and C.J. Leaver (1993). Oxidative stimulation of glutathione synthesis in *Arabidopsis thaliana* suspension cultures. *Plant Physiology* 103: 621-627.

Mehlhorn, H., D.A. Cottam, P.W. Lucas, and A.R. Wellburn (1987). Induction of ascorbate peroxidase and glutathione reductase activities by interactions of mixtures of air pollutants. *Free Radical Research Communications* 3: 1-5.

Miedema, P (1983). The effects of low temperature on *Zea mays. Advances in Agronomy* 35: 93-129.

Mishra, N.P., R.K. Mishra, and G.S. Singhal (1993). Changes in the activities of antioxidant enzymes during exposure of intact wheat leaves to strong visible light at different temperatures in the presence of protein synthesis inhibitors. *Plant Physiology* 102: 903-910.

Monk, L.S., K.V. Fagerstedt, and R.M.M. Crawford (1989). Oxygen toxicity and superoxide dismutase as an antioxidant in physiological stress. *Physiologia Plantarum* 76: 456-459.

Nakano, Y. and K. Asada (1981). Hydrogen peroxide is scavenged by ascorbate-specific peroxidase in spinach chloroplasts. *Plant and Cell Physiology* 22: 867-880.

Nakano, Y. and K. Asada (1987). Purification of ascorbate peroxidase in spinach chloroplasts: Its inactivation by ascorbate-depleted medium and reactivation by monodehydroascorbate radical. *Plant and Cell Physiology* 28: 131-140.

O'Kane, D., V. Grill, P. Boyd, and R.H. Burdon (1996). Chilling, oxidative stress and antioxidative responses in *Arabidopsis thaliana* callus. *Planta* 198: 366-370.

Öquist, G. (1983). Effects of low temperature on photosynthesis. *Plant, Cell and Environment* 6: 281-300.

Packer, J.E., T.F. Slater, and R.I. Willson (1979). Direct observation of a free radical interaction between vitamin E and vitamin C. *Nature* 278: 737-738.

Palm, J.M., M. Garrido, M.I. Rodriguez-Garcia, and L.A. del Rio (1991). Peroxisome proliferation and oxidative stress mediated by activated oxygen species in plant peroxisomes. *Archives of Biochemistry and Biophysics* 287: 68-74.

Parkin, K.L., A. Marangoni, R.L. Jackman, R.Y. Yada, and D.W. Stanely (1989). Chilling injury: A review of possible mechanisms. *Journal of Food Biochemistry* 13: 127-153.

Pastori, G.M. and V.S. Trippi (1992). Oxidative stress induces high rate of glutathione reductase synthesis in a drought-resistant maize strain. *Plant and Cell Physiology* 33: 957-961.

Patterson, B.D. and D. Graham (1987). Temperature and metabolism. In *The Biochemistry of Plants*, Volume 12. New York: Academic Press Inc., pp. 153-199.

Perl-Treves, R. and E. Galun (1991). The tomato Cu,Zn superoxide dismutase genes are developmentally regulated and respond to light and stress. *Plant Molecular Biology* 17: 745-760.

Pinhero, R.G., M.V. Rao, G. Paliyath, D.P. Murr, and R.A. Fletcher (1997). Changes in activities of antioxidant enzymes and their relationship to genetic and paclobutrazol-induced chilling tolerance of maize seedlings. *Plant Physiology* 114: 695-704.

Poskuta, J., M. Mikulska, M. Faltynowicz, B. Bielak, and B. Wroblewska (1974). Chloroplast development. *Zeitschrift für. Pflanzenphysiologie* 73: 387-393.

Prasad, T. (1996). Mechanisms of chilling-induced oxidative stress injury and tolerance in developing maize seedlings: Changes in antioxidant system, oxidation of proteins and lipids, and protease activities. *The Plant Journal* 10: 1017-1026.

Prasad, T.K., M.D. Anderson, B.A. Martin, and C.R. Stewart (1994). Evidence for chilling-induced oxidative stress in maize seedlings and a regulatory role for hydrogen peroxide. *Plant Cell* 6: 65-74.

Price, A.H., A. Taylor, S.J. Ripley, A. Griffiths, A.J. Trewavas, and M.R. Knight (1994). Oxidative signals in tobacco increase cytosolic calcium. *Plant Cell* 6: 1301-1310.

Prinsze, C., T.M.A.R. Bubbleman, and J. Van Steveninck (1990). Protein damage, induced by small amounts of photodynamically generated singlet oxygen or hydroxyl radicals. *Biochimica et Biophysica Acta* 37: 152-157.

Rabinowitch, H.D. and I. Fridovich (1983). Superoxide radicals, superoxide dismutases and oxygen toxicity in plants. *Photochemistry and Photobiology* 37: 679-690.

Salin, M.L. (1988). Toxic oxygen species protective systems of the chloroplast. *Physiologia Plantarum* 72: 681-689.

Sánchez-Casa, P. and D.F. Klessig (1994). A salicylic acid-binding activity and a salicyclic acid-inhibitable catalase activity are present in a variety of plant species. *Plant Physiology* 106: 1675-1679.

Saran, M., C. Michel, and W. Bors (1988). Reactivities of free radicals. In *Air Pollution and Plant Metabolism,* eds. S. Schulte-Hostede, N.M. Darrall, L.W. Blank, and A.R. Wellburn. London, England: Elsevier Applied Science, pp. 76-92.

Saruyama, H. and M. Tanida (1995). Effect of chilling on activated oxygen-scavenging enzymes in low temperature-sensitive and -tolerant cultivars of rice (*Oryza sativa* L.). *Plant Science* 109: 105-111.

Scandalios, J.G. (1993). Oxygen stress and superoxide dismutases. *Plant Physiology* 101: 7-12.

Schmidt, A. and K.J. Kunert (1986). Lipid peroxidation in higher plants: The role of glutathione reductase. *Plant Physiology* 82: 700-702.

Schöner, S. and G.H. Krause (1990). Protective systems against active oxygen species in spinach: Response to cold acclimation in excess light. *Planta* 180: 383-389.

Sen Gupta, A., J.L. Heinen, A.S. Holaday, J.L. Burke, and R.D. Allen (1993). Increased resistance to oxidative stress in transgenic plants that overexpress chloroplastic Cu/Zn superoxide dismutase. *Proceedings of the National Academy of Sciences USA* 90: 1629-1633.

Sen Gupta, A., R.P. Webb, A.S. Holaday, and R.D. Allen (1993). Overexpression of superoxide dismutase protects plants from oxidative stress. *Plant Physiology* 103: 1067-1073.

Senaratna, T., B.D. McKersie, and A. Borochov (1987). Dessication and free radical mediated changes in plant membranes. *Journal of Experimental Botany* 38: 2005-2014.

Slooten, L., K. Capiau, W. Van Camp, M. Van Montagu, C. Sybesma, and D. Inzé (1995). Factors affecting the enhancement of oxidative stress tolerance in transgenic tobacco overexpressing manganese superoxide dismutase in the chloroplasts. *Plant Physiology* 107: 737-750.

Smith, I.K. (1985). Stimulation of glutathione synthesis in photorespiring plants by catalase inhibitors. *Plant Physiology* 79: 1044-1047.

Sonoike, K. and I. Terashima (1994). Mechanisms of photosystem-I photoinhibition in leaves of *Cucumis sativus* L. *Planta* 194: 287-293.

Stamp, P. (1984). Chilling tolerance of young plants demonstrated on the example of maize (*Zea mays* L.). *Advances in Agronomy and Crop Science* 7.

Tamura, H. and A. Yamagami (1994). Antioxidative activity of monoacylated anthocyanins isolated from Muscat Bailey A grape. *Journal of Agriculture and Food Chemistry* 42: 1612-1615.

Tepperman, J.M. and P. Dunsmuir (1990). Transformed plants with elevated levels of chloroplastic SOD are not more resistant to superoxide toxicity. *Plant Molecular Biology* 14: 501-511.

Terashima, I., S. Funayama, and K. Sonoike (1994). The site of photoinhibition of *Cucumis sativus* L. at low temperatures is photosystem I, not photosystem II. *Planta* 193: 300-306.

Thompson, J.E., J.H. Brown, G. Paliyath, J.F. Todd, and K. Yao (1991). Membrane phospholipid catabolism primes the production of activated oxygen in senescing tissues. In *Active Oxygen/Oxidative Stress and Plant Metabolism, Volume 6, Current Topics in Plant Physiology,* eds. E.J. Pell and K.L. Steffen. Rockville, MD: American Society of Plant Physiologists, pp. 57-66.

Tsang, E.W.T., C. Bowler, D. Hérouart, W. Van Camp, R. Villaruel, C. Genetello, M. Van Montagu, and D. Inzé (1991). Differential regulation of superoxide dismutase in plants exposed to environmental stress. *Plant Cell* 3: 783-792.

Ushimaru, T., M. Shibaska, and H. Tsuji (1992). Development of the $O_2 -$ detoxification system during adaption to air of submerged rice seedlings. *Plant Cell Physiology* 33: 1065-1071.

Van Camp, W., K. Capiau, M. Van Montagu, D. Inzé, and L. Slooten (1996). Enhancement of oxidative stress tolerance in transgenic tobacco plants overproducing Fe-superoxide dismutase in chloroplasts. *Plant Physiology* 112: 1703-1714.

Volk, S. and J. Feierabend (1989). Photoinactivation of catalase at low temperature and its relevance to photosynthetic and peroxide metabolism in leaves. *Plant, Cell and Environment* 12: 701-712.

Walker, M.A. and B.D. McKersie (1993). Role of the ascorbate-glutathione antioxidant system in chilling resistance of tomato. J. *Plant Physiology* 141: 234-239.

Wang, H., G. Cao, and R.L. Prior (1996). Total antioxidant capacity of fruits. *Journal of Agriculture and Food Chemistry* 44: 701-705.

Wang, H., G. Cao, and R.L. Prior (1997). Oxygen radical absorbing capacity of anthocyanins. *Journal of Agriculture and Food Chemistry* 45: 304-309.

Wildi, B. and C. Lütz (1996). Antioxidant composition of selected high alpine plant species from differing altitudes. *Plant, Cell and Environment* 19: 138-146.

Wilson, D. (1976). The mechanism of chill- and drought-hardening of *Phaseolus vulgaris* leaves. *New Phytologist* 76: 257-270.

Wise, R.R. (1995). Chilling-enhanced photooxidation: The production, action and study of reactive oxygen species produced during chilling in the light. *Photosynthesis Research* 45: 79-97.

Wise, R.R. and A.W. Naylor (1987). Chilling-enhanced photooxidation: The peroxidative destruction of lipids during chilling injury to photosynthesis and ultrastructure. *Plant Physiology* 83: 272-277.

Yamasaki, H., H. Uefuji, and Y. Sakihama (1996). Bleaching of the red anthocyanin induced by superoxide radical. *Archives of Biochemistry and Biophysics* 332: 183-186.

Zhang, J., S. Cui, J. Li, J. Wei, and M.B. Kirkham (1995). Protoplasmic factors, antioxidant responses, and chilling resistance in maize. *Plant Physiology and Biochemistry* 33: 567-575.

Zheng, R. and Z. Yang (1991). Lipid peroxidation and antioxidative defense systems in early leaf growth. *Journal of Plant Growth Regulation* 10: 187-199.

Chapter 3

Root System Functions During Chilling Temperatures: Injury and Acclimation

Pamela L. Sanders
Albert H. Markhart III

INTRODUCTION

Chilling temperatures for plants are below optimal, but above freezing (no ice formation). The focus of this chapter is root system responses to chilling temperatures. We will discuss processes unique or specific to roots, leaving discussion of whole-plant subjects, such as respiration and cold-induced genes, to others. Though not always recognized, low soil temperatures can affect root systems and the whole plant's physiology in many agronomic and horticultural situations. For crop plants, the earliest spring plantings are the most economically rewarding for the grower, but the potential for yield loss due to chilling injury increases. An average soil temperature of $15°C$ prior to maize shoot emergence caused a developmental delay of more than one week at tasseling stage compared to $25°C$ soil temperature (Bollero, Bullock, and Hollinger, 1996). Root zone temperature is critical during germination and early stand establishment. For example, the rapidly growing apical tissue of cucumber radicles is the most chilling-sensitive part of the seedling (Rab and Saltveit, 1996). Soil temperature may even remain below optimal for much of the season, as it does for cereal crops in northern Europe, plants at higher elevations, and for cotton in the United States, for example (Clarkson and Warner, 1979; DeLucia, Day, and Öquist, 1991; McMichael and Burke, 1994). Diurnally, the soil may be slow to warm in the morning, causing differential root and shoot conditions; cold soil could limit photosynthesis and growth for part of the day. There are more questions than answers about chilled root systems. We hope to stimulate further interest and research in this intriguing area. Most important, we will demonstrate the significance of roots during chilling and highlight the ability of some root processes to acclimate.

The most important root system functions affected by chilling are growth, water uptake, and mineral uptake. Inhibition of any of these three functions can have serious implications for crop performance (see Figure 3.1). For many chilling-sensitive plants of tropical origins (beans, maize, tomato, pepper, cotton), air or soil temperatures of 10 to 15°C are near the limit for growth (Dickson and Boettger, 1984; Scott and Jones, 1986; Bradow, 1990; Bollero, Bullock, and Hollinger, 1996). Cotton roots grow into deeper zones of the soil only as the temperature of each zone warms to 20 to 26°C (Bland, 1993). Even chilling-tolerant species such as *Brassica napus* have reduced root growth at 5 to 10°C (MacDuff et al., 1986).

Low root temperature often causes water stress in transpiring plants because of the reduction in water movement through the roots (Kramer, 1940). Transient or permanent wilting can occur when plants are exposed to cold (e.g., see McWilliam, Kramer, and Musser, 1982; Bradow, 1990). We will discuss root system control of water status during chilling and its role in shoot-water relations.

Mineral uptake is often reduced by chilling-induced changes in roots. Over the short-term, ion uptake per se may be inhibited, while over the long term, decreased root elongation may become more significant (Moorby and Nye, 1984; Marschner, 1995). Crops may suffer from nutrient deficiencies during long cold periods (Cumbus and Nye, 1985; Engels, Munkle, and Marschner, 1992; Engels and Marschner, 1996).

Overview of Mechanisms of Injury

Raison and Orr (1990) defined sensitivity/insensitivity in absolute terms: either plants are damaged below a critical temperature or they are not. The degree of sensitivity corresponds to the critical temperature, not to the time required for symptom development. Chilling *tolerance*, however, is not an all-or-nothing phenomenon; duration and temperature both contribute to the stress (dose response). In addition, light level, relative humidity, and time of day all affect the amount of stress perceived by the plant. The conditions during the rewarming period are also critical to crop survival. Little is known about how these factors affect chilled roots in particular.

Membrane lipids are a popular candidate for both temperature sensor and the primary site of chilling injury (Lyons, Graham, and Raison, 1979; Raison and Orr, 1990; Murata and Los, 1997). Recently, Murata and Los

FIGURE 3.1. The Effects of Chilling Root Systems on Root and Shoot Processes

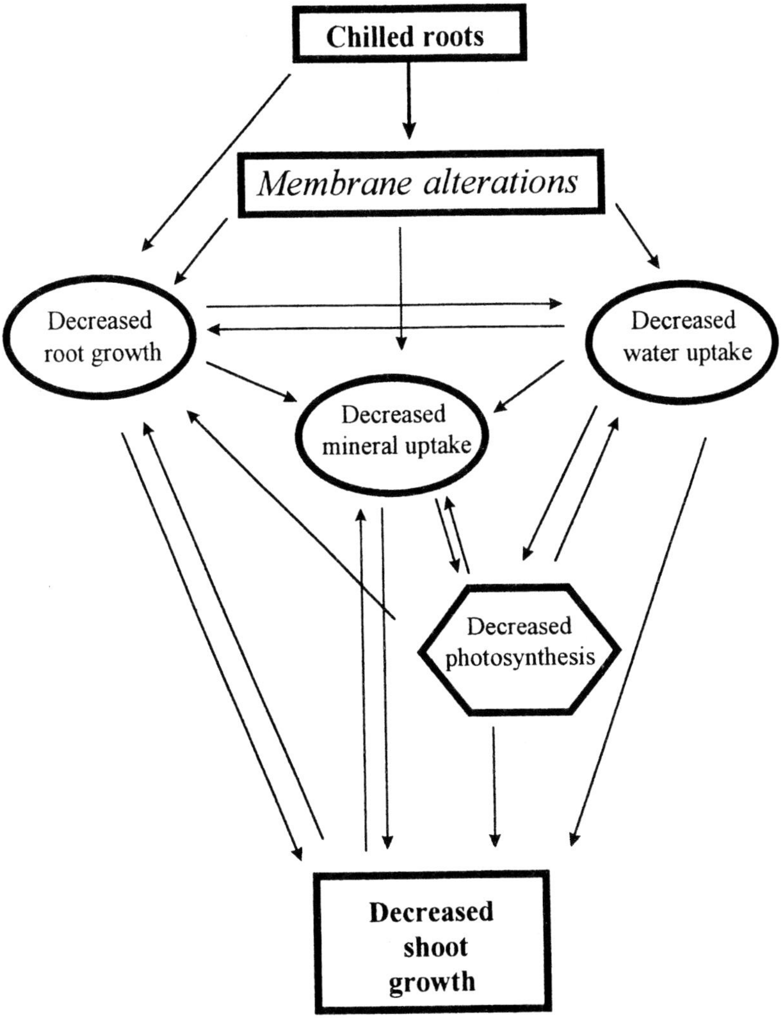

(1997) concluded that the role of membrane fluidity in temperature perception and in transduction of the signal is firmly established. Although membrane lipid composition may not provide the complete explanation of chilling sensitivity in all cases (e.g., Wu and Browse, 1995), many recent studies demonstrate the role of membrane lipids (e.g., Vigh et al., 1993; Kodama et al., 1995; Ishizaki-Nishizawa et al., 1996). Some other mechanisms of chilling injury include reduced water uptake, metabolic imbalances, loss of synchronization (Martino-Catt and Ort, 1992), oxidative stress (Prasad, Anderson, and Stewart, 1994), and toxic by-products. Our model of the mechanisms of root system chilling is based on a primary injury to the membranes, resulting in inhibition of root functions that then inhibits shoot functions (see Figure 3.1).

CHILLING INHIBITION OF ROOT GROWTH

Low soil temperature inhibits root system growth via elongation and biomass accumulation. Limitations of root growth, of course, usually limit shoot functions as well, even when shoots are at optimal temperature (Cumbus and Nye, 1985). Soil temperature is critical for radicle emergence during seed germination and stand establishment. Among twenty genotypes of *Phaseolus vulgaris*, radicle emergence occurred in four days at 12°C, but not until twenty-nine days at 8°C (Dickson and Boettger, 1984). Cucumber radicles were more sensitive as they grew longer, and the root tips were more sensitive than the basal regions (Rab and Saltveit, 1996).

Root elongation rates and lateral development are depressed in sensitive crops. Hydroponically grown maize roots at 20°C grew at 1.2 millimeters per hour (mm/h) over ten days, while those transferred to 5°C had elongation rates of only 0.02 mm/h (Pritchard et al., 1990). Upon rewarming, following a lag phase, growth rate fully recovered at 100 h. Cotton roots at 20°C (cold for cotton) developed no laterals whether shoots were at 20°C or 28°C; chilled taproot length was 60 percent of the warm roots (McMichael and Burke, 1994). Also in cotton, Bradow (1990) found that after seven days at 15°C, root length was 28 percent of the length of roots at 30°C. Scott and Jones (1986) evaluated cold tolerance of several tomato species: root elongation relative growth rate (RGR) at 10°C was less than one-third of the rate at 20°C. Root length was more limited by chilling than hypocotyl length (Scott and Jones, 1986). Even in the cold-tolerant *Brassica napus*, root extension was limited below 9 to 10°C, but lateral branching did not differ between warm and cold roots (Moorby and Nye, 1984; MacDuff et al., 1986).

Dry matter accumulation is also limited in chilled roots. Dry weight of squash and cucumber roots did not increase over four days at 10°C or below (Reyes and Jennings, 1994). After four days at 2 or 6°C, roots failed to regrow when returned to 26°C, indicating irreversible injury. *Phaseolus vulgaris* roots did not gain dry matter during seven days at

10°C, but growth rate fully recovered by the end of seven days of rewarming (Sanders, 1997).

Mechanisms of Growth Inhibition

Cell Expansion

Root tip elongation has been studied extensively as a model system for the biophysics of expansion (Pahlavanian and Silk, 1988; Pritchard et al., 1990; Pritchard, Wyn Jones, and Tomos, 1990). The components of cell expansion are cell turgor, cell wall extensibility, and the yield threshold. Cell wall extensibility limited growth of maize root cells at 5°C, not cell turgor pressure, which actually increased during the first 70 h of chilling (Pritchard et al., 1990). In wheat roots, chilling reduced the wall extensibility and increased the yield threshold; again, turgor was not limiting (Pritchard, Wyn Jones, and Tomos, 1990). In maize roots, cells in the elongating zone that stopped growing during the cold treatment did not grow again when warmed (Pritchard et al., 1990). Recovery of elongation required the production of new cells that would expand.

Maize seedlings grown at low temperatures (15/12°C) had primary roots of greater diameter than warm-grown plants; both stele and cortex were thicker. However, no correlation was found between cold-induced changes in root anatomy and shoot tolerance to chilling among genotypes (Kiel and Stamp, 1992).

Dry Matter Accumulation

The mechanisms by which low temperature inhibits growth in terms of dry matter accumulation are, of course, complex, and vary for leaves, roots, and seed/fruit. Carbon partitioning among source leaves and sinks is greatly affected by chilling, and not in a simple manner. Whether cold occurs during the light or dark period (Martino-Catt and Ort, 1992), the light intensity, the duration of cold stress, whether the roots or the whole plant is chilled, and the timing of onset of the cold period (King, Reid, and Patterson, 1982; King, Joyce, and Reid, 1988) affect photosynthesis and partitioning differently. Over several days, feedback can occur between sinks and source leaves that corrects metabolic imbalances but, for the researcher, masks the original limitation to growth.

Accumulation of carbohydrates in leaves is a common response to chilling. For example, in a tropical C4 grass during cold nights, glucose, sucrose, and starch remained in leaves; at warm temperatures the carbohydrates were transported from the leaves at night (Shatters and West, 1995). Carbohydrate accumulation could be caused by altered leaf carbohydrate metabolism (e.g., starch-degrading enzymes), reduced translocation (phloem loading/unloading), or low sink demand by cold roots. Crawford and Huxter (1977) found reduced soluble-sugar levels in chilled maize roots and noted that the addition of exogenous glucose stimulated root growth and respiration; they concluded that transport of sugars was limiting for chilled maize. In contrast, Paul and colleagues (1991) found increased soluble-carbohydrate levels in sunflower roots early in the chilling treatment, which they interpreted as evidence of reduced sink demand. After the first day of cold, feedback inhibition of photosynthesis partly corrected the source-sink imbalance. In their study, both low sink strength and inhibited translocation were implicated in carbohydrate accumulation in leaves. They also noted that accumulation of sugars in the sink could reduce the turgor pressure gradient that drives translocation. Both root and shoot responses are critical in determining whole-plant carbohydrate metabolism during chilling.

Chilled Roots Affect Shoot Growth

Low root zone temperatures accompanied by warm air temperatures occur in many situations. Differential root-shoot temperature treatments can clarify how root chilling affects shoot responses. For *Phaseolus vulgaris,* a root zone temperature of 10°C for nine days resulted in primary leaves 40 percent the size of leaves from plants with warm roots (Milligan and Dale, 1988). For bell pepper plants, root zone temperature of 12°C resulted in very little shoot growth over thirty-five days (Gosselin and Trudel, 1986). Pine and spruce root growth was almost completely inhibited at 8°C, but shoot growth was optimal when the root zone temperature was 12°C (Vapaavuori, Rikala, and Ryyppö, 1992). These reports emphasize the importance of root temperature to growth of the whole plant. The mechanisms underlying these responses are for the most part undefined; possibilities include water deficits, hormonal signals, or nutrient deficits.

WATER RELATIONS DURING CHILLING

Water Uptake

Water absorption is severely restricted by low temperatures (Kramer, 1940; Bagnall, Wolfe, and King, 1983; Markhart, 1986; Bolger, Upchurch, and McMichael, 1992; Cui and Nobel, 1994). For example, only minimal amounts of water were absorbed by winter wheat at any depth until the soil temperature at that depth reached 10°C (Wraith and Ferguson, 1994). Reduced water absorption by cold roots, coupled with moderate transpiration, can cause loss of leaf turgor (McWilliam, Kramer, and Musser, 1982; Fennell, Li, and Markhart, 1990). Water flux through a root system is limited by lipid membranes (Markhart et al., 1979a; Peterson, Murrmann, and Steudle, 1993). Plasma membranes become less fluid and are less permeable (conductive) to water when chilled (McElhaney, DeGier, and van der Neut-Kok, 1973). (They may become more permeable during freezing injury due to leakage; this is a different situation.) When water uptake is determined during a temperature decline, the effect is immediate: flow rates decrease smoothly as the temperature decreases. Once the target temperature is reached, flow may continue to decrease for another hour or so, then it stabilizes (Smit-Spinks, Swanson, and Markhart, 1984; Sanders, 1997).

The response to potential chilling-induced water deficit is a whole-plant characteristic. When cotton root zone temperature was lowered, leaf water potential and transpiration began to fall at 20 and 13°C, respectively, so that a decline in plant hydraulic conductance was evident around 15°C (Radin, 1990). Chilling tolerance may be due to root or stomatal factors, or both. Most plants usually need to maintain water uptake or to reduce water loss to offset reduced uptake. An interesting exception to this rule occurs in some cacti that require a reduction in cladode water content to survive cold periods, and the chilling-induced decrease in root hydraulic conductivity causes the reduction in water content (Cui and Nobel, 1994).

Several researchers have observed that when exposed to a decrease in root zone temperature, the water uptake and hydraulic conductance of the root system recovers from its initial precipitous decline (see Figure 3.2) (Smit-Spinks, Swanson, and Markhart, 1984; Fennell, 1985; Markhart, 1986; Sanders, 1997). These observations suggest that roots have the ability to acclimate rapidly to changes in temperature. Acclimation is discussed in more detail in the section Root System Acclimation.

Hydraulic and Nonhydraulic Factors

When roots are chilled and leaves are warm, shoot growth or physiology may be limited by hydraulic or nonhydraulic factors. For example, when bean roots are chilled, leaf expansion is inhibited. Whether hydraulic or nonhydraulic factors cause the inhibition is not clear. Expansion of primary leaves of *Phaseolus vulgaris* was inhibited within minutes of root chilling, before changes in leaf water potential could be detected with the pressure chamber (Sattin, Stacciarini Seraphin, and Dale, 1990). In addition, growth remained slow after leaf water status had recovered. When examined with

FIGURE 3.2. Acclimation of Water Uptake by Two Detopped *Phaseolus vulgaris* Root Systems at 10°C

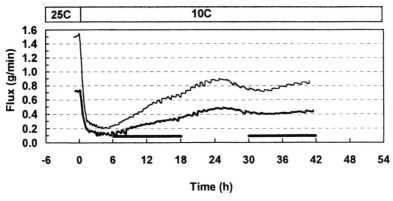

Source: Sanders, 1997, p. 29.

other methods, hydraulic factors appear to be sufficient to account for the inhibition of leaf growth (Malone, 1993; Pardossi, Pritchard, and Tomos, 1994). A similar question exists for photosynthesis in warm leaves with chilled roots: the roles of hydraulic and nonhydraulic signals can be complex. Photosynthesis decreased by more than 90 percent in Scots pine when the roots alone were chilled (DeLucia, Day, and Öquist, 1991). Stomata closed because of nonhydraulic signals or decreased *localized* water potentials, not lower bulk xylem water potential.

Stomatal Conductance

If impaired root function results in a chilling-induced water deficit, stomatal conductance response is crucial. At low temperatures, stomata of some chilling-sensitive species are sometimes slow to respond to the chilling-induced water deficit. Corn, cotton, bean, mung bean, and okra (McWilliam, Kramer, and Musser, 1982; Eamus, Fenton, and Wilson, 1983; Bagnall, Wolfe, and King, 1983; Eamus, 1986; Pardossi, Vernieri, and Tognoni, 1992) suffered from water stress for at least several hours before stomatal closure reduced transpiration sufficiently to restore turgor. Slow closure has been seen in osmotically stressed chilled epidermal strips as well as on whole plants (McWilliam, Kramer, and Musser, 1982). Stomata of mildly dehydrated or ABA-treated bean leaves opened rather than closed when the temperature was lowered from 27° to 10°C (Cornic and Ghashghaie, 1991).

Under optimal conditions, stomatal closure generally occurs within minutes, so this slow response has been called "stomatal locking open." The expectation is that, if possible, plants would reduce transpiration to offset reduced water uptake quickly enough to avoid wilting (McWilliam, Kramer, and Musser, 1982).

However, slow stomatal closure is not always evident in these species. In our experiments at 10°C, *P. vulgaris* stomata did respond to the chilling-induced water deficit with stomatal closure within 20 minutes (min) (Sanders, 1997). This response, however, was not enough to avoid wilting in most cases.

Chilling-tolerant species avoided wilting by closing stomata quickly (e.g., *Pisum sativum*, Eamus, Fenton, and Wilson, 1983) or by maintaining root conductance that was sufficient to offset transpiration without stomatal closure (e.g., *Brassica oleracea*, Markhart et al., 1979a; McWilliam, Kramer, and Musser, 1982). An interesting demonstration of the importance of the whole-plant response is evident in a recent paper describing amelioration of chilling-induced water stress by elevated carbon dioxide (CO_2) (Boese, Wolfe, and Melkonian, 1997). High CO_2 resulted in lower transpiration rates and higher photosynthetic rates during chilling.

Why do the stomata of chilling-sensitive plants not close in response to the chilling-induced water deficit? The major hypotheses focus on abscisic acid (ABA) concentrations in roots and leaves, or the effects of ABA on stomata at low temperatures (Pardossi, Vernieri, and Tognoni, 1992). An ABA pretreatment is expected to cause stomatal closure and ameliorate the chilling stress (McWilliam, Kramer, and Musser, 1982), but in some cases, stomata have opened in response to ABA at lower temperatures (Cornic and Ghashghaie, 1991). On the other hand, cold-induced reduction of the hydraulic conductivity of guard cell plasma membranes could slow the response to the signals that normally induce closure (slower transport of water and ions across the membrane) (McWilliam, Kramer, and Musser, 1982).

MINERAL UPTAKE

The third root system function injured by chilling is mineral uptake. The first day of chilling maize roots, phosphorus, potassium, and nitrogen uptake rates were 15, 20, and 35 percent of the warm rate, respectively, regardless of shoot temperature (Engels, Munkle, and Marschner, 1992). Under long-term (six weeks) exposure to several root zone temperatures, snapdragon growth and nutrient uptake were both optimal with 22°C roots, suggesting that at lower temperatures growth was limited by nutrient uptake (Hood and Mills, 1994). In the long term, shoot nutrient status may be limited by uptake per unit of root or by reduced root growth. (If shoot growth is also slowed by the cold, the reduced uptake rate could be sufficient to meet the demand.) Along with the net uptake rate, nutrient concentration and metabolism in various tissues and translocation rates should be considered when analyzing the effects of low temperatures on nutrient uptake and status. The temperature difference between root and shoot is an important factor in uptake rates and nutrient translocation, especially later in the cold period, as will be described in the Root System Acclimation section.

Nitrogen

Nitrogen uptake, metabolism, and partitioning is altered in chilled roots. In cold-tolerant *Brassica napus,* nitrate uptake by roots at 7°C was half that at 17°C, with shoots at 25°C (MacDuff, Jarvis, and Cockburn, 1994). However, slow growth of warm *Brassica* shoots when roots were at 10°C was not a result of reduced nitrogen uptake (Cumbus and Nye, 1982). When whole *Brassica* and rye plants were chilled, both nitrogen uptake and translocation to the shoots via the xylem were inhibited, causing an increase in root nitrogen concentration that could further inhibit uptake (Lainé et al., 1994). Partitioning of nitrate reductase activity between roots and leaves changes during chilling and affects leaf nitrate concentration (Lawlor et al., 1987). Nitrate reductase activity was greater in cold barley roots (with warm shoots) than in warm roots and remained high during the dark period when warm roots have negligible activity (Deane-Drummond, Clarkson, and Johnson, 1980). Low temperature increased the relative partitioning of nitrogen into roots of *Phaseolus vulgaris* that were dependent on fixing nitrogen (Thomas and Sprent, 1984).

Low soil temperatures alter the relative rates of absorption of ammonium and nitrate (MacDuff and Wild, 1989). Nitrate uptake is more sensitive to low temperatures than ammonium uptake in chilling-tolerant grasses (Clarkson and Warner, 1979). In chilling-sensitive tomato, however, ammonium, not nitrate, absorption was irreversibly inhibited by 4 h of chilling; uptake did not recover when rewarmed (Smart and Bloom, 1991). Neither ammonium nor nitrate uptake were injured in chilling-resistant tomato genotypes. See Smart and Bloom (1991) for an excellent discussion of possible mechanisms of chilling injury to ammonium absorption and the importance of nutrient status during growth in these types of experiments.

Phosphorus and Potassium

Phosphate deficiency can be induced by low soil temperature. In maize, phosphorus uptake was more sensitive to temperature than potassium uptake (greater Q_{10} at chilling temperatures) (Bravo and Uribe, 1981; Engels, Munkle, and Marschner, 1992). Phosphate *uptake* by chilling-tolerant *Brassica* was not affected until root temperature was below $10°C$ (Moorby and Nye, 1984; Cumbus and Nye, 1985). However, phosphate translocation or metabolism apparently was restricted: leaf phosphate concentrations were low, and deficiency symptoms were obvious when roots were chilled, regardless of shoot temperature (Cumbus and Nye, 1985).

Micronutrients

Less is known about micronutrient uptake in cold roots. In maize, translocation of manganese and zinc was reduced by root chilling and was not responsive to demand by warm shoots, leading Engels and Marschner (1996) to conclude that deficiencies could result from low soil temperature. Supply of iron and copper to shoots kept pace with shoot demand in spite of the cold roots (Engels and Marschner, 1996). In snapdragon, copper and molybdenum uptake were unaffected by root temperature, but uptake of other micronutrients declined below $15°C$ (Hood and Mills, 1994).

Nutrition and Chilling

Nutritional deficits can affect how plants respond to chilling stress. Transpiration of nitrogen- or phosphorus-deficient cotton plants began to decline at $25°C$ compared to $13°C$ for fully nourished plants (Radin, 1990). Although the response was specific to root temperature, it was apparently the stomatal conductance that differed between stressed and well-nourished plants.

CIRCADIAN RHYTHMS

Many of the processes important in chilling responses follow diurnal or circadian rhythms, or light/dark regulation. Injury could be caused by loss of synchronization among rhythmic physiological processes. In tomato leaves, low temperature suspends the normal diurnal expression pattern of some genes of carbon metabolism and interferes with turnover of the existing transcripts (Martino-Catt and Ort, 1992). When the plants are rewarmed, gene expression and protein levels are no longer synchronized with the light/dark cycle. We do not yet know if loss of synchronization occurs in chilled roots, or which processes would be affected. Loss of synchronization could cause injury if chilling interfered with the relationship between stomatal rhythms and root water uptake rhythms, transport of carbohydrates from leaves to root systems, nitrogen supply and demand, or hormone levels. Many fascinating questions remain about rhythms in water and nutrient uptake. Do the rhythms in water uptake stop in the cold? What effect might this have? Are there rhythms in enzyme activity or gene expression that could affect water uptake, for example, in desaturases or aquaporins?

Rhythms in Sensitivity

Shoots of many chilling-sensitive species are most sensitive when chilling begins at the end of the dark period (King, Reid, and Patterson, 1982). Diurnal sensitivity was evaluated by the length of time at 2°C required to kill half the plants. (Following Raison and Orr [1990], this should perhaps be described as *least tolerant* rather than *most sensitive*.) Removing the roots of tomato seedlings did not affect the degree or rhythm of chilling injury in tomato shoots (King and Reid, 1987). King and colleagues (1988) suggested that, for tomato shoots, the low carbohydrate levels found at the end of a dark period are critical in the diurnal sensitivity to cold. Whether root systems show a diurnal pattern of tolerance and sensitivity is unknown. Important basic issues yet to be understood include whether the critical temperature for root injury varies diurnally, and whether the time × temperature dose response varies diurnally. If root tolerance/sensitivity follows a rhythm, do similar mechanisms determine rhythms in roots and shoots? Significant interactions occur between roots and shoots during chilling (see Figure 3.1), which suggests many practical implications of rhythms in root sensitivity. For example, are diurnal responses of roots and shoots synchronized in a manner that benefits or harms the whole plant? Given the variety of genotypic responses to chilling, one would expect variation in how rhythms in root sensitivity are integrated into the whole-plant chilling and acclimation responses. For example, if the root system carbohydrate levels were involved in timing of sensitivity, timing of root sensitivity would be expected to differ among species that partition carbon to roots differently.

HORMONES IN CHILLED ROOTS

Abscisic Acid

Increased abscisic acid (ABA) levels have been correlated with chilling tolerance, as well as other environmental stresses, in many tissues, but the exact roles of ABA in tolerance are unknown. Induction of stomatal closure, growth inhibition, and effects on gene expression are all potential mechanisms (Kurkela and Franck, 1990; Anderson et al., 1994; Gusta, Wilen, and Fu, 1996; Janowiak and Dörffling, 1996). For example, endogenous ABA levels increased within hours of chilling in roots, xylem sap, and leaves of a tolerant rice cultivar, but not the more sensitive genotype (Lee, Lur, and Chu, 1993). They suggested that the increase in xylem sap ABA concentration resulted in rapid stomatal closure in the more tolerant cultivar. Janowiak and Dörffling (1996) also concluded that chilling tolerance in maize genotypes involves the ability to accumulate ABA quickly, whether or not a chilling-induced water deficit is involved.

ABA applied to root systems may increase water uptake and hydraulic conductivity at low temperatures in chilling-sensitive species (Markhart et al., 1979b; BassiriRad and Radin, 1992). When roots pretreated with ABA were chilled (with warm shoots), leaf turgor was maintained in spite of *increased* stomatal conductance and lower leaf water potential (Markhart, 1984). Leaf water potential of ABA-treated plants was not as low as nontreated plants. How ABA is involved in root system tolerance or acclimation remains unresolved.

Ethylene

In shoots, and in flower/fruit postharvest physiology, ethylene is a critical hormone. Yet little is known about its role in roots. Ethylene production by leaves, fruits, and flowers often increases following chilling or, in some cases, during chilling (Tong and Yang, 1987; Field and Barrowclough, 1989; Guye, 1989). The magnitude of the increase in ethylene production depends on the duration of chilling and on secondary factors such as wounding or senescence (Field and Barrowclough, 1989). Cucumber radicles produced increasing amounts of ethylene after 48 to 144 h of chilling, with meristematically active apical tissue producing the most ethylene with the least chilling (Rab and Saltveit, 1996). However, the apical tissues lost the ability to produce ethylene after longer periods of chilling, which the authors interpreted as an indication of greater sensitivity.

ROOT SYSTEM ACCLIMATION

Acclimation to chilling temperatures is a potential mechanism of improving tolerance. When exposed to fluctuating temperatures beyond its optimal, a plant that can resume growth and normal functions quickly *at stressful temperatures* would have a selective advantage. The ability to acclimate to air or soil temperatures in the chilling range is especially important during field establishment in the spring and seed filling in late summer.

Little research has explicitly addressed the question of whether root system functions acclimate to chilling. If root growth, water uptake, or mineral uptake acclimated to low temperatures, the stress on the whole plant could be ameliorated. Acclimation of nutrient uptake has received the most attention. Acclimation by roots to other stresses has been described: maize root systems can acclimate to anoxia during pretreatment with hypoxia (Drew, 1997). Roots benefit from whole-plant acclimation during water stress: water uptake is improved by increased carbon partitioning to the roots (Geiger, Koch, and Shieh, 1996). In addition, roots play a part in acclimation of other organs: chilling *Brassica* root systems was as effective as chilling the shoots in inducing leaf acclimation to freezing (Kacperska and Szaniawski, 1993).

Acclimation of Root Growth

Although many papers describe the effects of temperature on root growth, few have looked explicitly at whether growth acclimates to reductions in temperature. Acclimation of growth rate would be indicated by reduced growth early in a chilling period, followed by a higher growth rate later in the chilling period. In *Phaseolus vulgaris,* neither root elongation rate nor dry-weight accumulation acclimated over a week at 10°C (Sanders, 1997). More research into the acclimation abilities of root growth of genotypes of varying cold tolerance is needed. Plants with root systems whose growth rate acclimates in cold soil would have an advantage in stand establishment. Acclimation of root growth could restore sink strength and correct root-shoot imbalances. Restored root growth would improve water and nutrient availability.

Acclimation of Mineral Uptake

After the root zone temperature is changed, nutrient uptake rates increase in temperature responsiveness (Q_{10} values > 2) (Deane-Drummond and Glass, 1983). After several hours or days, macronutrient uptake rates partly recover and become less affected by temperature. Whether this recovery (and reduction in Q_{10}) is acclimation has been debated (Siddiqi, Memon, and Glass, 1984; White, Clarkson, and Earnshaw, 1987; Clarkson et al., 1988; Engels, Munkle, and Marschner, 1992; MacDuff, Jarvis, and Cockburn, 1994; Engels and Marschner, 1996). Is recovery due to a root factor such as more ion transporters in the roots, or improved root hydraulic conductivity, or is it due to increased shoot demand? Shoot:root ratio is a critical factor in understanding the whole-plant response to chilling-induced changes in nutrient uptake. An increased shoot:root ratio can result in increased uptake per unit root due to increased shoot demand, not due to root characteristics (White, Clarkson, and Earnshaw, 1987; Engels, Munkle, and Marschner, 1992).

Evidence regarding acclimation is mixed. Under a limiting nitrate supply, maize plants previously given a 7°C root zone treatment had greater net nitrate uptake at 17°C than 17°C-grown plants, suggesting acclimation (MacDuff, Jarvis, and Cockburn, 1994). However, MacDuff and colleagues noted that the tissue nitrate status is an important factor in whether the apparent acclimation occurs. In another study, after three to five days of low root zone temperature, xylem exudation rates increased in maize plants with warm shoots (higher shoot demand) (Engels, Munkle, and Marschner, 1992). Nitrogen, calcium, and potassium uptake and translocation did not acclimate to low temperature per se, but increased in response to shoot demand (only when shoots were warm). Phosphorus concentration in the exudate and translocation continued to decrease over time when roots were cold, regardless of shoot temperature; phosphorus uptake and translocation did not acclimate to chilling or shoot demand.

Acclimation of Water Uptake and Root Hydraulic Conductance

One of the most intriguing aspects of a plant's responses to chilling is the observation that water uptake rates can partially recover after several hours of chilling (see Figure 3.2) (Smit-Spinks, Swanson, and Markhart, 1984; Fennell, 1985; Markhart, 1986; Sanders, 1997). Uptake increases while the roots are still cold, or, in other words, acclimates to the new environment. Acclimation of water uptake appears to be a general phenomenon in both chilling-tolerant and chilling-sensitive species. Earlier research described greater uptake on the second day at 5°C than on the first in tomato and sunflower, but not bean (Böhning and Lusanandana, 1952). They suggested that changes occurred in the tomato and sunflower tissues that improved water uptake (acclimation). Compared to warm-grown plants, cold-grown soybean and broccoli had higher hydraulic conductivity and/or lower activation energy for uptake when chilled (Markhart et al., 1979a). Acclimation was due to a change in conductance (and in the limiting membranes), not to a change in the driving force. Hydraulic conductance of mung bean plants was higher after five days of chilling than earlier in the cold treatment (Bagnall, Wolfe, and King, 1983). Acclimation has also been examined by measuring flow at warm temperatures after chilling. Rye, barley, and *Brassica rapa* plants given a cool root zone treatment for several days had greater rates of root-pressure-driven exudation at 20°C than warm-grown plants (Clarkson, 1976; Bigot and Boucaud, 1994). However, Bolger and colleagues (1992) found no evidence of acclimation in cotton roots.

Acclimation and Tolerance to Chilling

Is the ability to restore water uptake important to a plant's ability to tolerate chilling? Chilling-sensitive bean roots acclimated as much or more than chilling-tolerant canola (Sanders, 1997). Compared to beans, flux through zinnia roots was less affected by the cold, at first, but did not acclimate as much (Sanders, 1997). Uptake by spinach acclimated to 60 percent of the warm flux within 10 h of the temperature decrease (Fennell, 1985). Flux through Scots pine roots acclimated to 50 percent of the warm rate within 10 h of chilling (Smit-Spinks, Swanson, and Markhart, 1984). The spinach and pine experiments ended within 14 and 10 h, respectively, so we can not be certain that flux did not rise further, as it did in beans. As mentioned, cotton that is very sensitive did not acclimate (Bolger, Upchurch, and McMichael, 1992). Before drawing conclusions about acclimation among species, more plants need to be examined. So far, chilling tolerance of a species does not necessarily correlate with the species' ability to acclimate. During root chilling, a new steady-state water balance is established in the plant at the low temperature; both stomatal water loss and uptake are involved. Species with stomata that close more during chilling may not need to have roots that acclimate to as great a degree as species whose stomata remain more open. On the other hand, cold-tolerant *Brassica oleracea* did not reduce stomatal water loss upon chilling. Instead, the hydraulic conductance of the root system was adequate to maintain the water supply even when chilled (McWilliam, Kramer, and Musser, 1982).

Possible Mechanisms of Acclimation of Water Uptake

Both the uptake rate and the root system hydraulic conductivity acclimate to chilling (Markhart, 1986). The conductivity of the root system is determined for the most part by the structures that limit water flow through the radial pathway. Therefore, temperature-induced changes in conductivity are caused by changes in these structures, namely the root cell plasma membranes. The biochemical basis of acclimation may be an alteration of root membrane composition, such as unsaturation level, that changes the membrane fluid properties. We will come back to membrane lipids later. Other possibilities include root growth, abscisic acid effects on the membranes, or changes in activity of aquaporins.

Growth

Apparent acclimation might be explained if root systems grow significantly or if growth rate acclimates. Uptake might be restored by an increase in root size, and/or new root tissue produced at low temperature may be more conductive to water. When Markhart and colleagues (1980) found chilling-induced changes in lipid composition, it was the roots that grew during chilling that accounted for the change. Root growth could account for increased uptake at low temperatures over longer time periods, but in many crops, growth is probably too slow at low temperatures to account for much short-term acclimation.

Aquaporins

Water-selective channel proteins (aquaporins) are an additional path for water movement in plant tissues (Chrispeels, 1994; Steudle and Henzler, 1995; Maurel, 1997). Although the significance of these channels in overall water flux is still unknown, their presence in plants suggests the intriguing possibility of regulated water transport. The phenomenon of acclimation of water uptake by chilled root systems may be just the type of situation in which aquaporins could be important to the plant. The ability to open water channels when flux through the lipid bilayer decreased would reduce the chilling-induced water stress. Some aquaporins are regulated posttranslationally by phosphorylation (Maurel et al., 1995; Johansson et al., 1996). Aquaporin gene expression has responded to abscisic acid, blue light, salt stress, and desiccation (Yamaguchi-Shinozaki et al., 1992; Kaldenhoff, Kolling, and Richter, 1993; Yamada et al., 1995). Aquaporin channel activity and gene expression would be exciting areas for future research into acclimation mechanisms.

Membrane Unsaturation

Acclimation could occur by alterations in membrane lipids that would increase the permeability to water. Membranes composed of more unsaturated fatty acids (e.g., linolenic acid compared to oleic acid) are more fluid at lower temperatures (McElhaney, DeGier, and van der Neut-Kok, 1973). Unsaturated fatty acids pack less tightly together in the membrane, resulting in less resistance to the movement of water across the membrane (Träuble, 1971). An increase in unsaturation of root plasma membrane lipids would increase membrane fluidity and cause acclimation of root hydraulic conductivity, as well as acclimation of other physiological functions (Markhart, 1986).

Membrane Lipids of the Root System

Because of the potential importance of membrane lipids to root acclimation, we will review what is known about lipid composition and pathways in roots, where many aspects of lipid metabolism remain unclear. Most information about plant lipid synthetic pathways comes from leaf or oilseed studies. Lipid metabolism in nonphotosynthetic tissues has recently been reviewed (Sparace and Kleppinger-Sparace, 1993). Fatty acids are synthesized in plastids. Pea root plastids have all the necessary components of the pathway for fatty acid and glycerolipid biosynthesis from glucose, except adenosine triphosphate (ATP) (Qi, Kleppinger-Sparace, and Sparace, 1995). Root plastids differed from other nongreen plastids and chloroplasts in the amounts of various lipids (Xue et al., 1997). Only 25 percent of the total plastid lipid was MGDG (monogalactosyldiacylglycerol) and DGDG (diagalactosyldiacylglycerol). Another 25 percent was phospholipid, with phosphatidylcholine predominant. Surprisingly, neutral lipids (mainly TAG [triacylglycerol], DAG [diacylglycerol], and free fatty acids) made up 40 percent of the total. Mutant analysis suggests that the chloroplast (prokaryotic) desaturase genes are expressed in root plastids and can contribute 18:2 and 18:3 to both galactolipids and phospholipids (Browse et al., 1993). However, the eukaryotic (nonplastid, microsomal) pathway is the main producer of glycerolipids in roots and seeds (Somerville and Browse, 1991). Mitochondrial lipid metabolism may be important to lipid composition in nonphotosynthetic tissue where plastids are not dominant (Sparace and Kleppinger-Sparace, 1993). Information about how organelles in roots interact for lipid biosynthesis and regulation is particularly lacking (Sparace and Kleppinger-Sparace, 1993).

The phospholipid and fatty acid composition of bulk membranes from roots is well characterized. Phosphatidylcholine and phosphatidylethanolamine are the primary membrane lipids in roots of many species. They are the predominant (70 to 80 percent) lipids in wheat, soybean, winter rye, and *Arabidopsis* roots (Ashworth et al., 1981; Whitman and Travis, 1985; Kinney, Clarkson, and Loughman, 1987; Browse et al., 1993). Within these phospholipid classes, palmitic, linoleic, and linolenic acids (16:0, 18:2, 18:3) accounted for about 25, 50, and 20 percent, respectively, of the fatty acids in roots grown at warm temperatures (see Table 3.1). In *Arabidopsis* roots, less than 10 percent of the polar lipids are MGDG and DGDG (Browse et al., 1993).

Unsaturation, Desaturases, and Chilling

Root lipid unsaturation increased in roots exposed to chilling temperatures in many species (see Table 3.1). Although increased levels of unsaturation have been hypothesized to be involved in root system chilling

TABLE 3.1. Fatty Acid Composition of Roots at Warm and Cold Growth Temperatures (Percentage Composition)

	°C	16:0	16:1	18:0	18:1	18:2	18:3	Reference
Wheat[1]	25	27	nr	2	5	49	16	Ashworth et al., 1981
	10	24	nr	2	6	32	36	
Rye[1]	20	26	0	nr	2	48	23	Clarkson, Hall,
	8	24	0	0.4	3	33	39	and Roberts, 1980
Scots pine[3]	20	27	nr	2	7	58	8	Ryyppö et al., 1994
	5	25	nr	2	11	49	14	
Brassica[2]	25	30	7	14	19	17	14	Kwon, Sanders,
	10	24	9	6	13	11	38	and Markhart, 1996
Brassica[1]	25	16	3	7	10*	22	41	Smolenska and
	5	14	2	5	11*	15	54	Kuiper, 1977
Potato[1]	20	25	nr	nr	6	50	18	Diepenbrock, Müller-Rehbehn, and Sattelmacher, 1989
Arabidopsis[2]	22	22	3	2	4	37	32	Arondel et al., 1992
Arabidopsis[1]	22	22	0	2	16	37	24	Browse et al., 1993
Maize[1†]	nr	45	14	9	12	21	nr	Cowan et al., 1993
Soybean[4]	30	24	nr	7	3	30	36	Whitman and Travis, 1985
Tobacco[2]	25	26	nr	5	2	47	21	Ishizaki-Nishizawa et al., 1996

Note: [1] fatty acid composition of phosphatidylcholine; [2] fatty acid composition of total lipids; [3] fatty acid composition of total phospholipids; [4] plasma membrane-enriched fraction; *18:1 and 16:3; [†]cortex plasma membrane-enriched fraction; nr = not reported.

tolerance and acclimation, their significance is still unclear (Smolenska and Kuiper, 1977; Markhart et al., 1980). For example, linolenic acid levels increased with low temperature in wheat roots, but when the increase was blocked with an inhibitor, it proved not to be crucial to tolerance (Ashworth et al., 1981). We do not know whether the baseline unsaturation level or the ability to raise unsaturation is important. Lipids in chilling-tolerant broccoli roots were more saturated than in chilling-sensitive soybean, but fatty acid unsaturation in broccoli roots increased to a much greater extent than in soybean when chilled (Markhart et al., 1980). Bulk root system fatty acids have been examined more often than specific fractions, such as plasma membranes. The complexity of lipid metabolism means that it will be important to know which phospholipids in which membranes contain the increased unsaturation.

Desaturases

In glycerolipids of nongreen plant tissues, there are about three double bonds/molecules (Browse et al., 1993). As in leaves and seeds, the first fatty acid desaturation in the pathway is carried out by a soluble plastid stearoyl-ACP desaturase. Further desaturations are carried out on the fatty acids once they are incorporated into membrane glycerolipids in either the plastid membrane or the endoplasmic reticulum by membrane-bound desaturases (Roughan, Holland, and Slack, 1980). Much of this pathway has been elucidated by analyzing *Arabidopsis* mutants (Somerville and Browse, 1991; Miquel and Browse, 1992). The endoplasmic reticulum phospholipid desaturase genes *fad2* and *fad3* have been cloned in several species (Arondel et al., 1992; Yadav et al., 1993; Heppard, et al., 1996; Sakamoto and Bryant, 1997). In roots, the desaturase encoded by *fad3* accounts for most of the linolenate (18:3) present (Browse et al., 1993). The phenotype of a *fad3* mutation is strongly expressed in root tissue, and an increase in *fad3* gene dosage in the wild-type *Arabidopsis* increased the level of linolenate (Arondel et al., 1992).

The importance of unsaturation levels and desaturase gene expression in nonroot tissues during chilling has received a great deal of attention (Wada, Gombos, and Murata, 1990; Vigh et al., 1993; Gombos, Wada, and Murata, 1994; Murata and Los, 1997; Sakamoto and Bryant, 1997). Expression in transgenic plants or cyanobacteria of desaturase genes from more cold-tolerant species increased cold tolerance in several cases. For example, Kodama and colleagues (1994, 1995) transformed tobacco with the *Arabidopsis* chloroplast linoleate-palmitoleate desaturase, resulting in a higher molar percentage of trienoic fatty acids in leaves and greater leaf tolerance to chilling compared to wild type. These results suggest an important role for unsaturated lipids in cold tolerance. On the other hand, the *fab1* mutant of *Arabidopsis,* which contains a higher percentage of saturated leaf phosphatidylglycerol than many chilling-sensitive species, was not damaged by short-term chilling treatments (Wu and Browse, 1995). However, after three weeks at $2°C$, there was an extensive loss of chloroplasts and a nearly complete loss of photosynthesis (Wu et al., 1997). Wu and colleagues (1997) propose that unsaturation is critical in maintenance of chloroplasts.

The increase in root system fatty acid unsaturation that occurs after a decrease in temperature suggests that expression of a desaturase gene may be up-regulated by the change in temperature. The ability to increase the level of unsaturation in the membranes could be explained by cold-induced membrane-bound desaturases acting on newly formed or preexisting lipids. Several *fad* (fatty acid desaturase) genes are low-temperature induced (Los et al., 1993; Gibson et al., 1994; Sakamoto and Bryant, 1997). In *Arabidopsis,* a cold-regulated isozyme of the chloroplast ω-3 (*fad8*) has been found (Gibson et al., 1994). In contrast, other *fad* genes in *Arabidopsis* leaves and seeds are not low-temperature regulated (e.g., *fad7,* Iba et al., 1993; *fad2,* Okuley et al., 1994; Heppard et al., 1996). In *Brassica* roots, steady-state *fad3* mRNA level was about six times greater in cold than in warm roots in the dark (Sanders, 1997). However, it seems that the transcript level is fairly constant in the cold and decreases in the warm dark roots, suggesting that mRNA (messenger ribonucleic acid) stability is important, as it is in the cyanobacterium *Synechococcus* (Sakamoto and Bryant, 1997).

Increased unsaturation can result from many other mechanisms. First, increased desaturase activity has been found in leaves (Williams, Khan, and Wong, 1992) and seeds (Cheesbrough, 1990; Garces, Sarmiento, and Mancha, 1992). Second, a differential effect of temperature on the reaction rates of desaturases, fatty acid synthase, acyltransferases, or other enzymes in this complex pathway will also affect the unsaturation level. As Browse and colleagues (1981) have pointed out, if fatty acid synthesis slows and desaturation continues, the proportion of unsaturated fatty acids will increase. In leaves and seeds, desaturation continues at times when synthesis slows or stops, resulting in an increased percentage of unsaturated molecules without a specific increase in desaturase activity. The differential effect of temperature on the rate of fatty acid synthesis relative to the rate of oleate desaturation was enough to explain the temperature-induced changes in the proportion of unsaturated lipids in safflower seed (Browse and Slack, 1983). Third, overall unsaturation level may be altered by changing the proportion of certain lipid classes if the classes differ in fatty acid composition (Roughan, 1985). A fourth possibility is that different acyl transferases, which transfer fatty acid onto glycerolipids, may select for different fatty acids. Tobacco transformed with genes for glycerol-3-phosphate acyltransferases from different species has shown that these genes were critical in determining the unsaturation level of phosphatidylglycerol and tolerance in leaves (Murata et al., 1992). It is not yet known whether these potential mechanisms of increasing unsaturation levels occur in roots or are involved in root system tolerance or acclimation.

Other Lipid Changes in Chilled Roots

In most cases, phospholipid composition of roots does not change with growth temperatures (e.g., Ashworth et al., 1981). Total phospholipid content may increase in cold-grown roots (Kinney, Clarkson, and Loughman, 1987). Phospholipid turnover rates were similar at 5 and 20°C. Evidence from labeling experiments suggests that phospholipid synthesis increased during chilling (Kinney, Clarkson, and Loughman, 1987).

Sterols can have an impact on membrane fluidity, but little is known about their role, especially in roots. Kinney and colleagues (1987) found a decrease in the sterol content in cold-grown rye roots; they hypothesized that the resulting rise in the phospholipid to sterol ratio may help maintain membrane integrity. Chilled mung bean leaves had an increased sterol to phospholipid ratio because of a reduction in lipid content, which was associated with injury (Guye, 1989). In tomato fruit, sterol composition and sterol levels changed during chilling and rewarming (Whitaker, 1994).

Summary of Mechanisms of Acclimation of Water Uptake

Acclimation of water uptake and root hydraulic conductivity requires alterations in the membrane systems that limit water flow into roots. Potential mechanisms of acclimation are changes in membrane lipid fluidity or in water channel proteins (aquaporins). Unsaturation levels increase in chilled roots, and this should increase the root conductance to water. Unsaturation can be regulated at the gene or enzyme level, particularly by desaturases and acyltransferases. These enzymes are important in cold tolerance, and their potential roles in root system acclimation are being investigated.

CONCLUSION

Soil temperature is often below optimum for crop performance. Root growth, water uptake, and mineral uptake are quickly and significantly inhibited by low temperature. Localized injury, such as cessation of root elongation, can have far-reaching consequences, such as reduced nutrient uptake leading to nutrient deficiencies in the shoot. Inhibition of water uptake often causes leaf water deficit, reduced photosynthesis, and reduced shoot growth. As chilling continues over time, some root functions, especially water uptake, have the potential to acclimate, ameliorating the stress to the whole plant. Roots are an excellent system in which to examine the effects of chilling because low temperatures directly affect cell membrane fluidity and fluidity is central to several root functions.

REFERENCES

Anderson, M.D., T.K. Prasad, B.A. Martin, and C.R. Stewart (1994). Differential gene expression in chilling acclimated maize seedlings and evidence for the involvement of abscisic acid in chilling tolerance. *Plant Physiology* 105: 331-339.

Arondel, V., B. Lemieux, I. Hwang, S. Gibson, H.M. Goodman, and C. Somerville (1992). Map-based cloning of a gene controlling ω-3 fatty acid desaturation in Arabidopsis. *Science* 258: 1353-1355.

Ashworth, E.N., M.N. Christiansen, J.B. St. John, and G.W. Patterson (1981). Effect of temperature and BASF 13 338 on the lipid composition and respiration of wheat roots. *Plant Physiology* 67: 711-715.

Bagnall, D., J. Wolfe, and R.W. King (1983). Chill-induced wilting and hydraulic recovery in mung bean plants. *Plant, Cell and Environment* 6: 457-464.

BassiriRad, H. and J.W. Radin (1992). Temperature-dependent water and ion transport properties of barley and sorghum roots. *Plant Physiology* 99: 34-37.

Bigot, J. and J. Boucaud (1994). Low-temperature pretreatment of the root system of *Brassica rapa* L. plants: Effects on the xylem sap exudation and on the nitrate absorption rate. *Plant, Cell and Environment* 17: 721-729.

Bland, W.L. (1993). Cotton and soybean root system growth in three soil temperature regimes. *Agronomy Journal* 85: 906-911.

Boese, S.R., D.W. Wolfe, and J.J. Melkonian (1997). Elevated CO_2 mitigates chilling-induced water stress and photosynthetic reduction during chilling. *Plant, Cell and Environment* 20: 625-632.

Böhning, R.H. and B. Lusanandana (1952). A comparative study of gradual and abrupt changes in root temperature on water absorption. *Plant Physiology* 27: 475-488.

Bolger, T.P., D.R. Upchurch, and B.L. McMichael (1992). Temperature effects on cotton root hydraulic conductance. *Environmental and Experimental Botany* 32: 49-54.

Bollero, G.A., D.G. Bullock, and S.E. Hollinger (1996). Soil temperature and planting date effects on corn yield, leaf area, and plant development. *Agronomy Journal* 88: 385-390.

Bradow, J.M. (1990). Chilling sensitivity of photosynthetic oil seedlings. 1. Cotton and sunflower. *Journal of Experimental Botany* 41: 1585-1593.

Bravo, F.P. and E.G. Uribe (1981). Temperature dependence of the concentration kinetics of absorption of phosphate and potassium in corn roots. *Plant Physiology* 67: 815-819.

Browse, J., M. McConn, D. James, and M. Miquel (1993). Mutants of *Arabidopsis* deficient in the synthesis of α-linolenate: Biochemical and genetic characteristics of the endoplasmic reticulum linoleoyl desaturase. *Journal of Biological Chemistry* 268: 16345-16351.

Browse, J., P.G. Roughan, and C.R. Slack (1981). Light control of fatty acid synthesis and diurnal fluctuations of fatty acid composition of leaves. *Biochemistry Journal* 196: 347-354.

Browse, J. and C.R. Slack (1983). The effects of temperature and oxygen on the rates of fatty acid synthesis and oleate desaturation in safflower *(Carthamus tinctorius)* seed. *Biochimica et Biophysica Acta* 753: 145-152.

Cheesbrough, T.M. (1990). Decreased growth temperature increases soybean stearoyl-acyl carrier protein desaturase activity. *Plant Physiology* 93: 555-559.

Chrispeels, M.J. (1994). Aquaporins: The molecular basis of facilitated water movement through living plant cells. *Plant Physiology* 105: 9-13.

Clarkson, D.T (1976). The influence of temperature on the exudation of xylem sap from detached root systems of rye *(Secale cereale)* and barley *(Hordeum vulgare)*. *Planta* 132: 297-304.

Clarkson, D.T., M.J. Earnshaw, P.J. White, and H.D. Cooper (1988). Temperature dependent factors influencing nutrient uptake: An analysis of responses at different levels of organization. In *Plants and Temperature,* eds. S.P. Long and F.I. Woodward. Cambridge, UK: Company of Biologists, pp. 281-309.

Clarkson, D.T., K.C. Hall, and J.K.M. Roberts (1980). Phospholipid composition and fatty acid desaturation in the roots of rye during acclimatization of low temperature: Positional analysis of fatty acids. *Planta* 149: 464-471.

Clarkson, D.T. and A.J. Warner (1979). Relationships between root temperature and the transport of ammonium and nitrate ions by Italian and perennial ryegrass *(Lolium multiflorum* and *Lolium perenne)*. *Plant Physiology* 64: 557-561.

Cornic, G. and J. Ghashghaie (1991). Effect of temperature on net CO_2 assimilation and photosystem II quantum yield of electron transfer of French bean *(Phaseolus vulgaris* L.) leaves during drought stress. *Planta* 185: 255-260.

Cowan, D.S.C., D.T. Cooke, D.T. Clarkson, and J.L. Hall (1993). Lipid composition of plasma membranes isolated from stele and cortex of maize roots. *Journal of Experimental Botany* 44: 991-994.

Crawford, R.M.M. and T.J. Huxter (1977). Root growth and carbohydrate metabolism at low temperatures. *Journal of Experimental Botany* 28: 917-925.

Cui, M. and P.S. Nobel (1994). Water budgets and root hydraulic conductivity of opuntias shifted to low temperatures. *International Journal of Plant Science* 155: 167-172.

Cumbus, I.P. and P.H. Nye (1982). Root zone temperature effects on growth and nitrate absorption in rape *(Brassica napus* cv. Emerald). *Journal of Experimental Botany* 33: 1138-1146.

Cumbus, I.P. and P.H. Nye (1985). Root zone temperature effects on growth and phosphate absorption in rape (*Brassica napus* cv. Emerald). *Journal of Experimental Botany* 36: 219-227.

Deane-Drummond, C.E, D.T. Clarkson, and C.B. Johnson (1980). The effect of differential root and shoot temperature on the nitrate reductase activity, assayed *in vivo* and *in vitro* in roots of *Hordeum vulgare* (barley): Relationship with diurnal changes in endogenous malate and sugar. *Planta* 148: 455-461.

Deane-Drummond, C.E. and A.D.M. Glass (1983). Compensatory changes in ion fluxes into barley (*Hordeum vulgare* L. cv. Betzes) seedlings in response to differential root/shoot temperatures. *Journal of Experimental Botany* 34: 1711-1719.

DeLucia, E.H., T.A. Day, and G. Öquist (1991). The potential for photoinhibition of *Pinus sylvestris* L. seedlings exposed to high light and low soil temperature. *Journal of Experimental Botany* 42: 611-617.

Dickson, M.H. and M.A. Boettger (1984). Emergence, growth, and blossoming of bean (*Phaseolus vulgaris*) at suboptimal temperatures. *Journal of the American Society for Horticultural Science* 109: 257-260.

Diepenbrock, W., A. Müller-Rehbehn, and B. Sattelmacher (1989). Fatty acid composition of root membrane lipids from two potato (*Solanum tuberosum* L.) genotypes differing in heat tolerance as affected by supraoptimal root zone temperature. *Agrochimica* 33: 478-484.

Drew, M.C. (1997). Oxygen deficiency and root metabolism: Injury and acclimation under hypoxia and anoxia. *Annual Review of Plant Physiology and Plant Molecular Biology* 48: 223-250.

Eamus, D. (1986). The responses of leaf water potential and leaf diffusive resistance to abscisic acid, water stress and low temperature in *Hibiscus esculentus:* The effect of water stress and ABA pretreatments. *Journal of Experimental Botany* 37: 1854-1862.

Eamus, D., R. Fenton, and J.M. Wilson (1983). Stomatal behaviour and water relations of chilled *Phaseolus vulgaris* L. and *Pisum sativum* L. *Journal of Experimental Botany* 34: 434-441.

Engels, C. and H. Marschner (1996). Effects of suboptimal root zone temperatures and shoot demand on net translocation of micronutrients from the roots to the shoot of maize. *Plant and Soil* 186: 311-320.

Engels, C., L. Munkle, and H. Marschner (1992). Effect of root zone temperature and shoot demand on uptake and xylem transport of macronutrients in maize (*Zea mays* L.). *Journal of Experimental Botany* 43: 537-547.

Fennel, A (1985). Cold acclimation and water relations of *Spinacia oleracea* L. PhD Dissertation, University of Minnesota, St. Paul, MN.

Fennell, A., P.H. Li, and A.H. Markhart III (1990). Influence of air and soil temperature on water relations and freezing tolerance of spinach *(Spinacia oleracea)*. Physiologia Plantarum 78: 51-56.

Field, R.J. and P.M. Barrowclough (1989). Temperature-induced changes in ethylene and implications for post-harvest physiology. In: *Biochemical and Physiological Aspects of Ethylene Production in Lower and Higher Plants,* eds.

H. Clijsters, M. DeProft, and R. Marcelle. Dordrecht, The Netherlands: Kluwer, pp. 191-199.

Garces, R., C. Sarmiento, and M. Mancha (1992). Temperature regulation of oleate desaturase in sunflower (*Helianthus annuus* L.) seeds. *Planta* 186: 461-465.

Geiger, D.R., K.E. Koch, and W.-J. Shieh (1996). Effect of environmental factors on whole plant assimilate partitioning and associated gene expression. *Journal of Experimental Botany* 47: 1229-1238.

Gibson, S., V. Arondel, K. Iba, and C. Somerville (1994). Cloning of a temperature-regulated gene encoding a chloroplast ω-3 desaturase from *Arabidopsis thaliana*. *Plant Physiology* 106: 1615-1621.

Gombos, Z., H. Wada, and N. Murata (1994). The recovery of photosynthesis from low-temperature photoinhibition is accelerated by the unsaturation of membrane lipids: A mechanism of chilling tolerance. *Proceedings of the National Academy of Sciences, USA* 91: 8787-8791.

Gosselin, A. and M.-J. Trudel (1986). Root zone temperature effects on pepper. *Journal of the American Society for Horticultural Science* 111: 220-224.

Gusta, L.V., R.W. Wilen, and P. Fu (1996). Low-temperature stress tolerance—The role of abscisic acid, sugars, and heat-stable proteins. *HortScience* 31: 39-46.

Guye, M.G (1989). Phospholipid, sterol composition and ethylene production in relation to choline-induced chilling tolerance in mung bean (*Vigna radiata* L. Wilcz.) during a chill-warm cycle. *Journal of Experimental Botany* 40: 369-374.

Heppard, E.P., A.J. Kinney, K.L. Stecca, and G.-H. Miao (1996). Developmental and growth temperature regulation of two different microsomal ω-6 desaturase genes in soybeans. *Plant Physiology* 110: 311-319.

Hood, T.M. and H.A. Mills (1994). Root zone temperature affects nutrient uptake and growth of snapdragon. *Journal of Plant Nutrition* 17: 279-291.

Iba, K., S. Gibson, T. Nishiuchi, T. Fuse, M. Nishimura, V. Arondel, S. Hugly, and C. Somerville (1993). A gene encoding a chloroplast ω-3 fatty acid desaturase complements alterations in fatty acid desaturation and chloroplast copy number of the *fad7* mutant of *Arabidopsis thaliana*. *Journal of Biological Chemistry* 268: 24099-24105.

Ishizaki-Nishizawa, O., T. Fujii, M. Azuma, K. Sekiguchi, N. Murata, T. Ohtani, and T. Toguri (1996). Low-temperature resistance of higher plants is significantly enhanced by a nonspecific cyanobacterial desaturase. *Nature Biotechnology* 14: 1003-1006.

Janowiak, F. and K. Dörffling (1996). Chilling of maize seedlings: Changes in water status and abscisic acid content in ten genotypes differing in chilling tolerance. *Journal of Plant Physiology* 147: 582-588.

Johansson, I., C. Larsson, B. Ek, and P. Kjellbom (1996). The major integral proteins of spinach leaf plasma membranes are putative aquaporins and are phosphorylated in response to Ca^{2+} and apoplastic water potential. *Plant Cell* 8: 1181-1191.

Kacperska, A. and R.K. Szaniawski (1993). Frost resistance and water status of winter rape leaves as affected by differential shoot/root temperature. *Physiologia Plantarum* 89: 775-782.

Kaldenhoff, R., A. Kolling, and G. Richter (1993). A novel blue light- and abscisic acid-inducible gene of *Arabidopsis thaliana* encoding an intrinsic membrane protein. *Plant Molecular Biology* 23: 1187-1198.

Kiel, C. and P. Stamp (1992). Internal root anatomy of maize seedlings (*Zea mays* L.) as influenced by temperature and genotype. *Annals of Botany* 70: 125-128.

King, A.I., D.C. Joyce, and M.S. Reid (1988). Role of carbohydrates in diurnal chilling sensitivity of tomato seedlings. *Plant Physiology* 86: 764-768.

King, A.I. and M.S. Reid (1987). Diurnal chilling sensitivity and desiccation in seedlings of tomato. *Journal of the American Society for Horticultural Science* 112: 821-824.

King, A.I., M.S. Reid, and B.D. Patterson (1982). Diurnal changes in the chilling sensitivity of seedlings. *Plant Physiology* 70: 211-214.

Kinney, A.J., D.T. Clarkson, and B.C. Loughman (1987). Phospholipid metabolism and plasma membrane morphology of warm and cool rye roots. *Plant Physiology and Biochemistry* 25: 769-774.

Kodama, H., T. Hamada, G. Horiguchi, M. Nishimura, and K. Iba (1994). Genetic enhancement of cold tolerance by expression of a gene for chloroplast ω-3 fatty acid desaturase in transgenic tobacco. *Plant Physiology* 105: 601-605.

Kodama, H., G. Horiguchi, T. Nishiuchi, M. Nishimura, and K. Iba (1995). Fatty acid desaturation during chilling acclimation is one of the factors involved in conferring low-temperature tolerance to young tobacco leaves. *Plant Physiology* 107: 1177-1185.

Kramer, P.J. (1940). Root resistance as a cause of decreased water absorption by plants at low temperature. *Plant Physiology* 15: 63-79.

Kurkela, S. and M. Franck (1990). Cloning and characterization of a cold- and ABA-inducible *Arabidopsis* gene. *Plant Molecular Biology* 15: 137-144.

Kwon, S.W., P.L. Sanders, and A.H. Markhart III (1996). Fatty acid unsaturation and the expression of a fatty acid desaturase gene during root acclimation to chilling in *Brassica napus* (canola). *Acta Horticulturae* 434: 237-240.

Lainé, P., J. Bigot, A. Ourry, and J. Boucaud (1994). Effects of low temperature on nitrate uptake, and xylem and phloem flows of nitrogen, in *Secale cereale* L. and *Brassica napus* L. *New Phytologist* 127: 675-683.

Lawlor, D.W., F.A. Boyle, A.C. Kendall, and A.J. Keys (1987). Nitrate nutrition and temperature effects on wheat: Enzyme composition, nitrate and total amino acid content of leaves. *Journal of Experimental Botany* 38: 378-392.

Lee, T.-M., H.-S. Lur, and C. Chu (1993). Role of abscisic acid in chilling tolerance of rice (*Oryza sativa* L.) seedlings. I. Endogenous abscisic acid levels. *Plant, Cell and Environment* 16: 481-490.

Los, D., I. Horvath, L. Vigh, and N. Murata (1993). The temperature dependent expression of the desaturase gene *desA* in *Synechocystis* PCC6803. *FEBS Letters* 318: 57-60.

Lyons, J.M., D. Graham, and J.K. Raison, eds (1979). *Low Temperature Stress in Crop Plants: The Role of the Membrane.* New York: Academic Press.

MacDuff, J.H., S.C. Jarvis, and J.E. Cockburn (1994). Acclimation of NO_3^- fluxes to low root temperature by *Brassica napus* in relation to NO_3^- supply. *Journal of Experimental Botany* 45: 1045-1056.

MacDuff, J.H. and A. Wild (1989). Interactions between root temperature and nitrogen deficiency influence preferential uptake of NO_3^- and NH_4^+ by oilseed rape. *Journal of Experimental Botany* 40: 195-206.

MacDuff, J.H., A. Wild, M.J. Hopper, and M.S. Dhanoa (1986). Effects of temperature on parameters of root growth relevant to nutrient uptake: Measurements on oilseed rape and barley grown in flowing nutrient solution. *Plant and Soil* 94: 321-332.

Malone, M. (1993). Rapid inhibition of leaf growth by root cooling in wheat: Kinetics and mechanism. *Journal of Experimental Botany* 44: 1663-1669.

Markhart, A.H., III (1984). Amelioration of chilling-induced water stress by abscisic acid-induced changes in root hydraulic conductance. *Plant Physiology* 74: 81-83.

Markhart, A.H., III (1986). Chilling injury: A review of possible causes. *HortScience* 21: 1329-1333.

Markhart, A.H., III, E.L. Fiscus, A.W. Naylor, and P.J. Kramer (1979a). Effect of temperature on water and ion transport in soybean and broccoli systems. *Plant Physiology* 64: 83-87.

Markhart, A.H., III, E.L. Fiscus, A.W. Naylor, and P.J. Kramer (1979b). The effect of abscisic acid on root hydraulic conductivity. *Plant Physiology* 64: 611-614.

Markhart, A.H., III, M.M. Peet, N. Sionit, and P.J. Kramer (1980). Low temperature acclimation of root fatty acid composition, leaf water potential, gas exchange and growth of soybean seedlings. *Plant, Cell and Environment* 3: 435-441.

Marschner, H. (1995). *Mineral Nutrition of Higher Plants*. London: Academic Press.

Martino-Catt, S. and D.R. Ort (1992). Low temperature interrupts circadian regulation of transcriptional activity in chilling-sensitive plants. *Proceedings of the National Academy of Sciences, USA* 89: 3731-3735.

Maurel, C. (1997). Aquaporins and water permeability of plant membranes. *Annual Review of Plant Physiology and Plant Molecular Biology* 48: 399-429.

Maurel, C., R.T. Kado, J. Guern, and M.J. Chrispeels (1995). Phosphorylation regulates the water channel activity of the seed-specific aquaporin α-TIP. *EMBO Journal* 14: 3028-3035.

McElhaney, R.N., J. DeGier, and E.C.M. van der Neut-Kok (1973). The effect of alterations in fatty acid composition and cholesterol content on the nonelectrolyte permeability of *Acholeplasma laidlawii* B cells and derived liposomes. *Biochimica et Biophysica Acta* 298: 500-512.

McMichael, B.L. and J.J. Burke (1994). Metabolic activity of cotton roots in response to temperature. *Environmental and Experimental Botany* 34: 201-206.

McWilliam, J.R., P.J. Kramer, and R.L. Musser (1982). Temperature-induced water stress in chilling-sensitive plants. *Australian Journal of Plant Physiology* 9: 343-352.

Milligan, S.P. and J.E. Dale (1988). The effects of root treatments on growth of the primary leaves of *Phaseolus vulgaris* L.: General features. *New Phytologist* 108: 27-35.

Miquel, M. and J. Browse (1992). *Arabidopsis* mutants deficient in polyunsaturated fatty acid synthesis: Biochemical and genetic characterization of a plant oleoyl-phosphatidylcholine desaturase. *Journal of Biological Chemistry* 267: 1502-1509.

Moorby, H. and P.H. Nye (1984). The effect of temperature variation over the root system on root extension and phosphate uptake by rape. *Plant and Soil* 78: 283-293.

Murata, N., O. Ishizaki-Nishizawa, S. Higashi, H. Hayashi, Y. Tasaka, and I. Nishida (1992). Genetically engineered alteration in the chilling sensitivity of plants. *Nature* 356: 710-713.

Murata, N. and D.A. Los (1997). Membrane fluidity and temperature perception. *Plant Physiology* 115: 875-879.

Okuley, J., J. Lightner, K. Feldmann, N. Yadav, E. Lark, and J. Browse (1994). *Arabidopsis fad2* gene encodes the enzyme that is essential for polyunsaturated lipid synthesis. *The Plant Cell* 6: 147-158.

Pahlavanian, A.M. and W.K. Silk (1988). Effect of temperature on spatial and temporal aspects of growth in the primary maize root. *Plant Physiology* 87: 529-532.

Pardossi, A., J. Pritchard, and A.D. Tomos (1994). Leaf illumination and root cooling inhibit bean leaf expansion by decreasing turgor pressure. *Journal of Experimental Botany* 45: 415-422.

Pardossi, A., P. Vernieri, and F. Tognoni (1992). Involvement of abscisic acid in regulating water status in *Phaseolus vulgaris* L. during chilling. *Plant Physiology* 100: 1243-1250.

Paul, M.J., S.P. Driscoll, and D.W. Lawlor (1991). The effect of cooling on photosynthesis, amounts of carbohydrate and assimilate export in sunflower. *Journal of Experimental Botany* 42: 845-852.

Peterson, C.A., M. Murrmann, and E. Steudle (1993). Location of the major barriers to water and ion movement in young roots of *Zea mays* L. *Planta* 190: 127-136.

Prasad, T.K., M.D. Anderson, and C.R. Stewart (1994). Acclimation, hydrogen peroxide, and abscisic acid protect mitochondria against irreversible chilling injury in maize seedlings. *Plant Physiology* 105: 619-627.

Pritchard, J., P.W. Barlow, J.S. Adam, and A.D. Tomos (1990). Biophysics of the inhibition of the growth of maize roots by lowered temperature. *Plant Physiology* 93: 222-230.

Pritchard, J., R.G. Wyn Jones, and A.D. Tomos (1990). Measurement of yield threshold and cell wall extensibility of intact wheat roots under different ionic, osmotic and temperature treatments. *Journal of Experimental Botany* 41: 669-675.

Qi, Q., K.F. Kleppinger-Sparace, and S.A. Sparace (1995). The utilization of glycolytic intermediates as precursors for fatty acid biosynthesis by pea root plastids. *Plant Physiology* 107: 413-419.

Rab, A. and M.E. Saltveit (1996). Differential chilling sensitivity in cucumber (*Cucumis sativus*) seedlings. *Physiologia Plantarum* 96: 375-382.

Radin, J.W. (1990). Responses of transpiration and hydraulic conductance to root temperature in nitrogen- and phosphorus-deficient cotton seedlings. *Plant Physiology* 92: 855-857.

Raison, J.K. and G.R. Orr (1990). Proposals for a better understanding of the molecular basis of chilling injury. In *Chilling Injury of Horticultural Crops*, ed. C.Y. Wang. Boca Raton, FL: CRC Press, pp.145-164.

Reyes, F. and P.H. Jennings (1994). Response of cucumber (*Cucumis sativus* L.) and squash (*Cucurbita pepo* L. var. *melopepo*) roots to chilling stress during early stages of seedling development. *Journal of the American Society for Horticultural Science* 119: 964-970.

Roughan, P.G (1985). Phosphatidylglycerol and chilling sensitivity in plants. *Plant Physiology* 77: 740-746.

Roughan, P.G., R. Holland, and C.R. Slack (1980). The role of chloroplasts and microsomal fractions in polar lipid synthesis from [1-^{14}C] acetate by cell-free preparations from spinach *(Spinacia oleracea)* leaves. *Biochemistry Journal* 188: 17-24.

Ryyppö, A., E.M. Vapaavuori, R. Rikala, and M.-L. Sutinen (1994). Fatty acid composition of microsomal phospholipids and H$^+$-ATPase activity in the roots of Scots pine seedlings grown at different root temperatures during flushing. *Journal of Experimental Botany* 45: 1533-1539.

Sakamoto, T. and D.A. Bryant (1997). Temperature-regulated mRNA accumulation and stabilization for fatty acid desaturase genes in the cyanobacterium *Synechococcus* sp. strain PCC 7002. *Molecular Microbiology* 23: 1281-1292.

Sanders, P.L. (1997). Acclimation to chilling: Plant water relations, growth and linoleate desaturase gene expression. PhD Dissertation, University of Minnesota, St. Paul, MN.

Sattin, M., E. Stacciarini Seraphin, and J.E. Dale (1990). The effects of root cooling and the light-dark transition on growth of primary leaves of *Phaseolus vulgaris* L. *Journal of Experimental Botany* 41: 1319-1324.

Scott, S.J. and R.A. Jones (1986). Cold tolerance in tomato. II. Early seedling growth of *Lycopersicon* spp. *Physiologia Plantarum* 66: 659-663.

Shatters, R.G. Jr. and S.H. West (1995). Response of *Digitaria decumbens* leaf carbohydrate levels and glucan degrading enzymes to chilling night temperature. *Crop Science* 35: 516-523.

Siddiqi, M.Y., A.R. Memon, and A.D.M. Glass (1984). Regulation of K+ influx in barley. Effect of low temperature. *Plant Physiology* 74: 730-734.

Smart, D.R. and A.J. Bloom (1991) Influence of root NH$_4^+$ and NO$_3^-$ content on the temperature response of net NH$_4^+$ and NO$_3^-$ uptake in chilling sensitive and chilling resistance *Lycopersicon* taxa. *Journal of Experimental Botany* 42: 331-338.

Smit-Spinks, B., B.T. Swanson, and A.H. Markhart III (1984). Changes in water relations, water flux, and root exudate abscisic acid content with cold acclimation of *Pinus sylvestris* L. *Australian Journal of Plant Physiology* 11: 431-441.

Smolenska, G. and P.J.C. Kuiper (1977). Effect of low temperature upon lipid and fatty acid composition of roots and leaves of winter rape plants. *Physiologia Plantarum* 41: 29-35.

Somerville, C. and J. Browse (1991). Plant lipids: Metabolism, mutants and membranes. *Science* 252: 80-87.

Sparace, S.A. and K.F. Kleppinger-Sparace (1993). Metabolism in nonphotosynthetic, nonoilseed tissues. In *Lipid Metabolism in Plants,* ed. T.S. Moore Jr. Boca Raton, FL: CRC Press, pp. 569-589.

Steudle, E. and T. Henzler (1995). Water channels in plants: Do basic concepts of water transport change? *Journal of Experimental Botany* 46: 1067-1076.

Thomas, R.J. and J.I. Sprent (1984). The effects of temperature on vegetative and early reproductive growth of a cold-tolerant and a cold-sensitive line of *Phaseolus vulgaris* L. 1. Nodulation, growth and partitioning of dry matter, carbon, and nitrogen. *Annals of Botany* 53: 579-588.

Tong, C.B.S. and S.F. Yang (1987). Chilling-induced ethylene production by beans and peas. *Journal of Plant Growth Regulation* 6: 201-208.

Träuble, H (1971). The movement of molecules across lipid membranes: A molecular theory. *Journal of Membrane Biology* 4: 193-208.

Vapaavuori, E.M., R. Rikala, and A. Ryyppö (1992). Effects of root temperature on growth and photosynthesis in conifer seedlings during shoot elongation. *Tree Physiology* 10: 217-230.

Vigh, L., D.A. Los, I. Horvath, and N. Murata (1993). The primary signal in the biological perception of temperature: Pd-catalyzed hydrogenation of membrane lipids stimulated the expression of the *desA* gene in *Synechocystis* PCC6803. *Proceedings of the National Academy of Sciences, USA* 90: 9090-9094.

Wada, H., Z. Gombos, and N. Murata (1990). Enhancement of chilling tolerance of a cyano-bacterium by genetic manipulation of fatty acid desaturation. *Nature* 347: 200-203.

Whitaker, B.D (1994). Lipid changes in mature-green tomatoes during ripening, during chilling, and after rewarming subsequent to chilling. *Journal of the American Society for Horticultural Science* 119: 994-999.

White, P.J., D.T. Clarkson, and M.J. Earnshaw (1987). Acclimation of potassium influx in rye *(Secale cereale)* to low temperatures. *Planta* 171: 377-385.

Whitman, C.E. and R.L. Travis (1985). Phospholipid composition of a plasma membrane-enriched fraction from developing soybean roots. *Plant Physiology* 79: 494-498.

Williams, J.P., M.U. Khan, and D. Wong (1992). Low temperature-induced fatty acid desaturation in *Brassica napus:* Thermal deactivation and reactivation of the process. *Biochimica et Biophysica Acta* 1128: 275-279.

Wraith, J.M. and A.H. Ferguson (1994). Soil temperature limitation to water use by field-grown winter wheat. *Agronomy Journal* 86: 974-979.

Wu, J. and J. Browse (1995). Elevated levels of high-melting-point phosphatidylglycerols do not induce chilling sensitivity in a mutant of *Arabidopsis. Plant Cell* 7: 17-27.

Wu, J., J. Lightner, N. Warwick, and J. Browse (1997). Low-temperature damage and subsequent recovery of fab1 mutant *Arabidopsis* exposed to 2°C. *Plant Physiology* 113: 347-356.

Xue, L., L.M. McCune, K.F. Kleppinger-Sparace, M.J. Brown, M.K. Pomeroy, and S.A. Sparace (1997). Characterization of the glycerolipid composition and biosynthetic capacity of pea root plastids. *Plant Physiology* 113: 549-557.

Yadav, N.S., A. Wierzbicki, M. Aegerter, C.S. Caster, L. Perez-Grau, A.J. Kinney, W.D. Hitz, R. Booth Jr., B. Schweiger, K.L. Stecca, et al. (1993). Cloning of higher plant ω-3 fatty acid desaturases. *Plant Physiology* 103: 467-476.

Yamada, S., M. Katsuhara, W.B. Kelly, C.B. Michalowski, and H.J. Bohnert (1995). A family of transcripts encoding water channel proteins: Tissue-specific expression in the common ice plant. *Plant Cell* 7: 1129-1142.

Yamaguchi-Shinozaki, K., M. Koizumi, S. Urao, and K. Shinozaki (1992). Molecular cloning and characterization of 9 cDNAs for genes that are responsive to desiccation in *Arabidopsis thaliana*: Sequence analysis of one cDNA clone that encodes a putative transmembrane channel protein. *Plant Cell Physiology* 33: 217-224.

Chapter 4

Mechanisms of Cold Acclimation

Jean-Marc Ferullo
Marilyn Griffith

INTRODUCTION

To survive winter, plants that live in the temperate regions of the Earth undergo adaptive changes as early as the middle of summer. The most obvious changes are morphological and involve the partial or total loss of aerial organs, and the formation of specialized organs, such as buds and tubers. However, more subtle changes occur in the remaining, overwintering plant parts, allowing them to go from a cold-sensitive to a cold-tolerant state. This acquisition of low-temperature tolerance is usually referred to as cold acclimation. In regions where freezing temperatures last for prolonged periods, freezing avoidance strategies are inappropriate, and plants actually develop freezing tolerance, that is, a mechanism to withstand ice formation within their tissues. Intracellular ice formation is invariably lethal. Therefore, the freezing process within living organisms that tolerate the presence of ice within their tissues involves the formation of extracellular ice at a high subzero temperature. During freezing, as much as 70 to 80 percent of the liquid water in the tissue is frozen as extracellular ice, resulting in cellular dehydration (Levitt, 1980). The abilities of organisms to modify the growth of ice and to withstand extensive dehydration are key elements of cellular survival. In some cases, extracellular freezing is accompanied by a second mechanism known as deep supercooling, in which water is maintained in the liquid state, even at temperatures far below the heterogeneous nucleation point (Burke et al., 1976). Deep supercooling has only been observed in some woody angiosperms and is limited to certain cell types or organs, such as xylem ray parenchyma and floral buds.

It is difficult to determine the lowest temperature that a given plant can endure. This temperature is generally assessed by measuring the LT50 of a population, that is, the temperature at which 50 percent of the population dies in a freezing assay performed at a given cooling rate (typically $-1°C$/hour [h]). The majority of crop plants are freezing sensitive during the summer. As plants are exposed to low temperatures (2 to 5°C) and/or short daylengths, the LT50 progressively decreases during a few days or weeks until the plants reach their maximal freezing tolerance. At the same time, amino acids, carbohydrates, nucleic acids, lipids, and some phyto-hormones accumulate in plant tissues (Li, 1984). These modifications are accompanied by changes in enzyme activities, isozymic patterns (Levitt, 1980), and gene expression (Walbot and Hahn, 1989; Guy, 1990; Caston-guay, Nadeau, and Laberge, 1993). In some species, it has been demon-strated that subsequent exposure to mild freezing temperatures ($-2°C$) is necessary to reach the fully acclimated state (Kacperska, 1993). Albeit far less documented, this second "overacclimation" phase was recently shown to involve additional changes in gene expression (Castonguay, Nadeau, and Laberge, 1993). The significance of these phenomena with respect to the development of cold tolerance is still far from being totally under-stood.

It has proven difficult to establish causal relationships between the metabolic and molecular changes that occur during cold acclimation and the development of freezing tolerance. Most experiments attempt to corre-late changes in biochemical composition or gene expression with changes in the LT50 of the plants. However, in order to reach the LT50, plants first have to survive exposure to cold temperatures. Plants that are lethally injured by exposure to temperatures in the range of 0 to 10°C are consid-ered chilling sensitive. As the temperature is lowered below the freezing point, which is generally near 0°C, most of the liquid water in the plant freezes. Plants that are lethally injured by this initial freezing event are considered to be freezing sensitive. As the temperature is lowered further, additional water is drawn from the cells to the intercellular ice. Only the plants that survive the initial formation of ice are freezing tolerant, and their killing point (LT50) may be determined by the degree of dehydration the cells can withstand. Because the development of freezing tolerance involves resistance to events that occur at temperatures above the LT50, many of the components of freezing tolerance may not be directly corre-lated with LT50 and must be shown, instead, to have a proven function in influencing survival of the freezing process. A change at the molecular level may actually play a role in increasing freezing tolerance if it is proven to be involved in at least one of the following functions:

- Lowering the freezing temperature of tissues
- Increasing the deep supercooling ability of the tissues

- Promoting and/or modifying extracellular ice formation
- Limiting the extent of freeze-induced cell desiccation
- Preserving protein and membrane functionality at very low water potential or restoring it upon rehydration
- Adjusting cellular metabolism to low-temperature conditions

This chapter describes the molecular mechanisms associated with cold acclimation in relation to their possible role in the development of freezing survival strategies. In many cases, particularly in the case of cold-regulated genes, these roles are still hypothetical. Because conclusions are based on a correlation with LT50, one of the major difficulties faced by scientists is to obtain experimental evidence to establish whether the molecular changes observed during cold acclimation actually contribute to the development of cold and/or freezing tolerance or if they result from cellular disorders caused by low temperatures.

CONTROL OF WATER STATUS

In his review on the formation of ice in plant tissues, Ashworth (1992) concluded that the ability of plants to control water content and phase in overwintering tissues is probably the most important factor in freezing survival. The water status within plant tissues can be determined by several methods. The presence of supercooled water is detected by differential thermal analysis (DTA), which consists of gradually lowering the temperature until supercooled water freezes, causing the appearance of low-temperature exotherms (Quamme, 1991). The distribution of ice within plant organs is localized by low-temperature scanning electron microscopy (Pearce, 1988; Flinn and Ashworth, 1994). New, nondestructive approaches to visualize the growth of ice are making it possible to determine the spatial distribution and phase of water in different tissues, identify ice nucleation sites, and quantify the rate of ice propagation in vivo. These techniques include magnetic resonance imaging (Millard et al., 1995, 1996; Ishikawa et al., 1997) and infrared video thermography (Wisniewsk, Lindow, and Ashworth, 1997).

Extracellular Freezing

In plants or plant parts that exhibit extracellular freezing, ice first forms in apoplastic fluids, and intracellular water is gradually withdrawn from the cell due to lower vapor pressure over ice. Studies of ectothermic animals living in polar regions showed that extracellular freezing is under the control of two types of proteins, ice-nucleating proteins and antifreeze proteins, that accumulate in the bloodstream or hemolymph during winter (Duman et al., 1991). Ice nucleators initiate extracellular ice formation at temperatures between -6 and $-12°C$, while antifreeze proteins modify the rate of growth and the morphology of ice crystals.

In plant tissues, the process of extracellular freezing was first considered to be a spontaneous phenomenon because apoplastic fluids are generally more dilute and should freeze first. However, endogenous ice nucleation activity has now been observed in woody, succulent, and herbaceous plants. Woody *Prunus* spp. produce heterogeneous ice nucleators active at $-2°C$ that contain neither protein nor lipid components (Ashworth, Anderson, and Davis, 1985; Gross, Proebsting, and Maccrindle-Zimmerman, 1988). Cacti such as *Opuntia humifusa, O. ficus-indica,* and *O. streptacantha* also produce ice nucleators active at $-2°C$ that may be composed of polysaccharides (Goldstein and Nobel, 1991, 1994). On the other hand, the herbaceous monocot winter rye produces ice nucleators composed of proteins, lipids, and carbohydrates during cold acclimation (Brush, Griffith, and Mlynarz, 1994). These ice nucleators are active at temperatures below $-5°C$ (Marentes et al., 1993; Brush, Griffith, and Mlynarz, 1994).

The discovery of two cold-inducible genes in *Arabidopsis thaliana* that encode alanine-rich polypeptides similar to fish antifreeze proteins raised the possibility that plants may also accumulate antifreeze proteins during cold acclimation. More recent studies indicated that the products of *kin1* (Kurkela and Franck, 1990) and *cor6.6* (Gilmour et al., 1992) do not exhibit antifreeze activity; rather, these proteins may have cryoprotective activity (see *Cor15* and Homologues in the section Induction of Cryoprotective Proteins). Meanwhile, antifreeze proteins were discovered in apoplastic fluids of cold-acclimated winter rye leaves (Griffith et al., 1992). Six apoplastic antifreeze proteins with molecular masses of 19, 26, 32, 34, 36, and 36 kilodalton (kDa) were identified in extracts from cold-acclimated winter rye leaves (Hon et al., 1994). These rye proteins do not cross-react with antibodies raised against various animal antifreeze proteins; rather, the rye antifreeze proteins are all similar to pathogenesis-related proteins: two of them are endochitinases, two are thaumatin-like proteins, and two are β-1,3-endoglucanases (Hon et al., 1995). Because no antifreeze activity is detectable in pathogenesis-related proteins induced in winter rye by fungal infestations (Antikainen et al., 1996), it seems that the antifreeze activity of these cold-inducible pathogenesis-related proteins is due to subtle structural evolution that conferred the ability to bind to ice. As shown by immunolocalization, antifreeze proteins are present in leaves, crowns, and roots of cold-acclimated rye and preferentially located in the epidermis and in cells adjacent to vascular strands (Antikainen et al., 1996). Because of their apoplastic location, these proteins may provide barriers for secondary ice nucleation from external sources and from the xylem, prevent inoculative freezing of cells, and/or inhibit recrystallization of extracellular ice. Another cold-induced chitinase, COR27, was recently discovered in bermudagrass (Gatschet et al., 1996), but at this time, it is unknown whether this protein has antifreeze activity.

Antifreeze activity has been found in many overwintering plants (Duman and Olsen, 1993; Griffith and Ewart, 1995), although only one additional antifreeze protein, a 67 kDa glycoprotein, has been isolated (Duman, 1994). In a comparative study, apoplastic fluids from freezing-sensitive and freezing-tolerant cereals, as well as from freezing-sensitive and freezing-tolerant herbaceous dicots, were tested for antifreeze activity and antifreeze protein content (Antikainen and Griffith, 1997). Only cold-acclimated, freezing-tolerant cereals were found to exhibit detectable antifreeze activity and to possess antifreeze proteins similar to those found in winter rye. These observations indicate that the accumulation of antifreeze proteins in response to cold is limited to some species and suggest that the molecular mechanisms of freezing tolerance in monocots and dicots may be distinct. Another explanation is that antifreeze proteins of dicots could be intracellular or more tightly bound to cell walls or plasma membranes, so that they are not detected in apoplastic extracts.

Deep Supercooling

All herbaceous plants that survive prolonged exposure to temperatures below $-5°C$ resist freezing injury by extracellular freezing (Sakai and Larcher, 1987). Only some woody, deciduous species exhibit deep supercooling in certain tissues, namely xylem ray parenchyma and floral buds. For this reason, very little information is available on the cellular and molecular mechanisms of deep supercooling. It is believed, however, that anatomical and biochemical adaptations of cell walls and vascular systems limit the spread of ice-frozen tissues into supercooled areas (Quamme, Su, and Veto, 1995; Fujikawa, Kuroda, and Ohtani, 1997). For example, by using immunocytolocalization in peach tissues, Wisniewski and Davis (1995) showed that the degree of esterification of pectins was modified during cold acclimation in tissues that experience deep supercooling. Such modifications may change the pore size of the cell wall and inhibit ice penetration inside supercooled compartments.

The reason why certain tissues use deep-supercooling strategies rather than extracellular freezing is unclear. It is generally admitted that deep supercooling is an alternative survival strategy for cells unable to withstand dehydration. However, a recent study showed that tissues which survive winter by deep supercooling develop dehydration tolerance to the same extent as those which actually undergo extracellular freezing (Fujikawa, Kuroda, and Ohtani, 1997). It is possible that the mechanism by which a plant achieves desiccation tolerance also promotes deep supercooling.

CONTROL OF CELL DEHYDRATION

Excessive dehydration of protoplasts due to extracellular freezing is believed to be lethal. The accumulation of various osmotically active, small hydrophilic compounds in cold-acclimated plant cells has long been thought to limit protoplast dehydration by reducing the difference in vapor pressure between extracellular and intracellular spaces. These compounds may also function in a noncolligative manner to stabilize cellular structures during freezing-induced dehydration.

Alteration in Carbohydrate Composition and Metabolism

In early fall, all perennial plants accumulate carbohydrate reserves in the form of starch or fructan. By itself, this phenomenon seems to be an energetic necessity for resuming growth in the spring rather than a component of the mechanism of freezing tolerance. However, a subsequent conversion of carbohydrate polymer reserves into soluble sugars is observed at the beginning of the cold period (Levitt, 1980). The principal forms of sugars that appear are oligosaccharides such as sucrose, raffinose, and stachyose (Salerno and Pontis, 1989; Tognetti et al., 1990; Bachmann, Matile, and Keller, 1994; Olien and Clark, 1995; Hill et al., 1996).

The accumulation of soluble sugar is concomitant with an increase in several enzymatic activities involved in carbohydrate metabolism: amylases, sucrose synthase, sucrose phosphate synthase, fructose-1,6-bisphosphatase, sedoheptulose-1,7-bisphosphatase, and Rubisco (see Table 4.1). In fact, it appears that the accumulation of free sugars involves an enhancement of all photosynthetic pathways (Öquist, Hurry, and Huner, 1993; Hurry et al., 1994; Hurry, Tobieson, et al., 1995). The increase in some enzymatic activities, sucrose synthase, sucrose phosphate synthase, and amylase, is related to the appearance of new, cold-specific isozymes (Maraña, García-Olmedo, and Carbonero, 1990; Volonec, Boyce, and Hendershot, 1991; Hill et al., 1996; Nielsen, Deiting, and Stitt, 1997; Reimholz

TABLE 4.1. Carbohydrate Metabolism-Related Enzymes Induced by Low Temperatures

Enzyme	Species	References
Aldose reductase	Bromegrass	Lee and Chen, 1993
Amylases	Alfalfa, potato	Volonec, Boyce, and Hendershot, 1991; Hill et al., 1996; Nielsen, Deiting, and Stitt, 1997
Fructose-1,6-biphosphatase (stromal and cytosolic)	Barley, spinach, bean	Holoday et al., 1992; Hurry, Strand, et al., 1995
Glucose-6-phosphate dehydrogenase	*Lolium perenne*	Bredemeijer and Esselink, 1995
Glyceraldehyde-3-phosphate dehydrogenase	Alfalfa	Laberge, unpublished data
Phosphoenolpyruvate carboxykinase	*Brassica napus*	Sáez-Vásquez and Delseny, 1995
Rubisco	Bean, spinach, winter rye	Holoday et al., 1992; Hurry et al., 1994
Sedoheptulose 1,7-bisphosphatase	Spinach, bean	Holoday et al., 1992
Sucrose phosphate synthase	Wheat, barley, chlorella, spinach, winter rye, potato	Salerno and Pontis, 1989; Tognetti et al., 1990; Guy, Huber, and Huber, 1992; Holoday et al., 1992; Hurry et al., 1994; Hurry, Strand, et al., 1995; Reimholz et al., 1997
Sucrose synthase	Wheat, potato	Tognetti et al., 1990; Crespi et al., 1991; Hill et al., 1996; Reimholz et al., 1997

et al., 1997). However, in the case of glucose-6-phosphate dehydrogenase in *Lolium perenne*, the increase in activity appears to result from an accumulation of the same enzyme that is present under normal growth conditions (Bredemeijer and Esselink, 1995). In addition, an increase in steady-state levels of mRNAs (messenger ribonucleic acids) encoding sucrose synthase has been observed in sugar beet (Hesse and Wilmitzer, 1996), and sucrose phosphate synthase in potato and wheat (Maraña, García-Olmedo, and Carbonero, 1990; Crespi et al., 1991; Reimholz et al., 1997). Likewise, Sáez-Vásquez and Delseny (1995) observed that the level of an mRNA encoding phosphoenolpyruvate carboxykinase increased during cold acclimation. These observations indicate that the changes in carbohydrate metabolism in relation to cold acclimation are under transcriptional and/or posttranscriptional control. Other types of regulation might be involved as well. For example, Witt and colleagues (1995) suggested that the control of the conversion of starch into soluble sugars might involve temperature-dependent binding of alpha-amylase to starch granules.

Although the capacity to accumulate soluble sugars is commonly considered an important factor in freezing tolerance, it is difficult to determine to what extent this mechanism is responsible for plant winterhardiness. Varietal or interspecific differences in soluble-sugar contents are often positively correlated with the level of freezing tolerance (Tognetti et al., 1990; Olien and Clark, 1995). In some cases, a positive correlation is found between freezing tolerance and the level of some of the aforementioned enzymatic activities (Tognetti et al., 1990; Holoday et al., 1992; Bredemeijer and Esselink, 1995). However, sugar contents and enzymatic activities are not always correlated with the LT50. Although differences in sucrose concentration occur between alfalfa cultivars differing in winterhardiness during autumn acclimation (Duke and Doehlert, 1981), the levels do not differ significantly when plants are sampled in midwinter (Paquin, 1984). One explanation for the variable correlation between freezing tolerance and soluble-sugar contents lies in the different capacities of plants to photosynthesize at cold temperatures.

In comparisons of winter and spring cultivars of rape and wheat, Hurry, Strand, and colleagues (1995) demonstrated that only winter cultivars are capable of improving their photosynthetic efficiency at low temperatures, leading to higher production of sugars during cold acclimation. As a second explanation, Griffith and McIntyre (1993) proposed that degree of freezing tolerance is related not only to photosynthetic capacity in overwintering annuals such as winter rye but also to the partitioning of photoassimilates between processes involved in growth and freezing tolerance. Although exposure to low temperatures is an essential step in developing freezing tolerance, the total amount of light intercepted by the plants and the daylength are both important factors influencing the rate of growth and the level of freezing tolerance achieved by the plants. For example, rye plants raised at low temperature under short daylengths become very freezing tolerant, but they do not grow because they only have sufficient photoassimilates to support one sink. On the other hand, rye plants raised under continuous light produce sufficient photoassimilate to grow quickly and to develop a high degree of freezing tolerance. Other authors have reached similar conclusions about the influence of competing processes on the development and/or maintenance of freezing tolerance. For example, in their studies of premature dehardening of dormant pine and spruce seedlings, Ögren (1997) and Ögren and colleagues (1997) showed that two factors are important for winter survival when photoassimilate is limiting: (1) the level of soluble sugars attained during acclimation and (2) the ability to maintain sufficiently high levels of soluble sugars throughout winter by lowering the respiration rate.

A third explanation for the lack of a relationship between sugar content and freezing tolerance is that certain sugars may play specific roles in different compartments of the tissues. Although sugars are usually involved in metabolic and osmotic adjustments within the cell, Olien (1984) proposed that sugars accumulate in the apoplast following a freeze-thaw cycle and that these sugars prevent the adhesion of ice to critical tissues within the crowns of winter rye. Antikainen and Griffith (1997) confirmed that sugars accumulate in the apoplast of the leaves of rye, wheat, and barley during cold acclimation and hypothesized that the sugars may play a role complementary to antifreeze proteins in modifying the growth of ice. The mechanism for sugar accumulation has been examined only recently in winter oat during the second phase of hardening. After oat plants are exposed to subzero temperatures, the activities of the carbohydrate-degrading enzymes invertase and fructan exohydrolase increase in the apoplast of the crowns. Soluble sugars also accumulate in the apoplast. These sugars include not only glucose, fructose, and sucrose but also fructans with degrees of polymerization ranging from three to greater than seven (Livingston and Henson, 1998). Because the level of apoplastic sugars is too small to lower the freezing point of apoplastic fluids, they are envisioned to interact with ice and prevent adhesions. These sugars could also protect cell walls and membranes during freeze-thaw cycles (Livingston and Henson, 1998).

In addition to their role in osmoregulation, soluble sugars are suspected to have cryoprotective effects on biomembranes and proteins (Strauss and Hauser, 1986; Hoekstra, Crowe, and Crowe, 1989). For example, it was shown that interactions between sugars and hydrophilic extremities of phospholipids modify phase transition of lipid bilayers (Sun, Irving, and Leopold, 1994). Different sugar molecules seem to vary in their cryoprotective activity. Galactinol-containing oligosaccharides such as stachyose and raffinose are thought to provide enhanced cryoprotection due to their capacity to form a glassy state that slows down molecular motion under anhydrous conditions (Caffrey, Fonseca, and Leopold, 1988; Bruni and Leopold, 1991). This hypothesis is supported by several observations that have established a close relationship between the accumulation of sugars of the raffinose family and the development of cold (Weimken and Ineichen, 1993) and dehydration (Sun, Irving, and Leopold, 1994) tolerance. In a recent study, Castonguay and colleagues (1995) observed a strong correlation between the accumulation of stachyose and raffinose and winterhardiness of different alfalfa cultivars. In this species, even though sucrose accumulates at similar levels in cultivars differing in winterhardiness, a subsequent surge in stachyose and raffinose levels is observed only in the most hardy cultivars.

The role of sugars in freezing tolerance is expected to be directly assessed by the analysis of transgenic plants in which the levels of soluble sugars are artificially altered. Recently, a significant improvement in leaf freezing tolerance was observed in transgenic tobacco plants expressing heterologous pyrophosphatase or invertase (Hincha et al., 1996). However, no correlation was found between leaf osmolality and hardiness, suggesting that the accumulation of sugars by itself does not confer higher tolerance. Instead, high sugar contents may induce the onset of other mechanisms that are not activated in wild-type plants. Alternatively, the roles of individual sugars and sugar alcohols may be synergistic. The freezing tolerance of cell suspension cultures of a freezing-sensitive eucalyptus hybrid can be increased by incubating the cells in sucrose, raffinose, fructose, or mannitol (Travert et al., 1997). Cells from a freezing-tolerant eucalytus hybrid normally accumulate sucrose and fructose in response to low temperature, but their hardiness can be enhanced by even a small increase in cellular levels of mannitol. These results indicate that mannitol may act synergistically with the sugars involved in cryoprotection. One way to increase the freezing tolerance of eucalyptus hybrids may be to engineer an increase in cellular mannitol by transforming the plants with a gene for mannitol-1-P dehydrogenase (Tarczynski, Jensen, and Bohnert, 1993).

Accumulation of Amino Acids and Amines

The osmotic adjustment of dehydrated cells is not completely explained by the accumulation of soluble sugars (Guinchard et al., 1997). Other osmolytes, such as free amino acids, glycinebetaine (N-N-N trimethylglycine), and polyamines, are thought to contribute to lowering water potential within the protoplast during freeze-induced desiccation. This assumption is reinforced by the fact that these compounds often accumulate in response to water and osmotic stresses. Proline and polyamines are thought to have cryoprotective effects on the plasma membrane in a manner similar to free sugars.

An increase in proline content during cold acclimation has been found in a number of species (Koster and Lynch, 1992; Savouré et al., 1997), and in some cases, the level of proline has been positively correlated with freezing tolerance. In a recent study, Igarashi and colleagues (1997) showed that the mRNA encoding Δ1-pyrroline-5-carboxylate synthetase, an enzyme involved in the proline biosynthesis pathway, was induced by cold in *Oryza sativa*. Although rice is freezing sensitive, this observation suggests that proline production during cold acclimation could be controlled by transcriptional regulation of Δ1-pyrroline-5-carboxylate synthetase. Indeed, a slight induction of this mRNA in *Arabidopsis thaliana* had already been observed (Yoshiba et al., 1995) over a short period of cold treatment. Physiological evidence of the importance of proline in freezing tolerance is still lacking. Although proline may simply be a form of nitrogen storage, its role in osmoregulation has long been suspected (Delauney and Verma, 1993). Kavi Kishor and colleagues (1995) demonstrated that increasing proline production by overexpression of Δ1-pyrroline-5-carboxylate synthetase significantly enhanced osmotolerance of transgenic tobacco. A similar experimental approach may establish the causal relationship between proline synthesis and low-temperature tolerance.

In higher plants, the biosynthesis of glycinebetaine and polyamines (namely putrescine and spermidine) is stimulated by a number of environmental stresses, including cold (Nadeau, Delaney, and Chouinard, 1987; Koster and Lynch, 1992). Recently, two genes encoding S-adenosyl-L-methionine decarboxylase were found to be up-regulated by cold in a hardy *Solanum* species, *S. sogarandinum* (Rorat, Irzykowski, and Grygorowicz, 1997). This enzyme takes part in the biosynthesis of spermine and spermidine. In *Oryza sativa,* the increase in putrescine level during exposure to cold was shown to be due to an abscisic acid (ABA)-mediated increase in arginine decarboxylase activity (Lee, Lur, and Chu, 1997). As previously observed for proline biosynthesis, the importance of polyamines in the acquisition of freezing tolerance could be assessed by modulating the level of expression of arginine decarboxylase and/or S-adenosyl-L-methionine decarboxylase in transgenic plants.

In addition to the accumulation of free sugars, amino acids, and polyamines, a modulation of water and solute flows across membranes is likely to occur during cold acclimation. For example, Orr and colleagues (1995) observed in winter *Brassica napus* a neosynthesis of tonoplast H^+-ATPase. Although no evidence indicates a regulation of water transport during cold acclimation, it was shown that the activity of plant water channels is temperature dependent (Hertel and Steudle, 1997) and modulated by phosphorylation (Maurel et al., 1995). In the ice plant *(Craterostigma plantagineum),* a fluctuation in the content of mRNAs encoding water channel proteins was observed during salt stress (Yamada et al., 1995).

Adjustment of Mechanical Properties

In addition to the accumulation of solutes, it has been proposed that the adjustment of cell rigidity may play an important role in determining the extent of cellular dehydration during extracellular freezing (Rajashekar and Burke, 1996). During freezing, the formation of extracellular ice may result in a negative pressure that tends to cause a reduction in cell volume, contraction of the protoplasm, and, ultimately, collapse of the cell (Rajashekar and Burke, 1996; Fujikawa, Kuroda, and Ohtani, 1997). At equilibrium between extracellular and intracellular spaces, this negative pressure equals the difference in water potential and osmotic potential within the cell. The extent of negative pressure that a cell can withstand without collapsing depends on the rigidity of the surrounding structures (i.e., plasma membrane and cell wall). Typically, this pressure is expected to be very negative (i.e., have the largest absolute value) in cells that resist dehydration, and less negative in cells that undergo intense dehydration. As a consequence, the rigidity of the cell wall and the strength of cell wall–plasma membrane interaction are thought to play major roles in the control of freezing-induced dehydration.

Several observations indicate that cell wall proteins undergo adaptive changes during cold acclimation. In alfalfa, Laberge and colleagues (1993) isolated a cold-induced mRNA, *MsaCiA,* that encodes a putative cell wall glycine-rich protein. This class of proteins is characterized by the presence of both a leader peptide that promotes extracellular targeting and highly repeated glycine-rich domains that provide increased flexibility to the polypeptide secondary structure. Experimental evidence shows that this type of protein is integrated into cell walls, where it is thought to have a structural, though currently undefined, function (Schowalter, 1993).

The deduced amino acid sequence of *MsaCiA* contains 38 percent Gly (glycine), 9 percent His (histidine), 9 percent Asn (asparagine) and 7 percent Tyr (tyrosine) and lacks Pro (proline), Trp (tryptophan), Phe (phenylalanine), and Cys (cysteine). The glycine-rich domains are organized in repeats of the $G_{2-4}Y_{0-1}N_{0-1}H_{0-1}$ type. These repeats have been identified in several glycine-rich proteins, including a family of cold-inducible proteins of alfalfa (Luo et al., 1991, 1992). The presence of a typical leader peptide at the N-terminus and the absence of a sequence that causes retention in the endoplasmic reticulum (ER) indicate that MSACIA is likely to be secreted. In addition, cell wall glycine-rich proteins are typically rich in tyrosine, which is thought to participate in the irreversible binding of the polypeptide to the cell wall through the formation of di-tyrosine bonds. Recently, a 59 kDa protein immunologically related to MSACIA was detected in the cell wall of cold-acclimated alfalfa crowns (Ferullo et al., 1997). The level of accumulation of this protein is positively correlated with the freezing tolerance of various alfalfa cultivars, suggesting that this protein actually plays a role in freezing tolerance. The possibility that this protein contributes to limiting cell dehydration through modification of cell wall rigidity is under investigation.

Another class of proteins involved in cell mechanical properties are the cell wall–plasma membrane linkers. Two similar mRNAs, *MsaCiC* and *BnPRP*, encoding putative plasma membrane cell wall linkers are induced by cold in alfalfa (Castonguay et al., 1994) and *Brassica napus* (Goodwin, Pallas, and Jenkins, 1996), respectively. The polypeptides predicted by both nucleotide sequences are composed of three distinct domains: at the N-terminus, a leader peptide implicated in extracellular targeting; a central domain featuring high proline content (approximately 50 percent); and at the C-terminus, a conserved hydrophobic, membrane-spanning domain. Although the structures of the proline-rich domains of MSACIC and BNPRP are not typical of cell wall proline-rich proteins or extensins, it is believed that they have biochemical properties similar to extensins that allow them to associate with cell walls. In their mature forms, MSACIC and BNPRP might constitute extrinsic, bimodular proteins that form strong linkages between the cell wall and plasma membrane. During freezing, they might help prevent cell contraction and, consequently, excessive dehydration. These changes in cell wall properties are possibly accompanied by cytoskeletal reorganizations, as indicated by the shift in tubulin isotypes observed in cold-acclimated rye (Kerr and Carter, 1990). The verification of the roles of cell wall proteins and cell wall–plasma membrane linkers in modifying cellular responses to freezing-induced dehydration will need new experimental procedures that combine both biophysical and molecular approaches.

MEMBRANE STABILIZATION

Temperature and water availability affect both physical and biological properties of cell membranes. As a result, membranes, especially the plasma membrane, are regarded as key sites of freezing-induced injury. Steponkus and Webb (1992) demonstrated that the response of rye plasma membranes to low temperature is radically modified after cold acclimation. During freezing, protoplasts from unacclimated rye tissues undergo an irreversible contraction by formation of endocytotic vesicles, which leads to expansion-induced lysis upon thawing. In addition, freeze-induced dehydration leads to a transition from lamellar to hexagonal-II phases in regions where plasma membrane contacts subtending endomembranes. By contrast, acclimated protoplasts do not exhibit lamellar to hexagonal-II phase transitions and form exocytotic vesicles that prevent expansion-induced lysis. Cold-acclimated protoplasts can still be injured through the formation of so-called fracture-jump lesions, but this occurs at a much lower temperature. Similar observations have been made in intact tissues of rye (Steponkus and Webb, 1993), oat (Webb, Uemera, and Steponkus, 1994), and *Arabidopsis thaliana* (Uemera, Joseph, and Steponkus, 1995). Although these changes in plasma membrane properties occur during cold acclimation, it is unclear whether they function in intact plant tissues to promote survival of freezing and thawing.

In some plant species, changes in the lipid composition of the plasma membrane occur during cold acclimation (Palta, Whitaker, and Weiss, 1993). The more significant changes are an increase in the proportion of phospholipids and in the unsaturated to saturated fatty acid ratio (due to an increase in di-unsaturated classes of phosphatidylcholine and phosphatidylethanolamine), a decrease in the sterol to phospholipid ratio, and a decrease in the cerebroside content. Different experimental approaches, such as artificial modification of protoplast lipid compositions (Steponkus et al., 1988) and comparison of related species of contrasting freezing tolerance (Palta, Whitaker, and Weiss, 1993), have demonstrated that changes in lipid composition can account for the improvement of membrane cryostability and play an important role in cold acclimation.

Very little information is currently available on the cellular mechanisms responsible for the adjustment of plasma membrane lipid composition during cold acclimation. In barley, Hughes and colleagues (1992) have isolated a low-temperature-responsive gene that encodes a phospholipid transfer protein. Recent observations by Williams and colleagues (1997) suggest that cold-inducible chloroplastic desaturases are involved in the change in fatty acid composition. Because of the complexity of plant lipid metabolism, future progress in the knowledge of membrane cryostability will most probably come from the isolation of mutants impaired in their capacity to modify their plasma membrane lipid compositions upon acclimation.

INDUCTION OF CRYOPROTECTIVE PROTEINS

In the last decade, an impressive number of cold-regulated genes have been isolated from diverse plant species, principally by using differential hybridization of cDNA (complementary deoxyribonucleic acid) libraries. Briefly, mRNAs are isolated from the studied plant tissue and retrotranscripted into double-stranded cDNAs, which are cloned into phage or plasmidic libraries. The library is then consecutively screened with two cDNA probes synthesized with mRNAs from tissues taken at unacclimated and acclimated stages. The clones that exhibit differential hybridization signals identify mRNAs that are up- or down-regulated during cold acclimation. As noticed by Hughes and Dunn (1996), this approach is limited. For example, differential hybridization does not allow the discrimination between two genes that share high sequence homologuey, such as genes that encode isoenzymes. In fact, among all the genes that have been isolated from such differential screening experiments, only a few are related to the metabolic adaptations that have been previously identified, such as carbohydrate or lipid metabolism. Instead, novel gene classes were identified. The proteins corresponding to these genes exhibit a strong bias in their amino acid compositions and the presence of numerous repeated motifs. They are thought to participate to the stabilization of the diverse cellular structures under freeze-induced dehydration conditions.

Cor15 *and Homologues*

A family of homologueous cold-inducible genes was isolated from both *Arabidopsis thaliana* and *Brassica napus* (see Table 4.2). The corresponding deduced polypeptides constitute a class of highly hydrophilic proteins that remain soluble upon heating at 100°C. Their amino acid compositions are enriched in Ala (alanine), Gly, and Lys (lysine) and depleted in Cys, Trp, and Pro. Weretilnyk and colleagues (1993) observed the presence of mRNAs that cross-hybridize with the *cor15* homologue *BN115* in three cruciferous plants, but not in spinach, rye, or tobacco, indicating that this gene family might be specific to cruciferous species. Due to their high Ala content, a property that they share with flounder antifreeze protein, they were first proposed to have antifreeze activity (Kurkela and Franck, 1990). The *Arabidopsis* COR15a protein is the most thoroughly investigated. Although it has no antifreeze activity, it may act as a protein cryoprotectant (Lin and Thomashow, 1992b). However, the mature form of COR15a, COR15am, has only weak cryoprotective effect on lactate dehydrogenase (Artus et al., 1996). Moreover, COR15am accumulates in the chloroplast (Lin and Thomashow, 1992a), a compartment that does not experience freezing. Experiments using transgenic plants indicate that constitutive expression of the COR15a in *Arabidopsis thaliana* may increase the stability of thylakoids after exposure to freeze-induced dehydration (Artus et al., 1996). COR15a may act as a membrane cryoprotectant, but its mode of action is not known at this time. Further investigation is needed to determine whether all the members of the COR15 family share the same properties, or if they have distinct functions despite their sequence homologueies. For instance, kinetic studies of the accumulation of BN28 showed that this protein disappears during deacclimation before freezing tolerance

TABLE 4.2. Cor15 and Homologues

Species	Name	kDa	Amino acid composition	Other characteristics	References
Arabidopsis thaliana	Kin1 (C)	6.5	Ala: 22.7% Gly: 13.6% Lys: 13.6% Arg, Cys, His, Ile, Pro, Trp, Tyr: 0%		Kurkela and Franck, 1990
	Cor6.6 (kin2) (C)				Gilmour, Artus, and Thomashow, 1991
	Cor15a (cor15) (C, I)	14.6 15 (apparent) 9.4 (apparent, mature form)	Ala: 17.9% Lys: 14.3% Cys, Trp, Pro, Met, Arg, His: 0%	Chloroplastic	Lin and Thomashow, 1992a
	Cor15b (C)	15 9.6 (mature form)		77% identity with cor15a, putative chloroplast targeting peptide, homology with group III LEA	Wilhelm and Thomashow, 1993
Brassica napus	BN115 (C)	14.8	Ala + Gly: 26% Lys: 14% Cys, Hys, Pro, Trp: 0%	Putative transit signal	Weretilnyk et al., 1993
	BN26, BN19 (C)			Two homologues of BN115	Weretilnyk et al., 1993
	BN28 (C, I)	28 33 (apparent)	Ala + Gly: 34% Tyr, Trp, Cys, Pro: 0%		Orr et al., 1992; Boothe, de Beus, and Johnson-Flanagan, 1995

The letters below gene names indicate whether the data come from molecular cloning (C) or protein immunodetection (I), purification (P) or microsequencing (S). Unless indicated, the molecular weights of polypeptides are deduced from nucleotide sequences. In the column "amino acid composition" were listed abundant as well as absent amino acid residues.

is lost, suggesting that this protein is unlikely to be involved in maintaining membrane integrity or chloroplast function during a freeze-thaw cycle (Boothe, de Beus, and Johnson-Flanagan, 1995).

Dehydrin-like Proteins

Although dehydrin-like proteins were first identified as a subset (group II) of the late-embryogenesis-abundant (LEA) proteins, they are also induced by several stresses, including dehydration, salt stress, cold, and ABA-treatment (Close et al., 1993). The main features shared by these proteins are the presence of highly conserved amino acid motifs and the absence of tryptophan, cysteine, and phenylalanine residues. A stretch of seven to nine serine residues is generally found at the central part of the polypeptide, and a lysine-rich motif of the EKKGIMDKIKELPG-type is repeated between two and fifteen times. As observed previously with MSACIA and the COR15 family, dehydrin-like proteins are highly hydrophilic and remain soluble upon boiling.

The induction of dehydrin-like genes and/or proteins during cold acclimation has been reported in many species (see Table 4.3). In fact, the majority of the cold-induced genes described so far are dehydrin-like. All of them possess several imperfect repeats of the lysine-rich motif, even though the serine cluster is frequently absent or is sometimes localized at an unusual position. In addition, several cold-induced dehydrin-like proteins are rich in glycine (up to 27 mole [mol] percent) and threonine (up to 17 mol percent). In some of them, glycine and threonine residues are organized in numerous GT repeats. Another common feature of these proteins is the discrepancy between calculated and apparent molecular weights on SDS-PAGE: the protein (obtained directly from plant extracts or by in vitro translation or recombinant expression) always appears 20 to 40 percent larger than expected. Because dehydrins do not appear to undergo posttranslational modifications, a tentative explanation for their anomalous migration in SDS-PAGE is that their high hydrophilicity affects the binding of SDS (Weretilnyk and Hanson, 1990; Houde et al., 1992).

No definite function has been clearly determined for dehydrin-like proteins, but many observations indicate that they could play an important role in freezing tolerance. For example, comparing chilling-sensitive and freezing-tolerant gramineae, Danyluk and colleagues (1994) showed that all plants tested contained genes homologous to the dehydrin-like Wcor410, but these genes were expressed only in freezing-tolerant species. The accumulation of dehydrin-like proteins during stresses that cause dehydration

TABLE 4.3. Dehydrin-like Genes and Proteins Induced by Low Temperatures

Species	Name	kDa	Amino acid composition	Repeated motifs	Serine cluster	Other characteristics	References
Alfalfa	Cas17 (C)	16.7	Gly: 25% Thr: 16.9%	DKIKEKIPG GEKKGIMEKIKEKIPG $(GT)_n$ n>2	No	Homologue to Cas18	Wolfraim and Dhindsa, 1993
	Cas18 (C)	17.6	Gly: 21% Thr: 12.5%	(V/M)(G/D)KIKEKIP $(GT)_n$, n>2	No		Wolfraim et al., 1993
Arabidopsis thaliana	Rab18 (C)	18.5	Gly: 34% Cys, Trp, Val: 0%	GG(Q/G)GYG(T/S) (R/E)KKG(I/M)(T/M)(Q/D)KIKE KLPG	Yes		Lång and Palva, 1992
	Lti45 (C)	45		E/PEEKKGFMDKIKEKLPG	Yes	64% identity with COR47 and 65% with RAB18 and DHNX	Welin et al., 1994
	Lti30 (C)	30		K(V/I)(M/K)E(Q/K)LPG TNTGVVHHEKK	No	Structural similarities with spinach CAP85	Welin et al., 1994
	COR47 (C)	3.4 47 (apparent)	Glu: 20% Lys: 14% Cys, Trp: 0%	(E/P)E(E/D)KKG(L/I)(V/E)KIKE KLPG	Yes		Gilmour, Artus, and Thomashow, 1991
Barley	Dhn5 (I, C)	59 86 (apparent)	Rich in Gly and Lys	EKKGIMDKIKELPG	Yes		van Zee et al, 1995
Blueberry	(I, P, S)	65, 60, and 19 (apparent)			N.D.	Cross-react with an antibody raised against EKKGIMDKIKELPG	Muthalif and Rowland, 1994
Grapevine	(I)	27 (apparent)			N.D.	Cross-react with an antibody raised against EKKIKEKLPG	Salzman et al., 1996
Peach	PCA60 (I, P, S)	60 (apparent)	Rich in Gly, Glu/Gln, Asp/Asn, and Thr	R/KLPGGQ	N.D.	Accumulates at higher levels in deciduous than in evergreen genotypes	Arora and Wisniewski, 1994, 1996; Arora, Wisniewski, and Rowland, 1996

Species	Name	kDa	Amino acid composition	Repeated motifs	Serine cluster	Other characteristics	References
Potato	*Sscil2* (C)	N.D.	N.D.	N.D.	N.D.		Rorat, Irzykowski, and Grygorowicz, 1997
	Ci7 (C)	23.7	Charged amino acids: 52%	(E/K)KKG(F/L)(L/K)(E/D)KIKE KxxG	Yes		van Berkel, Salamini, and Gebhart, 1994
Poncirus trifoliata	COR19 (C)	19.8	Gly: 26% Gln: 12% Glu: 12% Lys: 12% His: 11%	KEGLVDKIKQ(K(Q)IPG(V/A)G HHG$_n$Y(R/H) n>3	Yes	Serine cluster positioned in C-terminus Putative nuclear targeting sequence	Cai, Moore, and Guy 1995
	COR11 (C)	11.4			Yes	Serine cluster positioned in C-terminus Putative nuclear targeting sequence Truncated version of COR19	Cai, Moore, and Guy 1995
Spinach	COR85	85	Asp/Asn: 17.8%	NKGGVFDKIKEKLPGQ	N.D.	Cytosolic; 350 kDa homo-oligomeric complex Cryoprotective activity	Kazuoka and Oeda, 1994
	(P, S)		Glu/Gln: 16% Lys: 9.6% His: 9.4% Gly: 8.9%				
	CAP85 (C)	85	Lysine-rich	LDKIKDKLPG HTQQLYPQSDHNYNTH EDKKNDYH			Neven et al., 1993
Wheat	COR39 (C)	39.1 48 (apparent)	Gly: 27% Thr: 16% His: 11% Cys, Trp: 0%	GEKK(G/S)(L/I/V)MENIK(D/E)K LPG(G/Q) G(A/H/Y)YGQQGHAG	No	Related to *Arabidopsis* cor49	Guo, Ward, and Thomashow, 1992

TABLE 4.3 (continued)

Species	Name	kDa	Amino acid composition	Repeated motifs	Serine cluster	Other characteristics	References
Wheat (continued)	Wcor410 (C)	28	rich in Glu		No	Acidic pl.	Danyluk et al., 1994
	Wcs120 (C, I)	39 50 (apparent)	Gly: 26.7% Thr: 16.7% His: 10.8% Cys, Phe, Trp: 0%	GEKKGVMENKEKLPGGHGDH QQ TGGTYGQQGHTGTT	No		Houde et al., 1992
	Wcs32 (I)	32 (apparent)				Immunologically related to WCS120	Dallaire et al., 1994
	Wcs66 (C, I)	46.8 66 (apparent)	Gly: 26.7% Thr: 17.1% His: 10.7%	GEKKGFVMENIKEKLPHHJHD HQQ TGGTYGQQGHTGTT	No	Immunologically related to WCS120	Chauvin, Houde, and Sarhan, 1994
	Wcs200 (P, I, S)	200 (apparent)	Ala: 22% Gly: 11.4% Thr: 13.3%	KGV(K/M)ENINDKLP(D/G)	N.D.	Immunologically related to WCS120	Ouelle, Houde, and Sarhan, 1993

Note: The letters below gene names indicate whether the data come from molecular cloning (C) or protein immunodetection (I), purification (P) or microsequencing (S). Unless indicated, the molecular weights of polypeptides are deduced from nucleotide sequences. In the column "amino acid composition" were listed abundant as well as absent amino acid residues. N.D. = not detected.

suggests that they are involved in a mechanism for tolerating low-water status. During freezing, water withdrawal and the resulting concentration of salts and other solutes in the cytosol create a physicochemical environment that is potentially denaturing for proteins, membranes, and enzymatic complexes. The highly repetitive structures of dehydrin-like proteins indicate that they have structural functions rather than enzymatic activities. It is thought that, due to their high content of hydrophilic residues (asp/asn, glu/gln, lys, thr), they might interact with other cell components to create a microenvironment that maintains them in an adequate solvation status. The high glycine content that is found in some dehydrins may confer relative flexibility to the polypeptide, allowing it to interact with a great variety of molecules. Indeed, two cold-associated dehydrin-like proteins, COR85 (Kazuoka and Oeda, 1994) and WCS120 (Houde et al., 1995), have been proven to significantly protect lactate dehydrogenase against denaturation upon freezing in vitro. Their cryoprotective power (PD50 = 15 and 10 micrograms per milliliter ($\mu g \bullet ml^{-1}$), respectively is similar to the cryoprotection afforded by 250 millimolar (mM) (85,575 $\mu g \bullet ml^{-1}$) sucrose. In the case of WCS120, the accumulation of the protein in vivo was estimated at 72 micrograms per gram ($\mu g \bullet g^{-1}$) fresh weight, which is compatible with a potential cryoprotective effect. However, such a function is only relevant if these proteins are localized in compartments that actually undergo dehydration during freezing. WCS120 is localized in developing vascular tissue in the crown (Houde et al., 1995). These cells are adjacent to the xylem where ice forms during freezing of the plant and may be subjected to intense freeze-induced dehydration.

HVA1, a cold-induced gene encoding another type of LEA (group 3), was isolated in barley (Sutton, Ding, and Kenefick, 1992). Transgenic rice expressing HVA1 exhibits higher growth rates than the wild-type rice plants under water deficit and salt stress (Xu et al., 1996). Likewise, Imai and colleagues (1996) reported that *Saccharomyces cerevisiae* constitutively expressing the tomato LEA-class gene *le25* grows more quickly in 1.2 molar (M) sodium chloride (NaCl) and survives freezing better. Although these data support the hypothesis that LEA proteins actually play a role in dehydration tolerance, equivalent studies are needed to confirm that dehydrin-like proteins function in a similar manner. In nature, dehydrin-like proteins do not accumulate equally in all cell types. In particular, it was observed that meristematic cells of acclimated wheat crowns were devoid of WCS120-type dehydrins (Kazuoka and Oeda, 1994). Consequently, it is likely that the acclimation of these cells relies on distinct molecular mechanisms. It would be of interest to determine to what extent an artificial increase in dehydrin-like protein content can enhance freezing tolerance in such tissues.

MOLECULAR CHANGES RELATED
TO PROTEIN SYNTHESIS

A number of molecular changes take place during cold acclimation that do not directly participate in the elaboration of freezing tolerance; rather, they are thought to maintain the efficiency of essential cellular processes under the biophysical constraints imposed by low temperatures, such as slower reaction rates and/or the effects of decreased water availability.

Protein Biosynthesis and Chaperones

In barley, an mRNA encoding a putative protein elongation factor, EF-1α, is induced by low temperature (Dunn et al., 1993). This factor is thought to bring about the alignment of aminoacyl-tRNAs (transfer RNAs) with the acceptor site of ribosomes. Similarly, cold-induced mRNAs encoding the transcription factors EF-1α and EF-G and a ribosomal protein L3 were recently discovered in potato (Rorat, Irzykowski, and Grygorowicz, 1997). These factors may improve the efficiency of the translational complex at low temperature, although no biochemical evidence is currently available to support this hypothesis.

Several authors have reported low-temperature induction of chaperone cognate proteins and/or corresponding mRNAs. This group of proteins, generally present at low basal levels in all plant tissues, plays an important role in protein posttranslational processing. It includes the family of heat shock proteins (HSPs), a ubiquitous group of chaperones that accumulate upon exposure to elevated temperatures. In cold-acclimated spinach, Neven and colleagues (1992) observed the accumulation of a polypeptide CAP-79 that shares structural and immunological homologuey with the 70 kDa heat shock protein HSP70. The induction by cold of HSP90 was also observed in *Brassica napus* (Krishna et al., 1995). In potato, one mRNA encoding the 22.3 kDa heat shock protein ci19 (van Berkel, Salamini, and Gebhardt, 1994) and two mRNAs encoding chloroplast chaperonins (Ssci3 and Ssci4) were found to be up-regulated by low temperature (Rorat, Irzykowski, and Grygorowicz, 1997).

Evidence supports the concept that the accumulation of chaperone proteins corresponds to an adjustment of processes involved in protein synthesis during the period of cold acclimation. Sabehat and colleagues (1996) demonstrated that the chilling resistance of tomato fruit is improved when the fruit accumulates HSPs following heat shock and returns to the initial level after the disappearance of HSPs. However, even though chaperones appear necessary for sustaining cellular homeostasis during the period of cold acclimation, they do not participate directly in the development of freezing tolerance. For example, CAP-79 accumulates in cold-treated tissues from both cold-tolerant (petunia and broccoli) and cold-sensitive (tomato and pepper) species (Neven et al., 1992). As a result, it is unlikely to play a determining role in the level of freezing tolerance attained by a given plant.

RNA-Binding Proteins

A class of glycine-rich proteins that have the ability to bind to RNA is known to be induced by various stresses in plants, including wounding, water stress, and ABA treatment. The primary structure of these proteins is composed of two distinct domains: an N-terminal moiety that contains two conserved RNA-binding motifs (RPN1 and RPN2), and a glycine-rich C-terminal moiety. Cold temperatures stimulate the accumulation of such proteins in bacteria (Goldstein, Pollitt, and Inouye, 1990; Sato, 1994; Mayo et al., 1997), mammals (Nishiyama et al., 1997), and plants (Carpenter, Kreps, and Simon, 1994). Although their cellular role is still unresolved, it is believed that they are most probably involved in mRNA metabolism. A recent study demonstrated that the cold-induced arrest of cell division in mammalian cell cultures was mediated by a glycine-rich RNA-binding protein (Nishiyama et al., 1997). In addition, two cold-regulated *Arabidopsis* genes that encode the glycine-rich RNA-binding proteins *Ccr1* and *Ccr2* were shown to follow an endogenous circadian rhythm of expression (Carpenter, Kreps, and Simon, 1994). These observations raise the hypothesis that the accumulation of glycine-rich RNA-binding proteins during cold acclimation might direct an adaptive modification of the cell cycle within plant tissues. In particular, their suppressive effect on mammalian cell growth underscores the idea that they might be involved in the inhibition of meristem activity during winter. However, an apparent contradiction exists between this hypothesis and observations by Rorat and colleagues (1997), who have reported the induction of *cdc48*, a gene involved in cell division and growth processes, during cold acclimation of potato. Further experiments will be needed to elucidate the role played by RNA-binding proteins in the cold acclimation process.

A second class of cold-induced mRNAs encoding the arginine-rich proteins pT59 and pA086 has been detected in barley (Cattivelli and Bartels,

1990). The corresponding proteins are suspected to act as stabilizers of specific mRNAs during cold stress.

Protein Phosphorylation

In all organisms, protein phosphorylation plays a central role in the regulation of numerous cellular functions, such as enzyme activity as well as protein-protein and protein-DNA interactions. Protein phosphorylation is controlled by specific protein kinases, which may themselves be activated by specific upstream kinases. Phosphorylation cascades are involved in the perception and intracellular transmission of informative signals.

Protein phosphorylation is a possible mediator of the molecular mechanisms of cold acclimation. In alfalfa, Monroy, Sarhan, and Dhindsa (1993) reported protein phosphorylation patterns change during the establishment of freezing tolerance. More recently, Kawczynski and Dhindsa (1996) identified cold-responsive phosphoproteins in alfalfa nuclei. In *Arabidopsis,* several transcripts encoding protein kinases and protein kinase regulators were shown to be up-regulated by cold, including a receptor-like protein kinase (Kenrick and Bishop, 1986), a mitogen-activated kinase (Mizoguchi et al., 1996), two S6 ribosomal protein kinases (Mizoguchi et al., 1995), and two 14-3-3 cognate kinase regulators (Jarillo et al., 1994). Another cold-inducible protein kinase-encoding mRNA, PKABA1, was found in wheat (Holappa and Walker-Simmons, 1995). Jonak and colleagues (1996) showed that the alfalfa mitogen-activated protein kinase p44MMK4, a protein involved in stress signal transduction, is activated in less than ten minutes after the beginning of a cold treatment. Moreover, a rapid accumulation of the corresponding mRNA was observed. Taken together, these results support the concept that not only the transduction of the cold signal but also the adjustment of some metabolic pathways to low temperatures (especially protein metabolism) involve multiple protein phosphorylation cascades.

COLD-INDUCED GENE PRODUCTS OF UNKNOWN FUNCTION

Even if the functions of the majority of the cold-induced genes cited previously are still hypothetical, there are others for which scientists do not have any clues concerning their possible roles. These are listed in Table 4.4. Indeed, the deduced amino acid sequences of some of them feature typical characteristics, such as high glycine content, absence of

TABLE 4.4. Other Cold-Induced Genes

Species	Name	kDa	Amino acid composition	Other characteristics	References
Alfalfa	Cas15 (MsaCiB) (C, I)*	14.5 23 (apparent)	Gly: 25.7% His: 15.4% Lys: 14.7% Glu: 11.8% Cys, Pro, Arg, Thr, Trp: 0%	GEQHGxyGzG (x, y, and Z are F, H, V, or none) repeated 4 times. Presence of a putative NTS, but immunolocalized in vacuoles (unpublished observation)	Monroy et al., 1993; Ferullo et al., 1997
Arabidopsis thaliana	Lti78 (lti140, cor78, rd29A) (C)	78 140 (apparent)	Hydroxyl groups or acidic residue: 53%	Five imperfect repeats of the VAEKL motif	Nordin, Heino, and Palva, 1991; Horvath, McLarney, and Thomashow, 1993; Nordin, Vahala, and Palva, 1993; Yamaguchi-Shinozaki and Shinozaki, 1994
Brassica napus	Blt14 (C)	9			Dunn et al., 1990
	BnC24A/B (C)	24 kDa (predicted and apparent)	Lys: 14% Arg+His: 27%	Homologous to the human tumor bbc1 (breast basic conserved) gene	Sáez-Vásquez et al., 1993
Potato	Ci21 (C, I)	12.4 (predicted and apparent)	Charged residue >50%	Homologous to ASR1, a water deficit- and ripening-induced gene from tomato	Schneider, Salamini, and Gebhardt, 1997
	PA13	26.7		Osmotin-like	Zhu, Chen, and Li, 1993
Wheat	AWPM-19 (C, P, S)	19 (predicted and apparent)	Ala: 21% Gly: 8%	Plasma membrane protein	Koike et al., 1997
	Wcs19 (C)	19 26 (apparent)	Lys: 8% Pro: 7%	Leaf specific, light dependent	Chauvin, Houde, and Sarhan, 1993

Note: The letters below gene names indicate whether the data come from molecular cloning (C) or protein immunodetection (I), purification (P) or microsequencing (S). Unless indicated, the molecular weights of polypeptides are deduced from nucleotide sequences. In the column "amino acid composition" were listed abundant as well as absent amino acid residues.

certain amino acids, or hydrophilicity, but these characteristics have been observed in numerous cold-induced polypeptides regardless of their putative cellular function. That many of them are induced by dehydration and osmotic stresses suggests that these genes could be important for the development of cold tolerance, but further experiments will be needed to discover their precise roles.

REGULATION OF MOLECULAR ADAPTATIONS TO COLD

As illustrated earlier, a tremendous number of molecular events occur during cold acclimation in plants. It is likely that a number of events have not been identified yet. Even though the individual roles of cold-induced genes involved in the development of freezing tolerance have not been clearly established, the final level of tolerance attained by the plant depends not only on the nature but also on the timing, the extent, and the spatial distribution of these events. To improve winter survival of plants, it will be essential to understand the mechanisms of regulation of responses to cold, from signal perception and transduction to the achievement of freezing tolerance. From the available data, cold acclimation is not controlled by a single, unique mechanism but rather by several, parallel pathways that act at the transcriptional, posttranscriptional, and posttranslational levels.

Evidence for Multiple Regulatory Pathways

Evidence for the existence of multiple regulatory pathways of the cold acclimation process has come from studies of the role of the phytohormone abscisic acid (ABA) in the development of freezing tolerance (Chandler and Robertson, 1994; Lång et al., 1994). ABA levels increase during cold acclimation in many, but not all, species. Moreover, exogenous application of ABA enhances the freezing tolerance of many plant tissues grown in warm conditions. Last, many genes associated with cold acclimation are induced by exogenous ABA. However, the use of inhibitors of ABA biosynthesis and of ABA-deficient and ABA-insensitive mutants has clearly demonstrated that cold acclimation involves both ABA-dependent and ABA-independent pathways (Nordin, Heino, and Palva, 1991; Nordin, Vahala, and Palva, 1993; Dallaire et al., 1994; Welin et al., 1994; Mäntylä, Lång, and Palva, 1995; Welbaum et al., 1997). Similarly, the comparison of cold with drought and osmotic stress has led to the identification of common as well as specific induction mechanisms. In particular, a number of cold-inducible genes are also induced by drought and/or osmotic stress, whereas others are strictly cold specific. Interestingly, three cold-inducible, but not drought-inducible, genes of *Arabidopsis thaliana* (*kin1, cor15b,* and *lti78*) possess a homologueous gene that is drought inducible (*kin2, cor15a,* and *lti65,* respectively) (Kurkela and Franck, 1992; Nordin, Vahala, and Palva, 1993; Wilhelm and Thomashow, 1993). These observations confirm that the various mechanisms of cold acclimation are regulated by both common and unique factors. For example, that some cold-inducible genes show organ specificity and even cell type specificity demonstrates that some intracellular factors must influence the regulation of cold responses at a local level (Ferullo et al., 1997).

Run-on transcription and time course experiments have demonstrated that a number of cold-induced transcripts are regulated at the posttranscriptional level (Horvath, McLarney, and Thomashow, 1993; Wolfraim et al., 1993). Although this type of regulation is known to involve special RNA secondary and tertiary structures (Klaff, Riesner, and Steger, 1996), the molecular basis of its dynamic adjustment to stress is not well understood. In studying the expression of nine low-temperature-responsive genes in barley, Dunn and colleagues (1994) showed that six of them are transcriptionally regulated, while the other three genes are regulated at the posttranscriptional level. Posttranscriptional regulation of one of these three genes, *blt14.0,* may involve a protein factor that could specifically stabilize the *blt14.0* transcript at low temperatures (Phillips, Dunn, and Hugues, 1997). Furthermore, this protein could be responsible for the organ-specific accumulation of *blt14.0* transcripts in that the absence of transcripts in certain tissues could be due to the absence of the stabilizing protein rather than the repression of transcriptional activity. Finally, the regulation of cold acclimation-related processes may also involve posttranslational factors. For example, immunodetection of the cold-induced protein MSACIA in alfalfa cell walls showed that mature proteins of different molecular weights accumulate in leaves and in crowns, which suggests that the proteins are modified posttranslationally (Ferullo et al., 1997).

Transcriptional Regulation of Cold-Inducible Genes

In the past few years, the promoters of several cold-induced genes from *Arabidopsis thaliana* and *Brassica napus* have been isolated and characterized (Yamaguchi-Shinozaki and Shinozaki, 1993; Baker, Wilhelm, and Thomashow, 1994; White et al., 1994; Yamaguchi-Shinozaki and Shinozaki, 1994; Wang et al., 1995). Functional analyses of these promoters identified regions responsible for ABA inducibility and other regions responsible for cold and drought inducibility. These data confirmed the coexistence of ABA-responsive and ABA-independent *cis*-elements capable of driving the expression of these genes. In addition, sequence comparison of the regions involved in cold and drought inducibility revealed the existence of C-repeat/DRE motifs that contain a conserved 5'CCGAC3' pentamere. Deletion studies proved that this motif is a *cis*-acting DNA regulatory element that imparts cold- and dehydration-dependent gene expression. C-repeat/DRE motifs have also been detected in the promoter sequence of the potato cold-regulated gene *ci7* (Kirch et al., 1997) and in the promoters of the alfalfa genes *MsaCiA* and *MsaCiB* (R. Desgagnés, personal communication), but their functionality has not yet been assessed. In *Arabidopsis thaliana*, Stockinger and colleagues (1997) isolated a gene encoding a 24 kDa protein, CBF1, that specifically binds to this DNA *cis*-acting element. Last, Jaglo-Ottosen and colleagues (1998) showed that the overexpression of CBF1 in transgenic plants leads to the accumulation of several cold-induced mRNAs and enhances freezing tolerance. Taken together, these data demonstrate that CBF1 is a *trans*-acting factor responsible for the regulation of at least part of the cold-induced genes. This important finding further supports the concept that the changes in gene expression associated with cold acclimation are part of a meaningful mechanism, tightly controlled by a cascade of molecular signals that involves interactions between *trans*-acting factors and *cis*-elements.

Regulatory elements have also been identified in the cold-inducible promoter of WCS120, a dehydrin-like protein from wheat (Vazquez-Tello, Ouellet, and Sarhan, 1998; Ouellet, Vazquez-Tello, and Sarhan, 1998). Although expression of *wcs120* is cold specific, the promoter region contains the low-temperature-responsive element GCCGAC, an unknown repressor element of GGGTATA, an ABA-responsive element CACCTGC, an element CACT-CAC that binds known transcriptional activators from yeast and *Drosophila*, and additional regions that enhance gene expression. This promoter is cold inducible in both freezing-sensitive and freezing-tolerant monocots and dicots, suggesting the existence of universal transcription factors responsive to low temperatures. Vazquez-Tello and colleagues (1998) have proposed that repressors bind to the *wcs120* promoter at warm temperatures and inhibit gene expression. At cold temperatures, these repressors are phosphorylated by Ca^{2+}-dependent protein kinases and released from the promoter region, allowing gene transcription to occur.

CONCLUSION

The study of the molecular mechanisms associated with cold acclimation has led to the characterization of numerous compounds and gene products that accumulate in plant tissues during the period of acclimation. The biochemical properties of these compounds, either actual or deduced from their chemical structures, are, in most cases, consistent with potential effects on water status and availability or on cryostabilization of membranes and/or proteins. However, the precise modes of action remain to be determined, and very little is known about their relative importance and interactions in the achievement of freezing tolerance. It is likely that many of these points will be addressed experimentally by the use of mutants and of transgenic plants in which a given mechanism is either impaired or enhanced. The localization of molecular events at the tissue and subcellular levels may also prove helpful in understanding their roles and in determining the spatial organization of the cold acclimation process.

Almost nothing is known of the metabolites and gene products that are down-regulated during cold acclimation. These may yet play an important role in the physicochemical adjustment of plant cells and in the ability of plants to reduce the accumulation of deleterious compounds that may inhibit the development of freezing tolerance.

Even though the study of cold acclimation at the molecular level has been an important source of information, this approach should not be considered as the only way to understand winter hardiness. Future efforts in integrating these data with anatomical, cytological, and biophysical studies are needed to draw a more accurate picture of the strategies used by plants to cope with winter.

REFERENCES

Antikainen, M., M. Griffith, J. Zhang, W.C. Hon, D.S.C. Yang, and K. Pihakaski-Maunbach (1996). Immunolocalization of antifreeze proteins in winter rye, leaves, crowns, and roots by tissue printing. *Plant Physiology* 110: 845-857.

Antikainen, M. and M. Griffith (1997). Antifreeze protein accumulation in freezing-tolerant cereals. *Physiologia Plantarum* 99: 423-432.

Arora, R. and M. Wisniewski (1994). Cold acclimation in genetically related (sibling) deciduous and evergreen peach (*Prunus persica* L. batsch). *Plant Physiology* 105: 95-101.

Arora, R. and M. Wisniewski (1996). Accumulation of a 60-kD dehydrin protein in peach xylem tissues and its relationship to cold acclimation. *HortScience* 31: 923-925.

Arora, R., M. Wisniewski, and L.J. Rowland (1996). Cold acclimation and alteration in dehydrin-like and bark storage proteins in the leaves of sibling deciduous and evergreen peach. *Journal of the American Society for Horticultural Science* 12: 915-919.

Artus, N.N., M. Uemera, P.L. Steponkus, S.J. Gilmour, and C. Lin (1996). Constitutive expression of the cold-regulated *Arabidopsis thaliana* COR15a gene affects both chloroplast and protoplast freezing tolerance. *Proceedings of the National Academy of Sciences, USA* 93: 13404-13409.

Ashworth, E.N. (1992). Formation and spread of ice in plant tissues. *Horticulture Review* 13: 215-255.

Ashworth, E.N., J.A. Anderson, and G.A. Davis (1985). Properties of ice nuclei associated with peach trees. *Journal of the American Society for Horticultural Science* 110: 287-291.

Bachmann, M., P. Matile, and F. Keller (1994). Metabolism of the raffinose family oligosaccharides in leaves of *Ajuga reptans* L. Cold acclimation, translocation, and sink to source transition: Discovery of chain elongation enzyme. *Plant Physiology* 105: 1335-1345.

Baker, S.S., K.S. Wilhelm, and M.F. Thomashow (1994). The 5′-region of *Arabidopsis thaliana cor15a* has *cis*-acting elements that confer cold-, drought- and ABA-regulated gene expression. *Plant Molecular Biology* 24: 701-713.

Boothe, J.G., M.D. de Beus, and A.M. Johnson-Flanagan (1995). Expression of a low-temperature-induced protein in *Brassica napus*. *Plant Physiology* 108: 795-803.

Bredemeijer, G.M.M. and G. Esselink (1995). Glucose 6-phosphate dehydrogenase during cold-hardening in *Lolium perenne*. *Journal of Plant Physiology* 145: 565-569.

Bruni, F. and A.C. Leopold (1991). Glass transitions in soybean seed—Relevance to anhydrous biology. *Plant Physiology* 96: 660-663.

Brush, R.A., M. Griffith, and A. Mlynarz (1994). Characterization and quantification of intrinsic ice nucleators in winter rye (*Secale cereale*) leaves. *Plant Physiology* 104: 725-735.

Burke, M., L.V. Gusta, H.A Quamme, C.J. Weiser, and P.H. Li (1976). Freezing and injury in plants. *Annual Review of Plant Physiology* 27: 507-528.

Caffrey, M., V. Fonseca, and A.C. Leopold (1988). Lipid-sugar interactions. Relevance to anhydrous biology. *Plant Physiology* 89: 754-758.

Cai, Q., G.A. Moore, and C.L. Guy (1995). An unusual group 2 LEA gene family in citrus responsive to low temperature. *Plant Molecular Biology* 29: 11-23.

Carpenter, C.D., J.A. Kreps, and A.E. Simon (1994). Genes encoding glycine-rich *Arabidopsis thaliana* proteins with RNA-binding motifs are influenced by cold treatment and an endogenous circadian rhythm. *Plant Physiology* 104: 1015-1025.

Castonguay, Y., S. Laberge, P. Nadeau, and L.-P. Vézina (1994). A cold-induced gene from *Medicago sativa* encodes a bimodular protein similar to development regulated proteins. *Plant Molecular Biology* 24: 799-804.

Castonguay, Y., P. Nadeau, and S. Laberge (1993). Freezing tolerance and alteration of translatable mRNAs in alfalfa (*Medicago sativa* L.) hardened at subzero temperatures. *Plant and Cell Physiology* 34: 31-38.

Castonguay, Y., P. Nadeau, P. Lechasseur, and L. Chouinard (1995). Differential accumulation of carbohydrates in alfalfa cultivars of contrasting winterhardiness. *Crop Science* 35: 509-516.

Cattivelli, L. and D. Bartels (1990). Molecular cloning and characterization of cold-regulated genes in barley. *Plant Physiology* 93: 1504-1510.

Chandler, P. M. and M. Robertson (1994). Gene expression regulated by abscissic acid and its relation to stress tolerance. *Annual Review of Plant Physiology and Molecular Biology* 45: 113-141.

Chauvin, L.P., M. Houde, and F. Sarhan (1993). A leaf-specific gene stimulated by light during wheat acclimation to low temperature. *Plant Molecular Biology* 23: 255-265.

Chauvin, L.P., M. Houde, and F. Sarhan (1994). Nucleotide sequence of a new member of the freezing tolerance-associated protein family in wheat. *Plant Physiology* 105: 1017-1018.

Close, T.J., R.D. Fenton, A. Yang, R. Asghar, D.A. DeMason, D.E. Crone, N.C. Meyer, and F. Moonan (1993) Dehydrin : The protein. *Current Topics in Plant Physiology* 10: 104-118.

Crespi, M.D., E.J. Zabaleta, H.G. Pontis, and G.L. Salerno (1991). Sucrose synthase expression during cold acclimation in wheat. *Plant Physiology* 96: 887-891.

Dallaire, S., M. Houde, Y. Gagné, H.S. Saini, S. Boileau, N. Chevrier, and F. Sarhan (1994). ABA and low temperature induce freezing tolerance via distinct regulatory pathways in wheat. *Plant and Cell Physiology* 35: 1-9.

Danyluk, J., M. Houde, E. Rassart, and F. Sarhan (1994). Differential expression of a gene encoding an acidic dehydrin in chilling-sensitive and freezing-tolerant gramineae species. *FEBS Letters* 344: 20-24.

Delauney, A.J. and D.P. Verma (1993). Proline biosynthesis osmoregulation in plants. *Plant Journal* 4: 215-223.

Duke, S.H. and D. C. Doehlert (1981). Root respiration, nodulation, and enzyme activities in alfalfa during cold acclimation. *Crop Science* 21: 489-494.

Duman, J.G (1994). Purification and characterization of a thermal hysteresis protein from a plant, the bittersweet nightshade *Solanum dulcamara*. *Biochimica et Biophysica Acta* 1206: 129-135.

Duman, J.G. and T.M. Olsen (1993). Thermal hysteresis protein activity in bacteria, fungi and phylogenetically diverse plants. *Cryobiology* 30: 322-328.

Duman, J.G., L. Xu, L.G. Neven, D. Tursman, and D.W. Wu (1991). Hemolymph proteins involved in insect subzero-temperature tolerance: Ice nucleators and antifreeze proteins. In *Insects at Low Temperature*, eds. R.E. Lee and D.L. Deninger. New York: Chapman and Hall, pp. 94-127.

Dunn, M.A., N.J. Goddard, L. Zhang, R.S. Pearce, and M.A. Hugues (1994). Low-temperature-responsive barley genes have different control mechanisms. *Plant Molecular Biology* 24: 879-888.

Dunn, M.A., M.A. Hugues, R.S. Pearce, and P.L. Jack (1990). Molecular characterization of a barley gene induced by cold treatment. *Journal of Experimental Botany* 232: 1405-1413.

Dunn, M.A., A. Morris, P L. Jack, and M.A. Hugues (1993). A low-temperature-responsive translation elongation factor 1α from barley (*Hordeum vulgare* L.). *Plant Molecular Biology* 23: 221-225.

Ferullo, J.-M., L.-P. Vézina, J. Rail, S. Laberge, P. Nadeau, and Y. Castonguay (1997). Differential accumulation of two glycine-rich proteins during cold-acclimation of alfalfa. *Plant Molecular Biology* 33: 625-633.

Flinn, C.L. and E.N. Ashworth (1994). Seasonal changes ice distribution and xylem development in blueberry flower buds. *Journal of the American Society for Horticultural Science* 119: 1176-1184.

Fujikawa, S., K. Kuroda, and J. Ohtani (1997). Seasonal changes in dehydration tolerance of xylem ray parenchyma cells of *Stylax obassia* twigs that survive freezing temperatures by deep supercooling. *Protoplasma* 197: 34-44.

Gatschet, M.J., C.M. Taliaferro, D.R. Porter, M.P. Anderson, J.A. Anderson, and K.W. Jackson (1996). A cold-regulated protein from bermudagrass crowns is a chitinase. *Crop Science* 36: 712-718.

Gilmour, S.J., N.N. Artus, and M.F. Thomashow (1991). cDNA sequence analysis and expression of two cold-regulated genes of *Arabidopsis thaliana*. *Plant Molecular Biology* 18: 13-21.

Gilmour, S.J., N.N. Artus, and M.F. Thomashow (1992). cDNA sequence analysis and expression of two cold-regulated genes of *Arabidopsis thaliana*. *Plant Molecular Biology* 18: 13-21.

Goldstein, G. and P.S. Nobel (1991). Changes in osmotic pressure and mucilage during low-temperature acclimation of *Opuntia ficus-indica*. *Plant Physiology* 97: 954-961.

Goldstein, G. and P.S. Nobel (1994). Water relations and low-temperature acclimation for cactus species varying in freezing tolerance. *Plant Physiology* 104: 675-681.

Goldstein, J., N.S. Pollitt, and M. Inouye (1990). Major cold shock proteins of *Escherichia coli*. *Proceedings of the National Academy of Sciences, USA* 87: 283-287.

Goodwin, W., J. Pallas, and G.I. Jenkins (1996). Transcripts of a gene encoding a putative cell wall-plasma membrane linker protein are specifically cold-induced in *Brassica napus*. *Plant Molecular Biology* 31: 771-781.

Griffith, M., P. Ala, D.S.C. Yang, W.C. Hon, and B.A. Moffat (1992). Antifreeze protein produced endogenously in winter rye leaves. *Plant Physiology* 100: 593-596.

Griffith, M. and K.V. Ewart (1995). Antifreeze proteins and their potential use in frozen foods. *Biotechnology Advances* 13: 375-402.

Griffith, M. and H.C.H. McIntyre (1993). The interrelationship between growth and frost tolerance in winter rye. *Physiologia Plantarum* 87: 335-344.

Gross, D.C., E.L. Proebsting Jr., and H. Maccrindle-Zimmerman (1988). Development, distribution, and characteristics of intrinsic, nonbacterial ice nuclei in *Prunus* wood. *Plant Physiology* 88: 915-922.

Guinchard, M.P., C. Robin, P. Grieu, and A. Guckert (1997). Cold acclimation in white clover subjected to chilling and frost: Changes in water and carbohydrates status. *European Journal of Agronomy* 6: 225-233.

Guo, W., R.W. Ward, and M.F. Thomashow (1992). Characterization of a cold-regulated wheat gene related to *Arabidopsis cor47*. *Plant Physiology* 100: 915-922.

Guy, C.L. (1990). Cold acclimation and freezing stress tolerance: Role of protein metabolism. *Annual Review of Plant Physiology and Molecular Biology* 41: 187-223.

Guy, C.L., J.L.A. Huber, and S.C. Huber (1992). Sucrose phosphate synthase and sucrose accumulation at low temperature. *Plant Physiology* 100: 502-508.

Hertel, A. and E. Steudle (1997). The function of water channels in *Chara:* The temperature dependence of water and solute flows provides evidence for composite membrane transport and for a slippage of small organic solutes across water channels. *Planta* 202: 324-335.

Hesse, H. and L. Willmitzer (1996). Expression analysis of a sucrose synthase gene from sugar beet (*Beta vulgaris* L.). *Plant Molecular Biology* 30: 863-872.

Hill, L.M., R. Reimholz, R. Schröder, T.H. Nielsen, and M. Stitt (1996). The onset of sucrose accumulation in cold-stored potato tubers is caused by an increased rate of sucose synthesis and coincides with low levels of hexose-phosphates, an activation of sucrose phosphate synthase and the appearance of a new form of amylase. *Plant Cell and Environment* 19: 1223-1237.

Hincha, D.K., U. Sonnewald, L. Willmitzer, and J.M. Schmitt (1996). The role of sugar accumulation in leaf frost hardiness—Investigations with transgenic tobacco expressing a bacterial pyrophosphatase or a yeast invertase gene. *Journal of Plant Physiology* 147: 604-610.

Hoekstra, F.A., L.M. Crowe, and J.H. Crowe (1989). Differential desiccation sensitivity of corn and *Pennisetum* pollen linked to their sucrose content. *Plant Cell and Environment* 12: 83-91.

Holappa, L.D., and M.K. Walker-Simmons (1995). The wheat abscissic acid-responsive protein kinase mRNA, PKABA1, is up-regulated by dehydration, cold temperature, and osmotic stress. *Plant Physiology* 108: 1203-1210.

Holoday, A.S.,W. Martindale, R. Alred, A.L. Brooks, and R.C. Leegood (1992). Changes in activities of enzymes of carbon metabolism in leaves during exposure of plants to low temperature. *Plant Physiology* 98: 1105-1114.

Hon, W.C., M. Griffith, P. Chong, and D.S.C. Yang (1994). Extraction and isolation of antifreeze proteins from winter rye (*Secale cereale* L.) leaves. *Plant Physiology* 104: 971-980.

Hon, W.C., M. Griffith, A. Mlynarz, Y.C. Kwok, and D.S.C. Yang (1995). Antifreeze proteins in winter rye are similar to pathogenesis-related proteins. *Plant Physiology* 109: 879-889.

Horvath, D.P., B.K. McLarney, and M.F. Thomashow (1993). Regulation of *Arabidopsis thaliana* (L.) Heyn. *cor78* in response to low temperature. *Plant Physiology* 103: 1047-1053.

Houde, M., C. Daniel, M. Lachapelle, F. Allard, S. Laliberté, and F. Sarhan (1995). Immunolocalization of freezing-tolerance-associated proteins in the cytoplasm and nucleoplasm of wheat crown tissues. *Plant Journal* 8: 583-593.

Houde, M., J. Danyluk, J.F. Laliberté, E. Rassart, R. Dhindsa, and F. Sarhan (1992). Cloning, characterization, and expression of a cDNA encoding a 50-kilodalton protein specifically induced by cold acclimation in wheat. *Plant Physiology* 99: 1381-1387.

Hughes, M.A. and M.A. Dunn (1996). The molecular biology of plant acclimation to low temperature. *Journal of Experimental Botany* 47: 291-305.

Hughes, M.A., M.A. Dunn, R.S. Pearce, A.J. White, and L. Zhang (1992). An abscissic-acid-responsive, low temperature barley gene has homologuey with a maize phospholipid transfer protein. *Plant Cell and Environment* 15: 861-865.

Hurry, V.M., G. Malmberg, P. Gardeström, and G. Öquist (1994). Effects of a short-term shift to low temperature and of long-term cold hardening on photosynthesis and ribulose 1,5-biphosphate carboxylase/oxygenase and sucrose phosphate synthase activity in leaves of winter rye (*Secale cereale* L.). *Plant Physiology* 106: 983-990.

Hurry, V.M., Å. Strand, M. Tobieson, P. Gardeström, and G. Öquist (1995). Cold hardening of spring and winter wheat and rape results in differential effects on growth, carbon metabolism, and carbohydrate content. *Plant Physiology* 109: 697-706.

Hurry, V.M., M. Tobieson, H. Kodama, S. Krömer, P. Gardeström, and G. Öquist (1995). Mitochondria contribute to increased photosynthetic capacity of leaves of winter rye (*Secale cereale* L.) following cold hardening. *Plant Cell and Environment* 18: 69-76.

Igarashi, Y., Y. Yoshiba, Y. Sanada, K. Yamaguchi-Shinozaki, K. Wada, and K. Shinozaki (1997). Characterization of the gene for $\Delta 1$-pyrroline-5-carboxylate synthetase and correlation between the expression of the gene and salt tolerance in *Oryza sativa* L. *Plant Molecular Biology* 33: 857-865.

Imai, R., L. Chang, A. Ohta, E.A. Bray, and M. Takagi (1996). A lea-class gene of tomato confers salt and freezing tolerance when expressed in *Saccharomyces cerevisiae*. *Gene* 170: 243-248.

Ishikawa, M., W.S. Price, H. Ide, and Y. Arata (1997). Visualization of freezing behaviors in leaf and flower buds of full-moon maple by nuclear magnetic resonance microscopy. *Plant Physiology* 115: 1515-1524.

Jaglo-Ottosen, K.R., S. Gilmour, D.G. Zarka, O. Schabenberger, and M.F. Thomashow (1998). *Arabidopsis* CBF1 overexpression induces COR genes and enhances freezing tolerance. *Science* 280: 104-106.

Jarillo, J.A., J. Capel, A. Leyva, J.M. Martínez-Zapatar, and J. Salinas (1994). Two related low-temperature-inducible genes of *Arabidopsis* encode proteins showing high homologuey to 14-3-3 proteins, a family of putative kinase regulators. *Plant Molecular Biology* 25: 693-704.

Jonak, C., S. Kiegerl, W. Ligterink, P.J. Barker, N.S. Huskisson, and H. Hirt (1996). Stress-signaling in plants: A mitogen-activated protein kinase pathway is activated by cold and drought. *Proceedings of the National Academy of Sciences, USA* 93: 11274-11279.

Kacperska, A. (1993). Water potential alterations—A prerequisite or a triggering stimulus for the development of freezing tolerance in overwintering herbaceous plants? In *Advances in Plant Cold Hardiness*, eds. P.H. Li and L. Christersson. Boca Raton, FL: CRC Press, Inc., pp. 73-91.

Kavi Kishor, P.B., Z. Hong, G.-H. Miao, C.-A. Hu, and D.P. Verma (1995). Overexpression of Δ1-pyrroline-5-carboxylate synthetase increases proline production and confers osmotolerance in transgenic plants. *Plant Physiology* 108: 1387-1394.

Kawczynski, W. and R. Dhindsa (1996). Alfalfa nuclei contain cold-responsive phosphoproteins and accumulate heat-stable proteins during cold treatment of seedlings. *Plant and Cell Physiology* 37: 1204-1210.

Kazuoka, T. and K. Oeda (1994). Purification and characterization of COR85-oligomeric complex from cold-acclimated spinach. *Plant and Cell Physiology* 35: 601-611.

Kenrick, J.R. and D.G. Bishop (1986). The fatty acid composition of phosphatidylglycerol and sulphoquinovosyldiaclyglycerol of higher plants in relation to chilling sensitivity. *Plant Physiology* 81: 946-949.

Kerr, G.P. and J.V. Carter (1990). Tubulin isotypes in rye roots are altered during cold acclimation. *Plant Physiology* 93: 83-88.

Kirch, H.H., J. van Berkel, H. Glaczinski, F. Salamini, and C. Gebhardt (1997). Structural organization, expression and promoter activity of a cold-stress-inducible gene of potato (*Solanum tuberosum* L.). *Plant Molecular Biology* 33: 897-909.

Klaff, P., D. Riesner, and G. Steger (1996). RNA structure and the regulation of gene expression. *Plant Molecular Biology* 32: 89-106.

Koike, M., D. Takezawa, K. Arakawa, and S. Yoshida (1997). Accumulation of 19-kDa plasma membrane polypeptide during induction of freezing tolerance in wheat suspension-cultured cells by abscissic acid. *Plant and Cell Physiology* 38: 707-716.

Koster, K.L., and D.V. Lynch (1992). Solute accumulation and compartmentation during the cold acclimation of puma rye. *Plant Physiology* 98: 108-113.

Krishna, P., M. Sacco, J.F. Cherutti, and S. Hill (1995). Cold-induced accumulation of *hsp90* transcript in *Brassica napus*. *Plant Physiology* 107: 915-923.

Kurkela, S. and M. Franck (1990). Cloning and characterization of a cold- and ABA-inducible *Arabidopsis* gene. *Plant Molecular Biology* 15: 137-144.

Kurkela, S. and M. Franck (1992). Structure and expression of *kin2*, one of two cold- and ABA-induced genes of *Arabidopsis thaliana*. *Plant Molecular Biology* 19: 689-692.

Laberge, S., Y. Castonguay, and L.-P. Vézina (1993). New cold- and drought-regulated gene from *Medicago sativa*. *Plant Physiology* 101: 1411-1412.

Lång, V., E. Mäntylä, B. Welin, B. Sundberg, and E.T. Palva (1994). Alterations in water status, endogenous abscissic acid content and expression of *rab18* gene during the development of freezing tolerance in *Arabidopsis thaliana*. *Plant Physiology* 104: 1341-1349.

Lång, V. and E. T. Palva (1992). The expression of a rab-related gene, *rab18*, is induced by abscissic acid during the cold acclimation process of *Arabidopsis thaliana* (L.) Heynh. *Plant Molecular Biology* 20: 951-962.

Lee, S.P. and T.H.H. Chen (1993). Molecular cloning of abscissic acid-responsive mRNAs expressed during the induction of freezing tolerance in bromegrass (*Bromus inermis* Leyss) suspension culture. *Plant Physiology* 101: 1089-1096.

Lee, T.-M., H.-S. Lur, and C. Chu (1997). Role of abscissic acid in chilling tolerance of rice (*Oryza sativa* L.) seedlings. II. Modulation of free polyamine levels. *Plant Science* 126: 1-10.

Levitt, J. (1980). *Responses of plants to environmental stresses,* Vol I: *Chilling, freezing, and high temperature stresses.* New York: Academic Press.

Li, P.H. (1984). Subzero temperature stress physiology of herbaceous plants. *Horticulture Review* 6: 373-416.

Lin, C. and M.F. Thomashow (1992a). DNA sequence analysis of a complementary DNA for cold-regulated *Arabidopsis* gene *cor15* and characterization of the COR15 polypeptide. *Plant Physiology* 99: 519-525.

Lin, C. and M.F. Thomashow (1992b). A cold-regulated *Arabidopsis* gene encodes a polypeptide having potent cryoprotective activity. *Biochemical and Biophysical Research Communications* 183: 1103-1108.

Livingston, D.P., III, and C.A. Henson (1998). Apoplastic sugars, fructan, fructan hydrolase, and invertase in winter oat: Responses to second-phase cold hardening. *Plant Physiology* 116: 403-408.

Luo, M., L. Lin, R.D. Hill, and S.S. Mohapatra (1991). Primary structure of an environmental stress and abscissic acid-inducible alfalfa protein. *Plant Molecular Biology* 17: 1267-1269.

Luo, M., J.H. Liu, S. Mohapatra, R.D. Hill, and S.S. Mohapatra (1992). Characterization of a gene family encoding abscissic acid- and environmental stress-inducible proteins of alfalfa. *Journal of Biological Chemistry* 267: 15367-15374.

Mäntylä, E., V. Lång, and E.T. Palva (1995). Role of abscissic acid in drought-induced freezing tolerance, cold acclimation, and accumulation of LT178 and RAB 18 proteins in *Arabidopsis thaliana. Plant Physiology* 107: 141-148.

Maraña, C., F. García-Olmedo, and P. Carbonero (1990). Differential expression of two types of sucrose synthase-encoding genes in wheat in response to anaerobiosis, cold shock and light. *Gene* 85: 167-172.

Marentes, E., M. Griffith, A. Mlynarz, and R.A. Brush (1993). Proteins accumulate in the apoplast of winter rye leaves during cold acclimation. *Physiologia Plantarum* 87: 499-507.

Maurel, C., R.T. Kado, J. Guern, and M.J. Chrispeels (1995). Phosphorylation regulates the water channel activity of the seed-specific aquaporin α-TIP. *EMBO Journal* 14: 3028-3035.

Mayo, B., S. Derzelle, M. Fernández, C. Léonard, T. Ferain, P. Hols, J.E. Suárez, and J. Delcour (1997). Cloning and characterization of *cspL* and *cspP,* two cold-inducible genes from *Lactobacillus plantarum. Journal of Bacteriology* 179: 3039-3042.

Millard, M.M., O.B. Veisz, D.T. Krizek, and M. Line (1995). Magnetic resonance imaging (MRI) of water during cold acclimation and freezing in winter wheat. *Plant Cell and Environment* 18: 535-544.

Millard, M.M., O.B. Veisz, D.T. Krizek, and M. Line (1996). Thermodynamic analysis of the physical state of water during freezing in plant tissue, based on the temperature dependence of proton spin-spin relaxation. *Plant Cell and Environment* 19: 33-42.

Mizoguchi, T., N. Hayashida, K. Yamaguchi-Shinozaki, H. Kamada, and K. Shinozaki (1995). Two genes that encode ribosomal-protein S6 kinase homologues are induced by cold or salinity stress in *Arabidopsis thaliana*. *FEBS letters* 358: 199-204.

Mizoguchi, T., K. Irie, T. Hirayama, N. Hayashida, K. Yamaguchi-Shinozaki, K. Matsumoto, and K. Shinozaki (1996). A gene encoding a mitogen-activated protein kinase is induced simulatneously with genes for a mitogen-activated protein kinase and an S6 ribosomal protein kinase by touch, cold, and water stress in *Arabidopsis thaliana*. *Proceedings of the National Academy Sciences, USA* 93: 765-769.

Monroy, A.F., Y. Castonguay, S. Laberge, F. Sarhan, L.-P. Vézina, and R. Dhindsa (1993). A new cold-induced alfalfa gene is associated with enhanced hardening at subzero temperature. *Plant Physiology* 102: 873-879.

Monroy, A.F., F. Sarhan, and R. Dhindsa (1993). Cold-induced changes in freezing tolerance, protein phosphorylation, and gene expression. *Plant Physiology* 102: 1227-1235.

Muthalif, M.M. and L.J. Rowland (1994). Identification of dehydrin-like proteins responsive to chilling in floral buds of blueberry (*Vaccinum*, section *cyanococcus*). *Plant Physiology* 104: 1439-1447.

Nadeau, P., S. Delaney, and L. Chouinard (1987). Effects of cold hardening on the regulation of polyamine levels in wheat (*Triticum aestivum* L.) and alfalfa (*Medicago sativa* L.). *Plant Physiology* 84: 73-77.

Neven, L.G., D.W. Haskell, C.L. Guy, N. Denslow, P.A. Klein, L.G. Green, and A. Silverman (1992). Association of 70-kilodalton heat-shock cognate proteins with acclimation to cold. *Plant Physiology* 99: 1362-1369.

Neven, L.G., D.W. Haskell, A. Hofig, Q.-B. Li, and C.L. Guy (1993). Characterization of a spinach gene responsive to low temperature and water stress. *Plant Molecular Biology* 21: 291-305.

Nielsen, T.H., U. Deiting, and M. Stitt (1997). A β-amylase in potato tubers is induced by storage at low temperature. *Plant Physiology* 113: 503-510.

Nishiyama, H., K. Itoh, Y. Kaneko, M. Kishishita, and O. Yoshida (1997). A glycine-rich RNA-binding protein mediating cold-inducible supression of mammalian cell growth. *Journal of Cell Biology* 137: 899-908.

Nordin, K., P. Heino, and E.T. Palva (1991). Separate signal pathways regulate the expression of a low-temperature-induced gene in *Arabidopsis thaliana* (L.) Heynh. *Plant Molecular Biology* 16: 1061-1071.

Nordin, K., T. Vahala, and E.T. Palva (1993). Differential expression of two related, low-temperature-induced genes in *Arabidopsis thaliana* (L.) Heynh. *Plant Molecular Biology* 21: 641-653.

Ögren, E. (1997). Relationship between temperature, respiratory loss of sugar and premature dehardening in dormant Scot pine seedlings. *Tree Physiology* 17: 47-51.

Ögren, E., T. Nilsson, and L.-G. Sundblad (1997). Relationship between respiratory depletion of sugars and loss of cold hardiness in coniferous seedlings over-wintering at raised temperatures: Indications of different sensitivities of spruce and pine. *Plant Cell and Environment* 20: 247-253.

Olien, C.R. (1984). An adaptive response of rye to freezing. *Crop Science* 24: 51-54.

Olien, C.R. and J.L. Clark (1995). Freeze-induced changes in carbohydrates associated with hardiness of barley and rye. *Crop Science* 35: 496-502.

Öquist, G., V.M. Hurry, and N.P.A. Huner (1993). Low-temperature effects on photosynthesis and correlation with freezing tolerance in spring and winter cultivars of wheat and rye. *Plant Physiology* 101: 245-250.

Orr, W., B. Iu, T.C. White, L.S. Robert, and J. Singh (1992). Complementary DNA sequence of a low-temperature-induced *Brassica napus* gene with homologuey to the *Arabidopsis thaliana* kin1 gene. *Plant Physiology* 98: 1532-1534.

Orr, W., T.C. White, B. Iu, L. Robert, and J. Singh (1995). Characterization of a low-temperature-induced cDNA from winter *Brassica napus* encoding the 70 kDa subunit of tonoplast ATPase. *Plant Molecular Biology* 28: 943-948.

Ouellet, F., M. Houde, and F. Sarhan (1993). Purification, characterization and cDNA cloning of the 200 kDa protein induced by cold acclimation in wheat. *Plant and Cell Physiology* 34: 59-65.

Ouellet, F., A. Vazquez-Tello, and F. Sarhan (1998). The wheat wcs120 promoter is cold-inducible in both monocotyledonous and dicotyledonous species. *FEBS Letters* 423: 324-328.

Palta, J.P., B.D. Whitaker, and L.S. Weiss (1993). Plasma membrane lipids associated with genetic variability in freezing tolerance and cold acclimation of *Solanum* species. *Plant Physiology* 103: 793-803.

Paquin, R. (1984). Influence of the environment on cold hardening and winter survival of forage and cereal species with consideration of proline as metabolic marker of hardening. In *Being Alive on Land*, eds. N.S. Margaris, M. Arianoustou-Faraggitaki, and W.C. Oechel. Boston, MA: Dr. W. Junk, Kluwer Academic Publishers, pp. 137-154.

Pearce, R.S. (1988). Extracellular ice and cell shape in frost-stressed cereal leaves: A low-temperature scanning-electron-microscopy study. *Planta* 175: 313-324.

Phillips, J.R., M.A. Dunn, and M.A. Hugues (1997). mRNA stability and localization of the low-temperature-responsive barley gene family blt14. *Plant Molecular Biology* 33: 1013-1023.

Quamme, H.A. (1991). Application of thermal analysis to breeding fruit crops for increased cold hardiness. *HortScience* 26: 513-517.

Quamme, H.A., W.A. Su, and L.J. Veto (1995). Anatomical features facilitating supercooling of the flower within the dormant peach flower bud. *Journal of the American Society for Horticultural Science* 120: 814-822.

Rajashekar, C.B. and M.J. Burke (1996). Freezing characteristics of rigid plant tissues. *Plant Physiology* 111: 597-603.

Reimholz, R., M. Geiger, U. Deiting, K.P. Krause, U. Sonnewald, and M. Stitt (1997). Potato plants contain multiple forms of sucrose phosphate synthase, which differ in their tissue distribution, their levels during development, and their responses to low temperature. *Plant Cell and Environment* 20: 291-305.

Rorat, T., W. Irzykowski, and W.J. Grygorowicz (1997). Identification and expression of novel cold-induced genes in potato *(Solanum sogarandinum)*. *Plant Science* 124: 69-78.

Sabehat, A., D. Weiss, and S. Lurie (1996). The correlation between heat-shock protein accumulation and persistence and chilling tolerance in tomato fruit. *Plant Physiology* 110: 531-537.

Sáez-Vásquez, J. and M. Delseny (1995). A rapeseed cold-inducible transcript encodes a phosphenolpyruvate carboxykinase. *Plant Physiology* 109: 611-618.

Sáez-Vásquez, J., M. Raynal, L. Meza-Basso, and M. Delseny (1993). Two related, low-temperature-induced genes from *Brassica napus* are homologous to the human tumour *bbc1* (breast basic conserved) gene. *Plant Molecular Biology* 23: 1211-1221.

Sakai, A. and W. Larcher (1987). *Frost Survival of Plants.* New York: Springer-Verlag.

Salerno, G.L. and H.G. Pontis (1989). Raffinose synthesis in *Chlorella vulgaris* cultures after a cold shock. *Plant Physiology* 89: 648-651.

Salzman, R.A., R.A. Bressan, P.M. Hasegawa, E.N. Ashworth, and B.P. Bordelon (1996). Programmed accumulation of LEA-like proteins during desiccation and cold acclimation of overwintering grape buds. *Plant Cell and Environment* 19: 713-720.

Sato, N. (1994). A cold-regulated cyanobacterial gene cluster encodes RNA-binding protein and ribosomal protein S21. *Plant Molecular Biology* 24: 819-823.

Savouré, A., X.J. Hua, N. Bertauche, M. Van Montagu, and N. Verbruggen (1997). Abscissic acid-independent and abscissic acid-dependent regulation of proline biosynthesis following cold and osmotic stresses in *Arabidopsis thaliana*. *Molecular and General Genetics* 254: 104-109.

Schneider, A., F. Salamini, and C. Gebhardt (1997). Expression patterns and promoter activity of the cold-regulated gene *ci21A* of potato. *Plant Physiology* 113: 335-345.

Schowalter, A.M. (1993). Structure and function of plant cell wall proteins. *Plant Cell* 5: 9-23.

Steponkus, P.L., M. Uemera, R.A. Balsamo, T. Arvinte, and D.V. Lynch (1988). Transformation of the cryobehavior of rye protoplasts by modification of the plasma membrane lipid composition. *Proceedings of the National Academy of Sciences, USA* 85: 9026-9030.

Steponkus, P.L. and M.S. Webb (1992). Freeze-induced dehydration and membrane destabilization in plants. In *Water and Life: Comparative Analysis of Water Relationships at the Organismic, Cellular, and Molecular Level*, Eds. G.N. Somero, C.B. Osmond, and C.L. Bolis. New York: Springer-Verlag, pp.338-362.

Steponkus, P.L. and M.S. Webb (1993). Freeze-induced membrane ultrastructural alterations in rye (*Secale cereale*) leaves. *Plant Physiology* 101: 955-963.

Stockinger, E.J., S.J. Gilmour, and M.F. Thomashow (1997). *Arabidopsis thaliana* CBF1 encodes an AP2 domain-containing transcriptional activator that binds to the C-repeat/DRE, a cis-acting DNA regulatory element that stimulates transcription in response to low temperature and water deficit. *Proceedings of the National Academy of Sciences, USA* 94: 1035-1040.

Strauss, G. and H. Hauser (1986). Stabilization of lipid bilayer vesicles by sucrose during freezing. *Proceedings of the National Academy of Sciences, USA* 83: 2422-2426.

Sun, W. Q., T C. Irving, and A.C. Leopold (1994). The role of sugar, vitrification and membrane phase transition in seed desiccation tolerance. *Physiologia Plantarum* 90: 621-628.

Sutton, F., X. Ding, and D.G. Kenefick (1992). Group 3 LEA gene HVA1 regulation by cold acclimation and deacclimation in two barley cultivars with varying freeze resistance. *Plant Physiology* 99: 338-340.

Tarczynski, M.C., R.G. Jensen, and H.J. Bohnert (1993). Stress protection of transgenic tobacco by production of the osmolyte mannitol. *Science* 259: 508-510.

Tognetti, J.A., G.L. Salerno, M.D. Crespi, and H.G. Pontis (1990). Sucrose and fructan metabolism of different wheat cultivars at chilling temperatures. *Physiologia Plantarum* 78: 554-559.

Uemera, M., R.A. Joseph, and P.L. Steponkus (1995). Cold acclimation of *Arabidopsis thaliana*. *Plant Physiology* 109: 15-30.

van Berkel, J., F. Salamini, and C. Gebhardt (1994). Transcripts accumulating during cold storage of potato (*Solanum tuberosum* L.) tubers are sequence related to stress-responsive genes. *Plant Physiology* 104: 445-452.

van Zee, K., F.Q. Chen, P.M. Hayes, T.J. Close, and T.H.H. Chen (1995). Cold-specific induction of a dehydrin gene family member in barley. *Plant Physiology* 108: 1233-1239.

Vazquez-Tello, A., F. Ouellet, and F. Sarhan (1998). Low-temperature-stimulated phosphorylation regulates the binding of nuclear factors to the promoter of *wcs120*, a cold-specific gene in wheat. *Molecular and General Genetics* 257: 157-166.

Volonec, J.J., P.J. Boyce, and K.L. Hendershot (1991). Carbohydrate metabolism in taproots of *Medicago sativa* L. during winter adaptation and spring regrowth. *Plant Physiology* 96: 786-793.

Walbot, V. and M. Hahn (1989). Effects of cold treatment on protein synthesis and mRNA levels in rice leaves. *Plant Physiology* 91: 930-938.

Wang, H., R. Datla, F. Georges, M. Loewen, and A.J. Cutler (1995). Promoters from *kin1* and *cor6.6*, two homologueous *Arabidopsis thaliana* genes: Transcriptional regulation and gene expression induced by low temperature, ABA, osmoticum and dehydration. *Plant Molecular Biology* 28: 605-617.

Webb, M.S., M. Uemera, and P.L. Steponkus (1994). A comparison of freezing injury in oat and rye. Two cereals at the extremes of freezing tolerance. *Plant Physiology* 104: 467-478.

Weimken, V. and K. Ineichen (1993). Effect of temperature and photoperiod on the raffinose content of spruce roots. *Planta* 190: 387-392.

Welbaum, G.E., D. Bian, D.R. Hill, R.L. Grayson, and M.K. Gunatilaka (1997). Freezing tolerance, protein composition, and abscissic acid localization and content of pea epicotyl, shoot, and root tissue in response to temperature and water stress. *Journal of Experimental Botany* 48: 643-654.

Welin, B.V., Å. Olson, M. Nylander, and E.T. Palva (1994). Characterization and differential expression of dhn/lea/rab-like genes during cold-acclimation and drought stress in *Arabidopsis thaliana*. *Plant Molecular Biology* 26: 131-144.

Weretilnyk, E.A. and A.D. Hanson (1990). Molecular cloning of a plant betaine-aldehyde dehydrogenase, an enzyme implicated in adaptation to salinity and drought. *Proceedings of the National Academy of Sciences, USA* 87: 2745-2749.

Weretilnyk, E.A., W. Orr, T.C. White, and J. Singh (1993). Characterization of three related low-temperature-regulated cDNAs from winter *Brassica napus*. *Plant Physiology* 101: 171-177.

White, T.C., D. Simmonds, P. Donaldson, and J. Singh (1994). Regulation of *BN115*, a low-temperature-reponsive gene from winter *Brassica napus*. *Plant Physiology* 106: 917-928.

Wilhelm, K.S. and M.F. Thomashow (1993). *Arabidopsis thaliana cor15b*, an apparent homologueue of *cor15a*, is strongly responsive to cold and ABA, but not drought. *Plant Molecular Biology* 23: 1073-1077.

Williams, J.P., O. Francisco, and M.U. Khan (1997). A low-temperature-sensitive mutant of *Brassica napus* that reduces the level of trienoic fatty acid in chloroplast lipids. *Plant Physiology* 35: 257-264.

Wisniewski, M. and G. Davis (1995). Immunogold localization of pectins and glycoproteins in tissues of peach with reference to deep supercooling. *Trees* 9: 253-260.

Wisniewski, M., S.E. Lindow, and E.N. Ashworth (1997). Observations of ice nucleation and propagation in plants using infrared video thermography. *Plant Physiology* 113: 327-334.

Witt, W., A. Buchholz, and J.J. Sauter (1995). Binding of endoamylase to native starch grains from poplar wood. *Journal of Experimental Botany* 46: 1761-1769.

Wolfraim, L.A. and R. Dhindsa (1993). Cloning and sequencing of the cDNA for *cas17*, a cold acclimation-specific gene of alfalfa. *Plant Physiology* 103: 667-668.

Wolfraim, L.A., R. Langis, H. Tyson, and R. Dhindsa (1993). cDNA sequence, expression, and transcript stability of a cold acclimation-specific gene, *cas18*, of alfalfa (*Medicago falcata*) cells. *Plant Physiology* 101: 1275-1282.

Xu, D., X. Duan, B. Wang, B. Hong, T.-H. D. Ho, and R. Wu (1996). Expression of a late embryogenesis abundant protein gene, HVA1, from barley confers tolerance to water deficit and salt stress in transgenic rice. *Plant Physiology* 110: 249-257.

Yamada, S., M. Katsuhara, W.B. Kelly, C.B. Michalowski, and H.J. Bohnert (1995). A family of transcripts encoding water channel proteins: Tissue specific expression in the common ice plant. *Plant Cell* 7: 1129-1142.

Yamaguchi-Shinozaki, K. and K. Shinozaki (1993). Characterization of the expression of a desiccation-responsive *rd29* gene of *Arabidopsis thaliana* and analysis of its promoter in transgenic plants. *Molecular and General Genetics* 236: 331-340.

Yamaguchi-Shinozaki, K. and K. Shinozaki (1994). A novel *cis*-acting element in an *Arabidopsis* gene is involved in responsiveness to drought, low-temperature, or high-salt stress. *Plant Cell* 6: 251-264.

Yoshiba, Y., T. Kiyosue, T. Katagiri, H. Ueda, T. Mizoguchi, K. Yamaguchi-Shinozaki, K. Wada, Y. Harada, and K. Shinozaki (1995). Correlation between the induction of a gene for Δ1-pyrroline-5-carboxylate synthetase and the accumulation of proline in *Arabidopsis thaliana* under osmotic stress. *Plant Journal* 7: 751-760.

Zhu, B., T.H.H. Chen, and P.H. Li (1993). Expression of an ABA responsive osmotin-like gene during the induction of freezing tolerance in *Solanum* commersonii. *Plant Molecular Biology* 21: 729-735.

Chapter 5

Signal Transduction Under Low-Temperature Stress

Sirpa Nuotio
Pekka Heino
E. Tapio Palva

INTRODUCTION

Plants are constantly exposed to, and need to survive, a changing and often unfavorable environment. This has led to the evolution of adaptive mechanisms that permit plant cells to sense environmental stimuli and to activate responses that increase their tolerance to environmental stresses. One of the major environmental challenges to plant growth and distribution is suboptimal temperature, which can cause serious damage to a number of plants. However, different plant species vary widely in their ability to tolerate low-temperature stress (Levitt, 1980; Sakai and Larcher, 1987). Chilling-sensitive tropical species can be irreparably damaged when temperatures drop below 10°C. Injuries are due to impairments in metabolic processes, such as alterations in membrane properties, interactions between macromolecules, and enzymatic reactions. Freezing-sensitive tissues are damaged at subzero temperatures as a result of ice formation. In contrast, freezing-tolerant temperate plants are able to survive considerable extracellular freezing. On a cellular level, freezing stress is mainly due to freeze-induced cellular dehydration and injuries in membranes (Steponkus, Uemura, and Webb, 1993). Overwintering woody plants, as well as dormant seeds of many plant species, exhibit extremes of low-temperature survival.

Many plant species native to regions commonly encountering subzero temperatures use the low nonfreezing temperatures as signals to increase their freezing tolerance. This adaptive process, known as cold acclimation, has been the focus of intensive studies since the beginning of this century. The

The work in the authors' laboratory is financed by the Academy of Finland, Biocentrum Helsinki, and EU BIO4-CT-960062.

acclimation process is correlated with a number of cellular and metabolic changes. Recent studies have started to define the molecular basis of these changes, which has led to the characterization of a large number of genes induced by low temperatures (for reviews, see Hughes and Dunn, 1996; Palva and Heino, 1997). However, little is known about the regulatory mechanisms controlling the low-temperature responses. Current attempts in several laboratories aim to clarify how plants perceive the low-temperature signal and transduce it to the nucleus to activate specific gene expression. This chapter will focus on discussing the current progress in the studies of cellular signaling that leads to increased freezing tolerance, as well as some exciting implications of that work.

Cold Acclimation

Plants need to adjust to both daily and seasonal fluctuations in temperature. Seasonal acclimation is operational in overwintering herbs and in woody plants. In woody species, acclimation is a two-step process in which the first phase of acclimation is triggered by shortening of the daylength and the second, by cooling temperatures. In both annual and overwintering herbaceous plants, freezing tolerance is not affected by daylength, but acclimation is triggered by low temperatures only.

Cold acclimation is a rather rapid process in many annual herbaceous plant species. A substantial increase in freezing tolerance of *Arabidopsis* can be observed already after twenty-four hours exposure to the acclimation temperatures (Thomashow, 1994). Upon return to normal growth temperatures, the level of freezing tolerance rapidly returns to that of nonacclimated plants. In addition to low nonfreezing temperatures, cold acclimation can be triggered by exposing the plants to mild water stress or exogenously applied abscisic acid (ABA) (Lång, Heino, and Palva, 1989; Guy et al., 1992; Mäntylä, Lång, and Palva, 1995).

Cold acclimation is associated with several physiological and biochemical alterations in the plants. The best-characterized changes include increases in soluble sugars, proteins, amino acids, and organic acids, as well as modification of membrane lipid composition and alterations in gene expression (for reviews, see Guy, 1990; Palva, 1994; Hughes and Dunn, 1996; Li and Chen, 1997; Palva and Heino 1997). Although the causal relationship of these changes to increased freezing tolerance is rather unclear, it has been demonstrated that new protein synthesis is essential for the acclimation process (Guy, 1990).

Genes and Proteins Induced by Low Temperatures

The ability to cold acclimate and tolerate freezing temperatures is a polygenic trait best characterized in overwintering cereals (Hughes and

Dunn, 1996; Sarhan, Ouellet, and Vazquez-Tello, 1997) and potato (Palta and Simon, 1993). Although only a limited number of genes may be crucial for the acclimation process (Palta and Simon, 1993), it has been estimated that the expression of hundreds of genes may be altered when plants are exposed to low temperatures (Guy, 1990). A large number of low-temperature-induced *(lti)* gene products are involved in the adjustment of plant metabolism for low-temperature conditions, while others are required for the increase in freezing tolerance.

Numerous low-temperature-responsive genes have been characterized from different plant species (more than forty genes so far in *Arabidopsis*), using differential screening, differential display, and subtractive cloning. Many of the deduced polypeptides have similarity to a wide variety of both structural and regulatory proteins, such as dehydrins and other late-embryo-genesis-abundant proteins, antifreeze proteins, lipid transfer proteins, proteases, catalases, alcohol dehydrogenases, protein kinases and kinase regulators, ribonucleic acid (RNA) binding proteins, transcription factors, and translation elongation factors (for reviews, see Thomashow, 1994; Hiilovaara-Teijo and Palva, 1998). However, in many cases, the predicted gene products are unique polypeptides without any similarity to previously characterized proteins. In most cases, the functions for the polypeptides corresponding to low-temperature-responsive genes are still unclear, and it has been very difficult to demonstrate whether and how a particular protein contributes to freezing tolerance. Some of the low-temperature-induced proteins are believed to act as cryoprotectants or molecular chaperones, or to protect cellular structures from dehydration (for reviews, see Hiilovaara-Teijo and Palva, 1998; Palva and Heino, 1997). Overexpression or antisense experiments with several *lti* genes have not shown the expected alteration in the freezing tolerance of transgenic plants. One exception in this regard is the ectopic expression of a cold-regulated gene, *cor15a,* that has been shown to increase the freezing tolerance of both chloroplasts frozen in situ and protoplasts frozen in vitro by about 1°C (Artus et al., 1996). The likely reasons for the lack of success in such experiments may lie in the fact that both cold acclimation and freezing tolerance are multigenic traits involving several genes that may have either redundant or additive effects. Therefore, overexpression of a single gene may bring about only minor enhancement of freezing tolerance.

Requirement of ABA

It has been suggested that low temperature triggers an increase of ABA content in plant cells, which in turn induces genes responsible for cold tolerance (Chen, Li, and Brenner, 1983). Elevated levels of endogenous ABA in response to low temperatures have been reported in the wild potato *Solanum commersonii* (Chen, Li, and Brenner, 1983), wheat (Lalk and Dörffling, 1985), spinach (Ryu and Li, 1994), and *Arabidopsis* (Lång et al., 1994). In *S. commersonii*, ABA is synthesized de novo during cold acclimation (Ryu and Li, 1994). In *Arabidopsis*, a threefold transient increase in ABA level was detected during the first day (12 to 24 hours [h]) of treatment, after which the ABA content rapidly returned to the noninduced level (Lång et al., 1994). Interestingly, the endogenous ABA level was increased in the ABA insensitive mutant *abi1* of *Arabidopsis* (see the following section) during the low-temperature treatment, but in contrast to wild-type plants, the ABA level remained high throughout the treatment (Lång et al., 1994). These results suggest that the *abi1* mutant might lack a negative feedback regulation of the ABA levels.

In addition to the transient increase of endogenous ABA content after cold treatment, other lines of evidence also demonstrate that ABA-controlled processes are required in signal transduction leading to development of freezing tolerance. Exogenous application of ABA to the plants at normal growth temperatures can substitute for the low-temperature treatment and induce an increase in freezing tolerance in several plant species (Chen, Li, and Brenner, 1983; Lång, Heino, and Palva, 1989). Several *lti* genes, whose expression correlates with increased freezing tolerance, are also responsive to exogenous ABA (Palva, 1994). Furthermore, both ABA-deficient and ABA-insensitive *Arabidopsis* mutants, *aba1* and *abi1*, respectively (Koornneef et al., 1982; Koornneef, Reuling, and Karssen, 1984) appear to be impaired in their ability to cold acclimate (Heino et al., 1990; Gilmour and Thomashow, 1991; Mäntylä, Lång, and Palva, 1995). Application of exogenous ABA does not induce cold acclimation in *abi1* mutant, in contrast to the wild-type and the *aba1* mutant plants (Mäntylä, Lång, and Palva, 1995).

Multiple Signal Pathways Control Gene Expression

Analysis of gene expression in ABA-deficient and ABA-insensitive mutants of *Arabidopsis* has demonstrated that some *lti* genes respond independently to low temperature and ABA. The *aba1* and *abi1* mutations had little, if any, effect on the low-temperature-responsive accumulation of transcripts for genes such as *lti78/rd29a, cor47, lti29,* and *lti30* (Nordin, Heino, and Palva, 1991; Gilmour and Thomashow, 1991; Welin et al., 1994), indicating that ABA and low temperature regulate these genes through separate signal transduction pathways (Nordin, Heino, and Palva, 1991; Gilmour and Thomashow, 1991). In contrast, no low-temperature induction of *rab18* gene was detected in these mutants (Lång and Palva, 1992), indicating that the expression of the gene in low temperatures is

ABA dependent. The existence of both ABA-independent and ABA-dependent signal cascades (see Figure 5.1) is supported by the fact that only a subset of the *lti* genes absolutely requires ABA for low-temperature induction (Lång and Palva, 1992). An additional group of *lti* genes appears not to be responsive to ABA at all (Carpenter, Kreps, and Simon, 1994; Jarillo et al., 1994), supporting an existence of a low-temperature-specific, ABA-independent pathway for gene activation. Interaction between the

FIGURE 5.1. Perception of Low Temperature and Related Stimuli, and Induction of Low-Temperature-Responsive Gene Expression

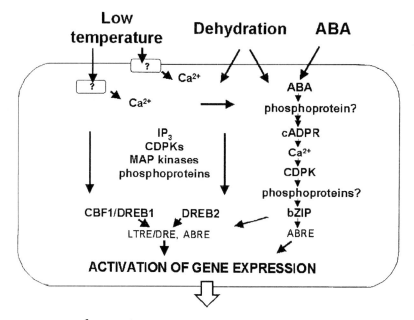

Low temperature response

Note: Temperature sensors may exist both in the plasma membrane and at intracellular locations. Signals are mediated by ABA-dependent and/or ABA-independent pathways. Calcium, calcium-dependent protein kinases (CDPKs), nitrogen-activated protein kinases (MAP-kinases), protein phosphates, and cyclic ADP-ribose are likely intermediates in the pathway. The signal flow continues into the nucleus where it results in induction of CBF1, bZIP, and other type transcription factors that subsequently bind to their cognate promoter elements and modulate transcription of *lti* genes. See text for details.

different pathways is obvious, and signaling cascades can be preferably regarded as a network. The balance of the flux through these pathways determines the specific response of the cell.

LOW-TEMPERATURE CONTROL
OF GENE EXPRESSION

Signal Perception and Sensors of Low-Temperature Stress

The nature of the primary sensor of low temperature is still a puzzle. Murata and Los (1997) suggested that a primary signal upon a change in temperature might be the change in membrane fluidity, which is one of the most rapid effects of temperature on the plasma membrane. Palladium (Pd)-catalyzed hydrogenation of the membrane lipids—a treatment expected to reduce membrane fluidity—rapidly induced expression of the *desA* gene (Vigh et al., 1993). The mechanisms by which reduction in membrane fluidity leads to gene activation are not understood. Murata and Los (1997) have proposed the relationship between membrane fluidity and calcium (Ca) channel activity at low temperature, but so far no such mediator has been detected.

Nordin-Henriksson and Palva (1998) have demonstrated that both the activation of low-temperature-responsive genes and the level of freezing tolerance obtained after acclimation are dependent on the acclimation temperature. Furthermore, a clear correlation between temperature and calcium influx to the cells has been demonstrated in *Arabidopsis* (K. Nordin-Henriksson, personal communication), suggesting that a temperature modulated Ca^{2+} channel could indeed be involved in the temperature sensing. Virtually all known receptors, including the identified plant receptors for ethylene, red light, blue light, and calcium, are located in membranes (for review, see Trewavas and Malhó, 1997). One class of well-characterized membrane receptors is formed by protein kinases, in which ligand binding triggers the kinase activity of the receptor and subsequent signal transduction occurs through protein kinase cascades. Mechanosensitive calcium channels exhibiting temperature-dependent modulation have been identified in plants, and it has been suggested that they may act as low-temperature sensors (Ding and Pickard, 1993). After exposure of *Arabidopsis* plants to low temperature, a minor decline in total water potential of the cells can be detected that may be sufficient for an osmosensor (Lång et al., 1994). In yeast, a two-component histidine kinase functions as osmosensor and monitors mechanical changes of the plasma membrane during osmotic stress

(Brewster et al., 1993; Wurgler-Murphy and Saito, 1997). An *Arabidopsis* SLN1 homologue, ATHK1, was recently shown to complement yeast *sln1* mutant and to function as an osmosensor in yeast (Shinozaki and Yamaguchi-Shinozaki, 1997).

It has been recently shown that temperature can cause conformational changes in macromolecular complexes (Powers and Noller, 1995). Different compartments of plant cells may sense temperature, not only indirectly through events generated at the plasma membrane, but also directly as their own temperature declines. This has been demonstrated by isolated alfalfa nuclei, which could respond to low temperature by changing their phosphorylation state (Kawczynski and Dhindsa, 1996). In addition to nuclei and plasma membrane, possible locations for low-temperature sensors are in vacuole and in organelles.

Gray and colleagues (1997) suggested that a photosynthetically generated redox signal might be the first component in a redox sensing/signaling pathway, acting synergistic to other signal transduction pathways that may trigger an integrated stress response. The accumulation of photosynthetic end products sucrose and fructans has been proposed to play a cryoprotective role, moderating the dehydration stress caused by water loss during extracellular ice formation, probably by stabilizing membranes. Therefore, any factor that chronically affects photosynthesis will ultimately influence the induction and maintenance of the cold-hardened state (Gray et al., 1997). Plants must constantly balance energy absorbed through the photosynthetic apparatus with energy utilized through metabolism. All environmental stresses have the potential to upset this balance and alter chloroplastic redox poise. Therefore, photosynthetic apparatus is not only involved in energy transduction but should also be considered an important "sensor," capable of detecting alteration in the prevailing environment through changes in chloroplastic redox poise (Anderson, Chow, and Park, 1995; Gray et al., 1997). A further trigger in stress signaling may be an oxidative burst, as suggested by Prasad and colleagues (1994).

Calcium As a Second Messenger

Calcium is frequently involved as a second messenger in plant responses to external stimuli (Bowler and Chua, 1994; Bush, 1995; Trewavas and Malhó, 1997). Several lines of evidence suggest that calcium is also acting as a second messenger in low-temperature signal transduction. Knight and colleagues (1991) showed a transient increase in the concentration of intracellular calcium as a result of cold shock. The increase in calcium is seen both in chilling-sensitive tobacco and in chilling-tolerant *Arabidopsis* (Knight, Trewavas, and Knight, 1996; Polisensky and Braam,

1996). Electrophysiological and chemical studies have shown that calcium channels activate as temperature declines, with a maximum activity at 6 to 7°C (Ding and Pickard, 1993; Monroy, Sarhan, and Dhindsa, 1993).

Monroy and Dhindsa (1995) have demonstrated that in alfalfa suspension cells, low temperature triggers an influx of apoplastic calcium into the cytosol. Treatment of cells with calcium chelators or calcium channel blockers prevented the calcium influx as well as the expression of low-temperature-responsive *cas15* gene and the development of freezing tolerance (Monroy, Sarhan, and Dhindsa, 1993; Monroy and Dhindsa, 1995). Thus, in alfalfa cells calcium involved in cold-triggered influx apparently comes from the cell wall. If calcium influx into the cytosol is artificially increased, *cas* genes are induced without low-temperature treatment. *Cas15* is expressed very rapidly at low temperature, independently of new protein synthesis (Monroy, Sarhan, and Dhindsa, 1993). In similar experiments with *Arabidopsis*, calcium was found to mediate the cold-induction of *kin* genes (Knight, Trewavas, and Knight, 1996; Tähtiharju et al., 1997) and *lti78/rd29a* and *rd30* genes (Olson, 1997). However, in these cases, at least some of the calcium influxes appear to occur from the intracellular stores. Both *lti78/rd29a* and *lti30* are induced by low temperature independently of ABA. On other hand, ABA induction of *rab18* gene, the expression of which is ABA dependent, was also calcium mediated, but only intracellular calcium stores were necessary (Olson, 1997). Thus, it is possible that influx of calcium from intracellular stores is involved in both ABA-dependent and ABA-independent signal pathways and might be a position where the two signal pathways interact. Extracellular calcium pools may be necessary in earlier steps of low-temperature signal perception.

cADPR and IP₃ Regulate Calcium Levels

Cyclic adenosine diphosphate (ADP)-ribose (cADPR) is a second messenger that in animal cells is known to induce calcium release from intracellular stores to the cytoplasm through a channel known as ryanodine receptor (RyR). Patch-clamp studies on beet taproot vacuoles have demonstrated cADPR-induced release of Ca^{2+} from plant endomembranes, indicating the presence of RyR-type channels also in plant cells (Allen, Muir, and Sanders, 1995). Wu and colleagues (1997) have recently presented evidence that cADPR is an intermediate in ABA signal transduction. The low-temperature-induced signal pathways mediated by ABA probably involve cADPR as well. Wu and colleagues (1997) microinjected tomato hypocotyl cells with either of the two stress-responsive promoters of *Arabidopsis*, *lti78/rd29a* or *kin2*, coupled to the reporter gene *uidA* (GUS), and monitored their activation in response to different treatments. External application of ABA or coinjections of Ca^{2+} turned on the two stress-responsive genes. cADPR by itself, in the absence of ABA, was able to activate the genes, whereas a competitive inhibitor of cADPR or inhibitors of calcium-ion release prevented the activation of the genes, even when ABA was present. In addition, Wu and colleagues (1997) monitored the expression of *kin2*-luciferase fusion in *Arabidopsis*. Luminescence was detected about 3 h after ABA addition, shortly after an increase in both cyclic ADP-ribose and Ca^{2+}-ion concentration.

A cDNA corresponding to a low-temperature-responsive gene encoding phospholipase C has been isolated from *Arabidopsis* (Hirayama et al., 1995). Analogous to animal cells, phospholipase C in plants is believed to hydrolyze phosphatidyinositol (1-4,5)-bisphosphate to generate two second messengers, diacylglycerol and inositol (1,4,5)-triphosphate (IP_3). IP_3 has been shown to induce Ca^{2+} release from intracellular sources, through channels that are distinct from cADPR gated channels. IP_3 content has been demonstrated to increase following hyperosmotic stress and changes in the levels of IP_3 precursor lipids and in the activity of enzymes involved in the metabolism of inositol phospholipids (Shinozaki and Yamaguchi-Shinozaki, 1997). The studies of Knight and colleagues (1996) with inhibitors neomycin and lithium suggest a role for IP_3-mediated calcium release from the vacuole after cold shock treatment.

Specificity of Calcium Signal

Cytosolic calcium levels have been found to change in response to a perplexing range of different stimuli, such as light, growth regulators, cold shock, pathogen attack, wind, and touch (Gilroy and Trewavas, 1994). The emerging question is, Where is the specificity in the signal? One answer could lie in information being encoded in the amplitude, frequency, and spatial localization of the changes in Ca^{2+} concentration in the cell (Gilroy and Trewavas, 1994; McAinsh and Hetherington, 1998). Changes in cytosolic Ca^{2+} levels exhibit enormous variability in amplitude and temporal and spatial distribution. For example, touch, wind, and cold shock all cause sharp spikes in cytosolic calcium levels in tobacco seedlings within 15 seconds (sec) (Knight et al., 1991; Knight, Trewavas, and Knight, 1996). Oxidative and salt stresses cause relatively low Ca^{2+} transients, lasting for several minutes (Price et al., 1994), and anoxia induces increases in Ca^{2+} level, lasting for several hours (Sedbrook et al., 1996). These differences may allow plant cells to distinguish one kind of stress from another and to induce distinct gene expression to adapt to a particular stress. Ratio and confocal imaging have indeed revealed spatially and temporally localized changes in calcium levels, implying that different parts of the cytoplasm may be regulated differently in response to a stimulus. Recent evidence suggests that plant cells can distinguish between different stimulus-induced increases in cellular calcium. The experiments of Gong and colleagues (1998) with transgenic, aequorin-expressing tobacco seedlings have demonstrated that a single heat shock initiated a refractory period in which additional heat shock signals failed to increase cytosolic Ca^{2+}. However, throughout this refractory period, cells retained a full responsiveness to other stimuli and, for example, responded to cold shock by a Ca^{2+} influx. Furthermore, the kinetics of cytosolic Ca^{2+} increase after a cold-shock was similar in both cold-sensitive tobacco and cold-tolerant *Arabidopsis,* but tobacco was able to recover its ability to respond fully to cold shock 30 minutes (min) after an initial cold shock, whereas *Arabidopsis* was not (Knight, Trewavas, and Knight, 1996). The authors suggest that this altered response to repeated cold stimulation is important in the cold acclimation process. Interestingly, hydrogen peroxide, which can mimic cold acclimation and endow chilling tolerance (Prasad et al., 1994), produced a similar response, suggesting that these two treatments act through a similar pathway.

Protein Phosphorylation

It is now well established that protein phosphorylation is involved in signal transduction during cold acclimation. Changes in the phosphorylation pattern of preexisting proteins have been observed in alfalfa cell suspension cultures during low-temperature treatment (Monroy, Sarhan, and Dhindsa, 1993). Inhibitors of protein kinases inhibit the development of freezing tolerance in *Arabidopsis* (Tähtiharju et al., 1997) and alfalfa (Monroy, Sarhan, and Dhindsa, 1993). Whether the cold-induced increase in protein phosphorylation is a result of increased kinase activity or an inhibition of protein phosphatase activity is not clear. In alfalfa cells, protein phosphatase inhibitor okadaic acid induced *cas15* gene at 25°C but had no effect on its induction at 4°C (Monroy, Sangwan, and Dhindsa, 1998). The protein kinase inhibitor staurosporine, on the other hand, had no effect on the inactive state of *cas15* at 25°C but prevented its induction by low temperature. In addition, protein phosphatase 2A activity was almost completely inhibited at 4°C (Monroy, Sangwan, and Dhindsa, 1998). Monroy and colleagues (1998) suggested that the cold activation of *cas15* could be mediated by derepression. Since the protein kinase inhibitor staurosporine prevents cold induction of *cas15*, the putative repressor protein is likely to be inactivated by phosphorylation. A decrease in temperature inhibits a protein phosphatase (PP2A), which acts on a repressor protein kinase. Through increased phosphorylation, the kinase becomes active and phosphorylates and inactivates the repressor.

As discussed previously, ABA-insensitive mutants of *Arabidopsis* are impaired in cold acclimation capacity. The *abi1* gene as well as the non-allelic *abi2* gene have been shown to encode proteins related to type 2C serine/threonine protein phosphatases (PP2C) (Leung et al., 1994; Meyer, Leube, and Grill, 1994). In addition, microinjection studies of Wu and colleagues (1997) support the hypothesis that phosporylation and dephosporylation events are involved in activation of stress-inducible genes. The phosphatase inhibitor okadaic acid activated gene expression, indicating that a negatively regulating phosphatase is also involved in ABA signaling. Okadaic acid-activated gene expression was blocked by EGTA and cADPR inhibitor, placing the phosphatase at an earlier stage than calcium and cADPR production in the signaling chain. Additionally, the protein kinase inhibitor K252a blocked ABA-induced gene expression, as well as the gene expression induced by cADPR, okadaic acid, and Ca^{2+} (Wu et al., 1997). This indicates that the affected kinases are downstream of the calcium signal.

CDPKs

Protein phosphorylation is apparently coupled to calcium signaling, and several calcium-dependent protein kinases (CDPKs) have been identified in *Arabidopsis* (Hrabak et al., 1996). These proteins have a serine/threonine protein kinase catalytic domain, an autoinhibitory domain, and, uniquely, a regulatory Ca^{2+}-binding domain. Thus, CDPKs in plants seem to be capable of both sensing increases in the cytoplasmic concentration of free calcium and mediating changes in cellular metabolism. Treatment of cell cultures with either a calcium channel blocker or with an inhibitor of calcium-dependent protein kinases (CDPKs) prevented cold-induced phosphorylation of proteins as well as cold acclimation and expression of low-temperature-responsive genes (Monroy, Sarhan, and Dhindsa, 1993). The involvement of CDPKs in cold acclimation was further demonstrated by showing low-temperature-responsive accumulation of a transcript encoding CDPK (Monroy and Dhindsa, 1995).

Sheen (1996) was first to connect specific CDPKs to stress signal pathways and also to link CDPK function to regulation of gene expression. The promoter of the stress-responsive barley gene *HVA1* was fused to green fluorescent protein or luciferace to monitor stress signaling in maize protoplasts. Expression of these reporter systems was induced, similar to *HVA1* gene in barley, by ABA and stress conditions such as drought, cold, and salinity. In absence of these stimuli, expression could be induced by increasing the cellular Ca^{2+} level. Interestingly, coexpression of catalytic domains of two CDPKs, ATCDPK1, and ATCDPK1a activated the *HVA1* promoter, bypassing the stress signals. Six other plant protein kinases, including two distinct CDPKs, failed to mimic this stress signaling. Furthermore, the activation was diminished by protein phosphatase 2C that is capable of blocking responses to ABA (Sheen, 1996). It remains to be seen whether the protein phosphorylated by ATCDPK1 is a transcription factor or a more upstream component in the pathway. Urao and colleagues (1994) showed that the messenger RNA (mRNA) encoding ATCDPK was present at a very low level in control plants, but the amount of the transcript increased rapidly in response to stress conditions by an ABA-independent manner.

Other Kinases

In addition to CDPKs, other protein kinases have also been suggested to be involved in the signal transduction pathway leading to increased freezing tolerance. Holappa and Walker-Simmons (1995) have shown that a wheat gene, *PKABA1*, encoding an ABA-inducible protein kinase is also activated by low temperature. Similarly, transcription levels of the *Arabidopsis ATPK19* and *ATPK6* genes, encoding proteins with similarity to serine/threonine kinases, rapidly and markedly increased when plants were subjected to cold or high-salt stresses (Mizoguchi et al., 1995). Mitogen-activated protein (MAP) kinases appear to be involved in the transduction of signals in response to many different stimuli in plants, for example, hormones (ABA, auxin, ethylene), wounding, drought, touch, and cold (for review, see Mizoguchi, Ichimura, and Shinozaki, 1997). Low-temperature-responsive components of the MAP kinase cascades have been characterized from *Arabidopsis.* These include the complementary deoxyribonucleic acids (cDNAs) of a MAP kinase, a MAP kinase kinase kinase, and and S6 ribosomal protein kinase (Mizoguchi et al., 1996). The transcription of these genes is induced simultaneously by low temperature, dehydration, salt stress, and touch. An alfalfa MAP kinase, MMK4, has been isolated by Jonak and colleagues (1996). The activity of MAP kinases and the transcript level of MMK4 were observed to increase after exposure to cold or drought stress. Surprisingly, the protein levels remained the same as in control. Bögre and colleagues (1997) showed that MMK4 was activated by wounding, independently of transcription or translation, whereas inactivation of the kinase was dependent on both transcription and translation.

A cDNA encoding a receptor-like protein kinase (RPK1) was isolated from *Arabidopsis* (Hong et al., 1997). The deduced polypeptide contained characteristic domains of receptor kinases: extracellular leucine-rich repeat, single membrane-spanning domain, and cytoplasmic protein kinase domain. The gene is induced within 1 h of treatment with ABA and environmental stresses such as dehydration, low temperature, and high salt, suggesting that the gene product is involved in a general stress response. The dehydration-induced expression is not impaired in *aba* and *abi* mutants, suggesting that dehydration-induced expression of the gene is ABA independent.

Two low-temperature-responsive *Arabidopsis* genes, *RCI1A* and *RCI1B*, were recently shown to encode polypeptides similar to the 14-3-3 family of proteins (Jarillo et al., 1994). Analogous to mammalian and yeast systems, the 14-3-3 proteins have been suggested to act as regulators of multifunctional protein kinases and thus regulate signal transduction cascades and other phosphorylation reactions. However, these *RCI* genes are activated relatively late (after three days) in cold acclimation and are thus not likely to participate in the initial signal transduction events. Instead, they might modulate the low-temperature signal or act in pathways that lead to adaptation to prolonged exposure to acclimating temperatures. Expression of these genes seems to be specific to low temperature, since they are not induced by water stress or ABA. The relatively low levels of transcripts also suggest a regulatory function.

Activation of Gene Expression

Both transcriptional and posttranscriptional controls have been shown to be involved in the expression of low-temperature-responsive genes (Hughes and Dunn, 1996). Temporal patterns of low-temperature-inducible gene activation are complex. In *Arabidopsis,* many of the low-temperature-responsive transcripts are detectable after 1 to 2 h of low-temperature (4°C) exposure, and the maximum induction is observed after 4 to 12 h (Palva, 1994). In most cases, the transcript levels remain high as long as the plants are kept in low temperatures and rapidly return to low basal levels upon return to normal growth temperatures (Palva, 1994). Some of the cold-induced genes/polypeptides are only present in the early and middle phases of cold acclimation and not when plants reach their maximum hardiness.

One approach to understanding gene regulation is definition of *cis*-elements mediating the low-temperature responses of the *lti* genes and characterization of *trans*-acting protein factors that bind to these sequences. Deletion analyses of promoters have demonstrated that different *cis*-elements mediate the low-temperature and ABA responses. This is consistent with the genetic studies indicating that low temperature and ABA can affect gene expression through independent pathways (Gilmour and Thomashow, 1991; Nordin, Heino, and Palva, 1991). A putative 5 base pair (bp) *cis*-element, CCGAC, has been suggested to function as a low-temperature response element (LTRE) in the *Arabidopsis* genes *lti78/rd29a* and *lti65/rd29b* (Nordin, Vahala, and Palva, 1993), and similar elements (referred to as the C-repeat) are present in other low-temperature-responsive genes of *Arabidopsis* (Baker, Wilhelm, and Thomashow, 1994) and barley (White et al., 1994). The same core sequence of drought response element (DRE, TACCGACAT) (Yamaguchi-Shinozaki and Shinozaki, 1994) was shown to mediate drought and low-temperature responses in the *Arabidopsis* genes *lti78/rd29a* and *lti65/rd29b* (Yamaguchi-Shinozaki and Shinozaki, 1994) and in *Brassica* gene *BN115* (White et al., 1994). This element does not seem to mediate ABA-regulated gene expression (Yamaguchi-Shinozaki and Shinozaki, 1994; Stockinger, Gilmour, and Thomashow, 1997). Mutation of the CCGAC sequence in the promoter of *lti78/rd29a* resulted in loss of response to the previous stresses (Yamaguchi-Shinozaki and Shinozaki, 1994). Mutation experiments of the two LTRE/DRE/C-repeat in 5′ proximal region of *Brassica napus BN115* gene showed that the element is critical to the low-temperature response (Jiang, Iu, and Singh, 1996). However, not all the genes induced by low temperature contain this element in their 5′ regions, for example, the *TCH* genes (Polisensky and Braam, 1996), indicating that other *cis*-elements are also mediating low-temperature responses.

Stockinger and colleagues (1997) have recently isolated an *Arabidopsis* cDNA clone that encodes the first transcription factor mediating cold responses. This factor, CBF1, was shown to bind to the C-repeat/DRE/LTRE element. The analysis of the deduced amino acid sequence indicated that this protein has a nuclear localization signal, a possible transcriptional activation domain, and a DNA-binding domain similar to that present in *Arabidopsis* APETALA2 and EREBP (ethylene response, element-binding protein) gene products. Expression of CBF1 in yeast activates transcription of reporter genes containing a LTRE/DRE/C-repeat as an upstream activator sequence, but not mutant versions of the element. Liu and colleagues (1998) have shown that at least three CBF1 homologues (DREB1 genes) are induced by low-temperature stress, whereas expression of another family of the DREB transcription factors (DREB2 genes), containing a similar DNA-binding domain, is induced by dehydration.

The responsiveness of several *lti* enes to ABA also suggests that *cis*-elements conferring ABA responsiveness should exist in their promoter regions. Indeed, several *Arabidopsis lti* genes carry in their promoters sequence elements closely resembling the ABA response elements (ABREs) shown to be involved in both ABA- and drought-regulated expression of a number of genes (Guiltinan, Marcotte, and Quatrano, 1990; Chandler and Robertson, 1994). An ABRE contains the motif CACGTG with the ACGT core sequence. ACGT-like elements have been observed in a large number of plant genes regulated by diverse environmental and physiological factors. cDNAs encoding DNA-binding proteins that specifically bind ABREs have been isolated and shown to contain the basic domain/leucine zipper (bZIP) structure (Guiltinan, Marcotte, and Quatrano, 1990). Shen and colleagues (1996) have demonstrated that an ABRE alone is not sufficient for induction, but an additional element (coupling element) must be present to allow transcriptional activation in response to ABA. Different types of coupling elements appear to be present in different genes. In *Adh* gene of *Arabidopsis,* a GT/GC element, in addition to an ABRE-like element, is required for inducibility in environmental stresses, including low oxygen, dehydration, and low temperature (de Bruxelles et al., 1996).

MYC and MYB types of transcription factors have been described to function in ABA and desiccation stress-responsive gene expression (Abe et al., 1997). The transcripts for these factors are also induced by ABA and dehydration stress, but not by cold treatment (Urao et al., 1993), suggesting that this may be one point at which dehydration and cold stresses differ.

Some low-temperature-responsive genes also appear to be regulated at the posttranscriptional level. Results of Phillips and colleagues (1997) indicate that mRNA stability is modulated by a low-temperature-dependent protein factor. Several RNA-binding proteins, which might stabilize or activate mRNA, are induced at low temperature in *Arabidopsis* and barley (Carpenter, Kreps, and Simon, 1994; Dunn et al., 1996). In barley, gene expression can also be controlled at the translational level. A low-temperature-responsive gene encoding an elongation factor 1α, involved in translation machinery, has been cloned from barley (Dunn et al., 1993), and a ribosomal S6 kinase, from *Arabidopsis* (Mizoguchi et al., 1995). Protein modification by phosphorylation, as discussed earlier, appears to be a frequent mechanism in regulation of gene expression at the posttranslational level.

GENETIC DISSECTION OF SIGNAL PATHWAYS

Many of the cellular components and genes involved in low-temperature perception and downstream signal transduction have been isolated, and their expression correlates with low-temperature stress. To elucidate their functional role, acquisition of more detailed biochemical and physiological information is necessary. One approach to identify essential components in low-temperature signaling involves the isolation and characterization of mutants with altered response to low temperature. At the same time, mutants' phenotypes could suggest specific functions for the cognate genes. Extensive screens for such mutants are now in progress in a number of laboratories.

Xin and Browse (1998) have developed a Petri dish freezing assay that allows rapid screening for mutants that are constitutively freezing tolerant. Over twenty independent mutant lines that are freezing tolerant without cold acclimation have been obtained. One interesting mutant is *Esk1*, which has LT_{50} of $-11.5°C$, even in the absence of acclimation. The phenotype of this mutant is also affected: it is smaller, darker green, and accumulates anthocyanin, proline, and soluble sugars (Xin and Browse, 1998). The authors suggest that the wild-type *ESK1* gene may encode a negative regulator that represses the genetic programs required to withstand freezing at normal growth conditions. Warren and colleagues (1996) have isolated several *sfr* (sensitivity to freezing) mutants of *Arabidopsis* that are not able to develop freezing tolerance after cold acclimation treatment (McKown, Kuroki, and Warren, 1996; Warren et al., 1996). Most of the mutations showed pleiotropic effects affecting anthocyanin accumulation, fatty acid composition, and elevation of sucrose and/or glucose levels during or after cold acclimation. These mutants may represent genes in the signaling pathway and others that are directly required for the development of freezing tolerance.

The power of the mutant approach was recently demonstrated by Ishitani and colleagues (1997). They utilized the firefly luciferase reporter system under control of the stress-responsive *lti78/rd29a* promoter to follow the alterations of gene activation in mutagenized plants. Hundreds of *Arabidopsis* mutants were obtained, which were defective in the regulation of *lti78/rd29a* gene expression following exposure either to cold, osmotic stress, or ABA. The mutants could be divided in several categories based on the expression pattern of the reporter gene. Mutants exhibited either constitutive, low, or high expression to one or all of the signals. Based on the recovered mutant class types, the authors suggest that ABA-dependent and ABA-independent stress signaling pathways act in a nonparallel manner and that pathways cross-talk and converge to activate stress gene expression. The characterization of the obtained mutants will undoubtedly have a major impact in understanding the stress signaling in *Arabidopsis*.

In our laboratory, a similar approach has been initiated to genetically dissect the signal transduction pathways that lead to activation of low-temperature-responsive genes and a subsequent increase in freezing tolerance. We are utilizing the *lti78/rd29a* promoter fused to the coding region of the maize anthocyanin regulatory gene *lc* as the reporter system. *Lc*-dependent anthocyanin production has been shown to function as a visual, nondestructive reporter in several plants (Lloyd, Walbot, and Davis, 1992). We have utilized the *ttg* mutant of *Arabidopsis*, which is defective in anthocyanin biosynthesis due to a mutation in the gene corresponding to the *lc* gene, and generated transgenic *Arabidopsis* lines harboring the *lti78/rd29a-lc* construct in this background. The *lc* gene complements the mutation in *ttg* plants, and, consequently, the transgenic lines exhibit low-temperature-dependent anthocyanin synthesis. Mutagenesis of the transgenic line has resulted in isolation of a number of mutants exhibiting constitutive anthocyanin production (Luoma et al., unpublished data). In addition to the *lti78/rd29a* gene, other low-temperature-responsive genes also appear to be constitutively expressed in the mutants. Work is currently in progress to identify and characterize the mutated genes. In addition, we use T-DNA mediated activation tagging approach to identify genes whose activation leads to alterations in the low-temperature responsiveness of the *lti78/rd29a* promoter (Aspegren et al., unpublished data). Multiple copies of the enhancer sequence from the CaMV 35S promoter are inserted in random locations in the genome of the transgenic *ttg* mutant. The use of T-DNA insertions facilitates an easy characterization of the mutated genes because the inserted DNA allows the use of homology-based isolation methods.

ENGINEERING FREEZING-TOLERANT CROPS

Although great progress has been achieved in understanding the molecular basis for plant cold acclimation during the last decade, the complexity of the systems hampers genetic engineering of plant freezing tolerance. As discussed previously, little is known about the exact function of the proteins corresponding to the low-temperature-induced genes. It is, therefore, essential to elucidate the function and importance of these genes, and their gene products, in tolerance before a general strategy for engineering plants with increased frost tolerance is available. Several transgenic plant lines over-expressing different low-temperature-responsive genes have been created, but only a marginal effect on freezing tolerance has been demonstrated (Artus et al., 1996). Similar results have been obtained in studies of water stress tolerance. However, there are a few examples in which increased drought tolerance was obtained by introducing simple metabolic traits from other organisms, such as production of trehalose (Holmström et al., 1996), proline (Kishor et al., 1995), mannitol (Tarczynski, Jensen, and Bohnert, 1993), and fructan (Pilon-Smits et al., 1995), into plants. Their possible effect also on tolerance of low-temperature stress is under study. Preliminary data from our laboratory suggest that transgenic *Arabidopsis* lines overproducing trehalose exhibit increased tolerance to freezing temperatures (Tamminen et al., unpublished data).

A completely different strategy to enhance plant stress tolerance involves manipulation of the signal transduction pathways leading to activation of stress-responsive genes. Expression of a big cluster of genes in the same regulatory pathway is expected to have greater effect on freezing tolerance than expression of an individual gene. Recently, Jaglo-Ottosen and colleagues (1998) succeeded in creating a new, cold-hardy strain of *Arabidopsis* using such an approach. In this pioneering work, transcriptional activator CBF1 that was earlier isolated and shown to regulate *cor15a* gene expression (Stockinger, Gilmour, and Thomashow, 1997) was overexpressed in *Arabidopsis*. The overexpression activated *cor15a* gene and at least four other cold-responsive genes and, more important, increased the freezing tolerance of nonacclimated *Arabidopsis* plants. Thus, a low-temperature response was obtained without low-temperature stimulus. Although the approach was used to modulate freezing tolerance in a plant fully capable of cold acclimation, it might be also extended to plants normally sensitive to freezing.

Many plants have the genetic potential for low-temperature and desiccation tolerance, as indicated by the ability of seeds to survive these stresses. The genes responsible for this tolerance are normally under strict developmental control and expressed only during embryo maturation. However, ectopic expression of seed-specific genes may confer to vegetative tissues the ability to accumulate seed-specific transcripts and influence responses in vegetative tissues, as shown by ectopic expression of *ABI3* gene in *Arabidopsis* (Parcy and Giraudat, 1997). In addition, it has recently been demonstrated that at least part of the presumed genetic potential for freezing tolerance is present also in plant species that do not cold acclimate (Baudo et al., 1996; Meza-Zepeda et al., 1998). Therefore, elucidation of the signal transduction pathways and identification of the regulators required for expression of the genes leading to cold acclimation may open new possibilities to improve stress tolerance in crop plants.

REFERENCES

Abe, H., K. Yamaguchi-Shinozaki, T. Urao, T. Iwasaki, D. Hosokawa, and K. Shinozaki (1997). Role of *Arabidopsis* MYC and MYB homologs in drought- and abscisic acid-regulated gene expression. *Plant Cell* 9: 1859-1868.

Allen, G.J., S.R. Muir, and D. Sanders (1995). Release of Ca^{2+} from individual plant vacuoles by both InsP$_3$ and cyclic ADP-ribose. *Science* 268: 735-737.

Anderson, J.M., W.S. Chow, and Y.I. Park (1995). The grand design of photosynthesis: Acclimation of the photosynthetic apparatus to environmental cues. *Photosynthesis Research* 46: 129-139.

Artus, N.N., M. Uemura, P.L. Steponkus, S.J. Gilmour, C. Lin, and M.F. Thomashow (1996). Constitutive expression of the cold-regulated *Arabidopsis thaliana COR15a* gene affects both chloroplast and protoplast freezing tolerance. *Proceedings of the National Academy of Sciences, USA* 93: 13404-13409.

Baker, S.S., K.S. Wilhelm, and M.F. Thomashow (1994). The 5'-region of *Arabidopsis thaliana cor15a* has *cis*-acting elements that confer cold-, drought- and ABA-regulated gene expression. *Plant Molecular Biology* 24: 701-713.

Baudo, M.M., L.A. Meza-Zepeda, E.T. Palva, and P. Heino (1996). Induction of homologous low temperature and ABA-responsive genes in frost resistant (*Solanum commersonii*) and frost-sensitive (*Solanum tuberosum* cv. Bintje) potato species. *Plant Molecular Biology* 30: 331-336.

Bögre, L., W. Ligterink, I. Meskiene, P.J. Barker, E. Heberle-Bors, N.S. Huskisson, and H. Hirt (1997). Wounding induces the rapid and transient activation of a specific MAP kinase pathway. *Plant Cell* 9: 75-83.

Bowler, C. and N.-H. Chua (1994). Emerging themes of plant signal transduction. *Plant Cell* 6: 1529-1541.

Brewster, J.L., T.D. Valoir, N.D. Dwyer, E. Winter, and M.C. Gustin (1993). An osmosensing signal transduction pathway in yeast. *Science* 259: 1760-1763.

Bush, D.S (1995). Calcium regulation in plant cells and its role in signaling. *Annual Review of Plant Physiology and Plant Molecular Biology* 46: 95-122.

Carpenter, C.D., J.A. Kreps, and A.E. Simon (1994). Genes encoding glycine-rich *Arabidopsis thaliana* proteins with RNA-binding motifs are influenced by cold treatment and an endogenous circadian rhythm. *Plant Physiology* 104: 1015-1025.

Chandler, P.M. and M. Robertson (1994). Gene expression regulated by abscisic acid and its relation to stress tolerance. *Annual Review of Plant Physiology and Plant Molecular Biology* 45: 113-141.

Chen, H.H., P.H. Li, and M.L. Brenner (1983). Involvement of abscisic acid in potato cold acclimation. *Plant Physiology* 71: 362-365.

de Bruxelles, G.L., W.J. Peacock, E.S. Dennis, and R. Dolferus (1996). Abscisic acid induces the alcohol dehydrogenase gene in *Arabidopsis*. *Plant Physiology* 111: 381-391.

Ding, J.P. and B.G. Pickard (1993). Modulation of mechanosensitive calcium-selective channels by temperature. *Plant Journal* 3: 713-720.

Dunn, M.A., K. Brown, R.L. Lightowlers, and M.A. Hughes (1996). A low-temperature-responsive gene from barley encodes a protein with single stranded nucleic acid binding activity which is phosphorylated *in vitro*. *Plant Molecular Biology* 30: 947-959.

Dunn, M.A., A. Morris, P.L. Jack, and M.A. Hughes (1993). A low-temperature-responsive translation elongation factor 1α from barley (*Hordeum vulgare* L.). *Plant Molecular Biology* 23: 221-225.

Gilmour, S.J. and M.F. Thomashow (1991). Cold acclimation and cold-regulated gene expression in ABA mutants of *Arabidopsis thaliana*. *Plant Molecular Biology* 17: 1233-1240.

Gilroy, S. and T. Trewavas (1994). A decade of plant signals. *BioEssays* 16: 677-682.

Gong, M., A.H. van der Luit, M.R. Knight, and A.J. Trewavas (1998). Heat-shock-induced changes in intracellular Ca^{2+} level in tobacco seedlings in relation to thermotolerance. *Plant Physiology* 116: 429-437.

Gray, G.R., L.-P. Chauvin, F. Sarhan, and N.P.A. Huner (1997). Cold acclimation and freezing tolerance. A complex interaction of light and temperature. *Plant Physiology* 114: 467-474.

Guiltinan, M.J., W.R. Marcotte, and R.S. Quatrano (1990). A plant leucine zipper protein that recognizes an abscisic acid response element. *Science* 250: 267-271.

Guy, C.L. (1990). Cold acclimation and freezing stress tolerance: Role of protein metabolism. *Annual Review of Plant Physiology and Plant Molecular Biology* 41: 187-223.

Guy, C., D. Haskell, L. Neven, P. Klein, and C. Smelser (1992). Hydration-state-responsive proteins link cold and drought stress in spinach. *Planta* 188: 265-270.

Heino, P., G. Sandman, V. Lång, K. Nordin, and E.T. Palva (1990). Abscisic acid deficiency prevents development of freezing tolerance in *Arabidopsis thaliana* (L.) Heynh. *Theoretical and Applied Genetics* 79: 801-806.

Hiilovaara-Teijo, M. and E.T. Palva (1998). Molecular responses in cold-adapted plants. In *Cold-Adapted Organisms—Fundamentals and Applications*, eds. R. Margesin and F. Schinner. Heidelberg: Springer-Verlag.

Hirayama, T., C. Ohto, T. Mizoguchi, and K. Shinozaki (1995). A gene encoding a phosphatidylinositol-specific phospholipase C is induced by dehydration and salt stress in *Arabidopsis thaliana*. *Proceedings of the National Academy of Sciences, USA* 92: 3903-3907.

Holappa, L.D. and M.K. Walker-Simmons (1995). The wheat abscisic acid-responsive protein kinase mRNA, PKABA1, is upregulated by dehydration, cold temperature and osmotic stress. *Plant Physiology* 108: 1203-1210.

Holmström, K.-O., E. Mäntylä, B. Welin, A. Mandal, E.T. Palva, O.E. Tunnela, and J. Londesborough (1996). Drought tolerance in tobacco. *Nature* 379: 683-684.

Hong, S.W., J.H. Jon, J.M. Kwak, and H.G. Nam (1997). Identification of a receptor-like protein kinase gene rapidly induced by abscisic acid, dehydration, high salt, and cold treatments in *Arabidopsis thaliana*. *Plant Physiology* 113: 1203-1212.

Hrabak, E.M., L.J. Dickmann, J.S. Satterlee, and M.R. Sussman (1996). Characterization of eight new members of the calmodulin-like domain protein kinase gene family from *Arabidopsis thaliana*. *Plant Molecular Biology* 31: 405-412.

Hughes, M.A. and M.A. Dunn (1996). The molecular biology of plant acclimation to low temperature. *Journal of Experimental Botany* 47: 291-305.

Ishitani, M., L. Xiong, B. Stevenson, and J.-K. Zhu (1997). Genetic analysis of osmotic and cold stress signal transduction in *Arabidopsis:* Interactions and convergence of abscisic acid-dependent and abscisic acid-independent pathways. *Plant Cell* 9: 1935-1949.

Jaglo-Ottosen, K.R., S.J. Gilmour, D.G. Zarka, O. Schabenberger, and M.F. Thomashow (1998). *Arabidopsis CBF1* overexpression induces *COR* genes and enhances freezing tolerance. *Science* 280: 104-106.

Jarillo, J.A., J. Capel, A. Leyva, J.M. Martinez-Zapater, and J. Salinas (1994). Two related low-temperature-inducible genes of *Arabidopsis* encode protein showing high homology to 14-3-3 proteins, a family of putative kinase regulators. *Plant Molecular Biology* 25: 693-704.

Jiang, C., B. Iu, and J. Singh (1996). Requirement of a CCGAC *cis*-acting element for cold induction of the *BN115* gene from winter *Brassica napus. Plant Molecular Biology* 30: 679-684.

Jonak, C., S. Kiegerl, W. Ligterink, P.J. Barker, N.S. Huskisson, and H. Hirt (1996). Stress signaling in plants: A mitogen-activated protein kinase pathway is activated by cold and drought. *Proceedings of the National Academy of Sciences, USA* 93: 11274-11279.

Kawczynski, W. and R.S. Dhindsa (1996). Alfalfa nuclei contain cold-responsive phosphoproteins and accumulate heat-stable proteins during cold treatment of seedlings. *Plant and Cell Physiology* 37: 1204-1210.

Kishor, P.B.K., Z. Hong, G.-H. Miao, C.-A.A. Hu, and D.P.S. Verma (1995). Overexpression of Δ^1-pyrroline-5-carboxylate synthetase increases proline production and confers osmotolerance in transgenic plants. *Plant Physiology* 108: 1387-1394.

Knight, H., A.J. Trewavas, and M.R. Knight (1996). Cold calcium signaling in *Arabidopsis* involves two cellular pools and a change in calcium signature after acclimation. *Plant Cell* 8: 489-503.

Knight, M.R., A.K. Cambell, S.M. Smith, and A.J. Trewavas (1991). Transgenic plant aequorin reports the effects of touch and cold-shock and elicitors on cytoplasmic calcium. *Nature* 352: 524-526.

Koornneef, M., M.L. Jorna, D.L.C. Brinkhorst-van der Swan, and C.M. Karssen (1982). The isolation of abscisic acid (ABA) deficient mutants by selection of induced revertants in non-germinating gibberellin sensitive lines of *Arabidopsis thaliana* (L.) Heynh. *Theoretical and Applied Genetics* 61: 385-393.

Koornneef, M., G. Reuling, and C.M. Karssen (1984). The isolation of abscisic acid-insensitive mutants of *Arabidopsis thaliana. Physiologia Plantarum* 61: 377-383.

Lalk, I. and K. Dörffling (1985). Hardening, abscisic acid, proline and freezing resistance in two winter wheat varieties. *Physiologia Plantarum* 63: 287-292.

Lång, V., P. Heino, and E.T. Palva (1989). Low temperature acclimation and treatment with exogenous abscisic acid induce common polypeptides in *Arabidopsis thaliana* (L.) Heynh. *Theoretical and Applied Genetics* 77: 729-734.

Lång, V., E. Mäntylä, B. Welin, B. Sundberg, and E.T. Palva (1994). Alterations in water status, endogenous abscisic acid content, and expression of *rab18* gene during the development of freezing tolerance in *Arabidopsis thaliana*. *Plant Physiology* 104: 1341-1349.

Lång, V. and E.T. Palva (1992). The expression of a *rab*-related gene, *rab18*, is induced by abscisic acid during the cold acclimation process of *Arabidopsis thaliana* (L.) Heynh. *Plant Molecular Biology* 20: 951-962.

Leung, J., M. Bouvier-Durand, P.-C. Morris, D. Guerrier, F. Chefdor, and J. Giraudat (1994). *Arabidopsis* ABA response gene *ABI1*: Features of a calcium-modulated protein phosphatase. *Science* 264: 1448-1452.

Levitt, J (1980). *Responses of Plants to Environmental Stresses: Chilling, Freezing and High Temperature Stresses*, Volume 1. New York: Academic Press.

Li, P.H. and T.H.H. Chen, eds (1997). *Plant Cold Hardiness*. New York: Plenum Press.

Liu, Q., M. Kasuga, Y. Sakuma, H. Abe, S. Miura, K. Yamaguchi-Shinozaki, and K. Shinozaki (1998). Two transcription factors, DREB1 and DREB2, with an EREBP/AP2 DNA binding domain separate two cellular signal transduction pathways in drought- and low-temperature-responsive gene expression, respectively, in *Arabidopsis*. *Plant Cell* 10: 1391-1406.

Lloyd, A.M., V. Walbot, and R.W. Davis (1992). *Arabidopsis* and *Nicotiana* anthocyanin production activated by maize regulators R and C1. *Science* 258: 1773-1775.

Mäntylä, E., V. Lång, and E.T. Palva (1995). Role of abscisic acid in drought-induced freezing tolerance, cold acclimation and accumulation of LTI78 and RAB18 proteins in *Arabidopsis thaliana*. *Plant Physiology* 107: 141-148.

McAinsh, M.R. and A.M. Hetherington (1998). Encoding specificity in Ca^{2+} signalling systems. *Trends in Plant Sciences* 3: 32-35.

McKown, R., G. Kuroki, and G. Warren (1996). Cold responses of *Arabidopsis* mutants impaired in freezing tolerance. *Journal of Experimental Botany* 47: 1919-1925.

Meyer, K., M.P. Leube, and E. Grill (1994). A protein phosphatase 2C involved in ABA signal transduction in *Arabidopsis thaliana*. *Science* 264: 1452-1455.

Meza-Zepeda, L.A., M.M. Baudo, E.T. Palva, and P. Heino (1998). Isolation and characterization of a cDNA clone corresponding to a stress activated cyclophilin gene in *Solanum commersonii*. *Journal of Experimental Botany* 49: 1451-1452.

Mizoguchi, T., N. Hayashida, K. Yamaguchi-Shinozaki, H. Kamada, and K. Shinozaki (1995). Two genes that encode ribosomal-protein S6 kinase homologs are induced by cold or salinity stress in *Arabidopsis thaliana*. *FEBS Letters* 358: 199-204.

Mizoguchi, T., K. Ichimura, and K. Shinozaki (1997). Environmental stress response in plants: The role of mitogen-activated protein kinases. *Trends in Biotechnology* 15: 15-19.

Mizoguchi, T., K. Irie, T. Hirayama, N. Hayashida, K. Yamaguchi-Shinozaki, K. Matsumoto, and K. Shinozaki (1996). A gene encoding a mitogen-activated protein kinase kinase kinase is induced simultaneously with genes for a mitogen-activated protein kinase and an S6 ribosomal protein kinase by touch, cold, and water stress in *Arabidopsis thaliana*. *Proceedings of the National Academy of Sciences, USA* 93: 765-769.

Monroy, A.F. and R.S. Dhindsa (1995). Low-temperature signal transduction: Induction of cold acclimation-specific genes of alfalfa by calcium at 25°C. *Plant Cell* 7: 321-331.

Monroy, A.F., V. Sangwan, and R.S. Dhindsa (1998). Low-temperature signal transduction during cold acclimation: Protein phosphatase 2A as an early target for cold-inactivation. *Plant Journal* 13: 653-660.

Monroy, A.F., F. Sarhan, and R.S. Dhindsa (1993). Cold-induced changes in freezing tolerance, protein phosphorylation, and gene expression: Evidence for a role of calcium. *Plant Physiology* 102: 1227-1235.

Murata, N. and D.A. Los (1997). Membrane fluidity and temperature perception. *Plant Physiology* 115: 875-879.

Nordin, K., P. Heino, and E.T. Palva (1991). Separate signal pathways regulate the expression of a low-temperature-induced gene in *Arabidopsis thaliana* (L.) Heynh. *Plant Molecular Biology* 16: 1061-1071.

Nordin, K., T. Vahala, and E.T. Palva (1993). Differential expression of two related, low temperature-induced genes in *Arabidopsis thaliana* (L.) Heynh. *Plant Molecular Biology* 21: 641-653.

Nordin-Henriksson, K. and E.T. Palva (1998). Temperature-dependent development of freezing tolerance and expression of low temperature-induced genes in *Arabidopsis thaliana*. *Journal of Experimental Botany*.

Olson, Å (1997). Molecular responses to cold stress. Regulation of low temperature-induced genes in *Arabidopsis thaliana*. PhD Thesis, Swedish University of Agricultural Sciences, Uppsala, Sweden.

Palta, J.P. and G. Simon (1993). Breeding potential for improvement of freezing stress resistance: Genetic separation of freezing tolerance, freezing avoidance, and capacity to cold acclimate. In *Advances in Plant Cold Hardiness*, eds. P.H. Li and L. Christersson. Boca Raton, FL: CRC Press, pp. 299-310.

Palva, E.T (1994). Gene expression under low temperature stress. In *Stress Induced Gene Expression in Plants*, ed. A.S. Basra. Chur, Switzerland: Harwood Academic Publishers, pp. 103-130.

Palva, E.T. and P. Heino (1997). Molecular mechanism of plant cold acclimation and freezing tolerance. In *Plant Cold Hardiness*, eds. P.H. Li and T.H.H. Chen. New York: Plenum Press, pp. 3-14.

Parcy, F. and J. Giraudat (1997). Interaction between the *ABI1* and the ectopically expressed *ABI3* genes in controlling abscisic acid responses in *Arabidopsis* vegetative tissues. *Plant Journal* 11: 693-702.

Phillips, J.R., M.A. Dunn, and M.A. Hughes (1997) mRNA stability and localization of the low-temperature-responsive barley gene family *blt14*. *Plant Molecular Biology* 33: 1013-1023.

Pilon-Smits, E.A.H., M.J.M. Ebskamp, M.J. Paul, M.J.W. Jeuken, P.J. Weisbeek, and S.C.M. Smeekens (1995). Improved performance of transgenic fructan-accumulating tobacco under drought stress. *Plant Physiology* 107: 125-130.

Polisensky, D.H. and J. Braam (1996). Cold-shock regulation of the *Arabidopsis TCH* genes and the effects modulating intracellular calcium levels. *Plant Physiology* 111: 1271-1279.

Powers, T. and H.F. Noller (1995). A temperature-dependent conformational rearrangement in the ribosomal protein S4 16S rRNA complex. *Journal of Biological Chemistry* 270: 1238-1242.

Prasad, T.K., M.D. Anderson, B.A. Martin, and C.R. Stewart (1994). Evidence for chilling-induced oxidative stress in maize seedlings and a regulatory role for hydrogen peroxide. *Plant Cell* 6: 65-74.

Price, A.H., A. Taylor, S.J. Ripley, A. Griffiths, A.J. Trewavas, and M.R. Knight (1994). Oxidative signals in tobacco increase cytosolic calcium. *Plant Cell* 6: 1301-1310.

Ryu, S.B. and P.H. Li (1994). Potato cold hardiness development and abscisic acid. II. *De novo* protein synthesis is required for the increase in free abscisic acid during potato (*Solanum commersonii*) cold acclimation. *Physiologia Plantarum* 90: 21-26.

Sakai, A. and W. Larcher (1987). *Frost Survival of Plants.* Berlin: Springer-Verlag.

Sarhan, F., F. Ouellet, and A. Vazquez-Tello (1997). The wheat *wcs120* gene family. A useful model to understand the molecular genetics of freezing tolerance in cereals. *Physiologia Plantarum* 101: 439-445.

Sedbrook, J.C., P.J. Kronebusch, G.G. Borisy, A.J. Trewavas, and P.H. Masson (1996). Transgenic *AEQUORIN* reveals organ-specific cytosolic Ca^{2+} responses to anoxia in *Arabidopsis thaliana* seedlings. *Plant Physiology* 111: 243-257.

Sheen, J. (1996). Ca^{2+}-dependent protein kinases and stress signal transduction in plants. *Science* 274: 1900-1902.

Shen, Q., P. Zhang, and T.-H.D. Ho (1996). Modular nature of abscisic acid (ABA) response complexes: Composite promoter units that are necessary and sufficient for ABA induction of gene expression in barley. *Plant Cell* 8: 1107-1119.

Shinozaki, K. and K. Yamaguchi-Shinozaki (1997). Gene expression and signal transduction in water-stress response. *Plant Physiology* 115: 327-334.

Steponkus, P.L., M. Uemura, and M.S. Webb (1993). A contrast of the cryostability of the plasmamembrane of winter rye and spring oat: Two species that widely differ in their freezing tolerance and plasma membrane lipid composition. In *Advances in Low Temperature Biology,* ed. P.L. Steponkus. London: JAI Press, pp. 211-313.

Stockinger, E.J., S.J. Gilmour, and M.F. Thomashow (1997). *Arabidopsis thaliana CBF1* encodes an AP2 domain-containing transcriptional activator that binds to the C-repeat/DRE, a *cis*-acting DNA regulatory element that stimulates transcription in response to low temperature and water deficit. *Proceedings of the National Academy of Sciences, USA* 94: 1035-1040.

Tähtiharju, S., V. Sangwan, A.F. Monroy. R.S. Dhindsa, and M. Borg (1997). The induction of *kin* genes in cold-acclimating *Arabidopsis thaliana*. Evidence of a role for calcium. *Planta* 203: 442-447.

Tarczynski, M.C., R.G. Jensen, and H.J. Bohnert (1993). Stress protection of transgenic tobacco by production of the osmolyte mannitol. *Science* 259: 508-510.

Thomashow, M.F (1994). *Arabidopsis thaliana* as a model for studying mechanisms of plant cold tolerance. In *Arabidopsis,* eds. E. Meyerowitz and C. Somerville. New York: Cold Spring Harbor Laboratory Press, pp. 807-834.

Trewavas, A.J. and R. Malhó (1997). Signal perception and transduction: The origin of the phenotype. *Plant Cell* 9: 1181-1195.

Urao, T., T. Katagiri, T. Mizoguchi, K. Yamaguchi-Shinozaki, N. Hayashida, and K. Shinozaki (1994). Two genes that encode Ca^{2+}-dependent protein kinases are induced by drought and high-salt stresses in *Arabidopsis thaliana*. *Molecular and General Genetics* 224: 331-340.

Urao, T., K. Yamaguchi-Shinozaki, S. Urao, and K. Shinozaki (1993). An *Arabidopsis myb* homolog is induced by dehydration stress and its gene product binds to the conserved MYB recognition sequence. *Plant Cell* 5: 1529-1539.

Vigh, L., D. Los, I. Horváth, and N. Murata (1993). The primary signal in the biological perception of temperature: Pd-catalyzed hydrogenation of membrane lipids stimulated the expression of the *desA* gene in *Synechocystis* PCC6803. *Proceedings of the National Academy of Sciences, USA* 90: 9090-9094.

Warren, G., R. McKown, A. Marin, and R. Teutonico (1996). Isolation of mutations affecting the development of freezing tolerance in *Arabidopsis thaliana* (L.) Heynh. *Plant Physiology* 111: 1011-1019.

Welin, B.V., Å. Olson, M. Nylander, and E.T. Palva (1994). Characterization and differential expression of *dhn/lea/rab*-like genes during cold acclimation and drought stress in *Arabidopsis thaliana*. *Plant Molecular Biology* 26: 131-144.

White, T.C., D. Simmonds, P. Donaldson, and J. Singh (1994). Regulation of *BN115*, a low-temperature-responsive gene from *Brassica napus. Plant Physiology* 106: 917-928.

Wu, Y., J. Kuzma, E. Maréchal, R. Graeff, H.C. Lee, R. Foster, and N.-H. Chua (1997). Abscisic acid signaling through cyclic ADP-ribose in plants. *Science* 278: 2126-2130.

Wurgler-Murphy, S.M. and H. Saito (1997). Two-component signal transducers and MAPK cascades. *Trends in Biochemical Sciences* 22: 172-176.

Xin, Z. and J.A. Browse (1998). *Eskimo1* mutants of *Arabidopsis* are constitutively freezing tolerant. *Proceedings of the National Academy of Sciences, USA* 95: 7799-7804.

Yamaguchi-Shinozaki, K. and K. Shinozaki (1994). A novel *cis*-acting element in an *Arabidopsis* gene is involved in responsiveness to drought, low temperature, or high-salt stress. *Plant Cell* 6: 251-264.

Chapter 6

Mechanisms
of Thermotolerance in Crops

Natalya Y. Klueva
Elena Maestri
Nelson Marmiroli
Henry T. Nguyen

INTRODUCTION

High-temperature stress is one of the most prominent abiotic stresses affecting crop productivity (Boyer, 1982). As much as 23 percent of the earth's land surface shows annual mean air temperatures above 40°C. In the temperate zone, soil surface temperatures of 60°C have been recorded, and the shoot apices of grasses may be exposed to temperatures around 40°C for several hours daily and to 50°C for shorter periods. Exposure to heat stress usually recurs on a daily basis (Pollock et al., 1993) and can be accompanied by other stresses, such as water deficits, low night temperatures, and soil metal contamination. It is estimated that heat and drought stresses affect 25 percent of the total arable lands. In the United States, average yields of major crops are three to seven times lower than the expected yields. Heat and drought stresses cause a major part of these losses.

The growing demand for human food supply makes breeding for high-yielding crops with built-in resistance against environmental constraints one of the most important challenges for plant breeders (Zhang, Klueva, and Nguyen, 1999). Significant increases in our understanding of the

The authors acknowledge support from the USDA-NRI grant no. 96-35100-3180 to H.T. Nguyen and the NATO grant no. CRG961159 to N. Marmiroli and H.T. Nguyen.

177

physiological basis of plant stress resistance and the advent of molecular technologies to crop breeding allow breeders to address these problems much more efficiently than in the past. As a result, in recent years, various abiotic stress resistance traits were successfully manipulated in several major crop species (reviewed in Klueva, Zhang, and Nguyen 1999). The overall aim of this chapter is to review the current knowledge on molecular biology of plant responses to high-temperature stress and the advances of biotechnology to improve high-temperature tolerance in crops.

IMPACT OF HIGH TEMPERATURE ON CROPS

Optimum and Stress

The extent of the damage caused by exposure to high temperature differs depending on the crop, the stage of growth, and the type of plant tissue. Certain stages of the plant's life cycle are more heat susceptible than others, and heat stress during these stages may be critical for the final yield determination, even when the temperature regime of the rest of the growing season is optimum. For example, in pearl millet *(Pennisetum americanum),* the seedling stage is the most vulnerable to heat due to high soil surface temperatures during emergence (Howarth, Pollack, and Peacock, 1997). Other critical phases are those related to reproductive functions such as seed and fruit maturation, which usually occur during the hottest months. Heat stress during these phases is directly reflected in reduced productivity and yield (Hall, 1992).

Crops must be grown in various geographic zones. Some crops (e.g., wheat) are well adapted to the cool environments, while others (e.g., rice, maize, and millet) grow in warm and hot environments (Monteith and Elston, 1993). For the purpose of crop production, the optimum temperature range for each species should be defined as a range of temperatures that, when maintained during the growth season, does not significantly impair crop productivity. It is desirable to select some indicator(s) that would allow the measurement of the amount of stress and the prediction of its bearing on end-point productivity.

Mahan and colleagues (1987) proposed using thermal dependencies of the apparent Michaelis constant (K_m) of plant enzymes to define the limits of thermal stress in various crop species. "Thermal kinetic window" (TKW) of an enzyme was defined as a temperature range in which K_m of this enzyme remained within 200 percent of the minimum (Mahan, Burke, and Orzech, 1987). Thermal kinetic window of several plant enzymes was shown to be species specific. In cotton, TKW for major enzymes lay within 23.5°C and 32°C, while for wheat it was within the 17.5°C to 23°C range (Burke, Mahan, and Hatfield, 1988). The time during which leaf temperatures were within the TKW was positively correlated with dry-matter accumulation in cotton and wheat (Burke, Mahan, and Hatfield 1988). The TKW concept is a useful practical tool allowing assessment of the level of stress in crop plants. An irrigation control system based upon the TKW predictions allowed the reduction of the stress levels and an increase of the yield in the field trials (Mahan, McMichael, and Wanjura, 1997).

The extent of the energy resource depletion in the cells under a heat stress can be another indicator of the stress level. Decrease of the intracellular adenosine triphosphate (ATP) concentration is a well-known consequence of the heat stress. The former results from the combined effects of metabolic disturbance and an increased metabolic cost (i.e., the amount of energy that plants have to spend for the activation of their stress resistance mechanisms). It can be quantitatively expressed as adenylate energy charge (AEC):

$$AEC = \frac{[ATP] + 0.5[ADP]}{[ATP] + [ADP] + [AMP]}$$

A significant deterioration of a plant's vitality is observed when AEC is below 0.6 (Ivanovici and Wiebe, 1981). It would be interesting to determine whether a significant correlation exists between the energy depletion factor and crop productivity under heat stress. The level of energy depletion may be a universal assessment of the damaging effect of heat stress in plants.

A few stress-induced compounds (discussed later) are sometimes considered molecular indicators of stress. These include heat shock proteins (HSPs) and other stress-induced metabolites (e.g., antioxidants). While their induction by heat stress in plant cells is ubiquitous, their relationships to plant productivity under high temperatures are not yet established.

Molecular Sites of Damage

At the whole-plant level, heat stress causes an acceleration of developmental stages, abortion of seed development, and impairment of grain filling. At the physiological level, high temperatures affect major plant cell functions, including photosynthesis, energy metabolism, and translocation of assimilates. Discussion of physiological effects of heat stress is beyond the scope of this chapter. For more information on this subject, readers are referred to several detailed reviews (Levitt, 1980; Berry and Bjorkman, 1980; Burke, 1990; Paulsen, 1994). Furthermore, this chapter focuses on the damage caused by high temperatures and the acquired thermotolerance phenomenon in plant somatic cells. Examples of major functions that have been shown to be impaired in plant cells under heat stress are discussed in the following material.

The critical site of heat injury in plant cells is proteins, in particular, enzymes. Heat stress may interfere with de novo protein biosynthesis, inhibit enzyme activity, and induce degradation of existing proteins. A heat-induced inhibition of the constitutive protein synthesis (Nover, 1991), together with an increased enzyme degradation rate, may deplete the major enzyme pools essential for cell functions during the heat stress and recovery period. Several plant enzyme activities are thermolabile, including ribulose-1,5-bisphosphate carboxylase (Weis, 1981), sucrose synthase (Rijven, 1986), catalase (Feierabend, Schaan, and Hertwig, 1992; Willekens et al., 1995), and superoxide dismutase (SOD) (Matters and Scandalios, 1986). Inhibition of ribulose-1,5-bisphosphate carboxylase was the major cause of inactivation of carbon dioxide (CO_2) fixation at high temperatures (Weis, 1981). Heat stress severely inhibited plant capacity to convert sucrose into starch in barley, wheat, and potato, indicating that one or more enzymes of the conversion pathway were specifically inhibited by heat (see discussion in Burke, 1990). Heat stress directly affected soluble-starch synthase activity of wheat endosperm both in vitro and in vivo, leading to an inhibition of starch accumulation in wheat grains (Rijven, 1986).

High temperatures inhibited catalase activity in several plant species, while other antioxidant enzyme activities were not affected (Feierabend, Schaan, and Hertwig, 1992). In rye, catalase activity was reversible and left no apparent damage upon termination of heat stress, whereas in cucumber the recovery of catalase activity was delayed and accompanied by bleaching of chlorophyll, indicating more significant oxidative damage (Feierabend, Schaan, and Hertwig, 1992). In maize seedlings grown at elevated temperatures, SOD activity was lower compared to those grown at lower temperatures (Matters and Scandalios, 1986). The activity of copper/zinc superoxide dismutase (Cu/ZnSOD) was optimum at 10°C and significantly reduced above 35°C (Burke and Oliver, 1992).

Heat stress impaired cell membrane integrity, leading to an increased membrane permeability to solutes and ions (Burke and Orzech, 1988). Simultaneously, functions of membrane integral enzymes essential for basic functions of photosynthesis, respiration, and translocation were disrupted. In barley aleurone layers, heat stress increased the degree of fatty-acid saturation in the endoplasmic reticulum-membrane phospholipids (Grindstaff, Fielding, and Brodl, 1996). Upon severe heat stress, membranes of endoplasmic reticulum were selectively damaged, resulting in the degradation of α-amylose mRNA (messenger ribonucleic acid), (Belanger, Brodl, and Ho, 1986). Heat-induced leakiness of cell membranes affects the redox potential of major cellular compartments which further impairs metabolic processes, leading to membrane disintegration, mixing of the contents of cell compartments, and, in the extreme case, cell death.

Oxidative stress was recently recognized as one of the major damaging factors of heat stress in many organisms, including plants. The impact of oxidative damage can be severe. In yeast, oxidative damage was the major cause of cell death under a lethal high temperature (Davidson et al., 1996). In plants, heat stress causes an imbalance between the amount of solar energy captured by the pigments and the electron transport through the photosynthetic cytochromes. This process is known as photoinhibition. The excess energy can be transferred to oxygen, resulting in a generation of reactive oxygen species (ROS). The critical intracellular sites of oxidative damage are chloroplasts and mitochondria, where the heat-induced disruption of electron transport takes place. In chloroplasts, heat stress caused photoinhibition of photosynthesis and inactivation of catalase, leading to an accumulation of ROS and bleaching of chlorophyll (Willekens et al., 1995). Photosystem (PS) II was shown to be the most heat-susceptible component of photosynthetic function in plant cells under heat stress (Schuster et al., 1988; Burke, 1990). Heat stress induced an alteration in the composition of chloroplast thylakoid membranes (Suss and Yordanov, 1986) and reversible disaggregation of the functional components in the PSII complex. This led to an uncoupling of the electron transport between PSI and PSII (Burke, Mahan, and Hatfield, 1988), an increase of the electron flow to molecular oxygen, and to the production of ROS. In a green alga, *Chlamydomonas reinhardtii*, heat stress under light caused an aggregation of thylakoid proteins associated with the PSII function (Schuster et al., 1988). Protein aggregates of high molecular weight were observed under denaturing sodium dodecylsulfate (SDS)-polyacrylamide gel electrophoresis (SDS-PAGE) conditions. This led to the suggestion that the thylakoid proteins were cross-linked by the ROS induced by the combined effects of heat and light.

Heat stress causes a profound modification of all aspects of cell and whole-plant metabolism. The main sites of injury of plant cells under heat stress are proteins (enzymes) and membranes. These are damaged by a direct effect of heat stress—elevation of intracellular temperature—and by the secondary stresses—oxidative damage and dehydration. The disfunction of photosynthetic enzymes and membranes leads to the inhibition of photosynthesis, which is the major cause of heat-stress-induced biomass loss in crop plants.

DEFENSE MECHANISMS
AGAINST HIGH-TEMPERATURE-INDUCED DAMAGE

Among all living organisms is a remarkable variation in the extent of their tolerance to high temperatures (see Table 6.1). All microorganism and plant species can be classified into three groups according to their thermotolerance rating (after Larcher, 1995):

1. *Heat-sensitive species:* comprises species injured at 39 to 40°C, mostly eukaryotic algae and submerged cormophytes, lichens in hydrated state, and soft-leaved terrestrial plants.
2. *Relatively heat-resistant eukaryotes:* plants from sunny and dry places, able to acquire hardiness to heat, can survive a 30 minute (min) exposure to 50 to 60°C and may tolerate shorter exposures to temperatures of up to 70°C.
3. *Heat-tolerant species:* mostly thermophilic prokaryotes that can endure temperatures of 75 to 90°C. Archaebacteria, unicellular eukaryotes, can survive even higher temperatures of up to 105°C after an acclimation treatment (Trent, 1996). The organisms adapted to extreme high temperatures possess thermostable cell components. A study of the molecular mechanisms of thermotolerance in thermotolerant species may be useful for an understanding of the molecular basis of high-temperature tolerance in plants.

Wild plants rarely or never die of overheating in natural environments and are able to disseminate even under harsh heat stress conditions. Their heat tolerance comes at the expense of productivity. The range of heat adaptation mechanisms in agricultural plants is narrow compared to that of wild species. For example, crop plants cannot utilize escape or many of the avoidance mechanisms used by their wild counterparts (Levitt, 1980). Simultaneously, high-temperature tolerance in many crop species has been compromised by the centuries-long selection for yield potential and by the human demand for a single crop species (although in numerous cultivars) that is ecologically ubiquitous.

Examples of the avoidance mechanisms utilized by crop species are transpirational cooling and evasion of sunlight through leaf movement. Leaf temperatures could be reduced by up to 10°C if the leaves in a canopy rotated relative to the sun (Hall, 1992). Thus, avoidance mechanisms may play a significant role in plant thermoresistance, and an effort needs to be made to study their molecular bases. We will not discuss these mechanisms in this chapter.

TABLE 6.1. Maximal Temperature Resistance of Microorganisms and Plants

Plant Group	Heat Injuries Above °C	
	Wet	Dry
Bacteria		
Archaebacteria	100-110	
Cyanobacteria	55-75	
Saprophytic bacteria	60-70	
Thermophilic bacteria	to 95	
Bacteria spores	80-120	up to 160
Fungi		
Phytopathogenic fungi	45-65 (70)	
Saprophytic fungi	40-60 (80)	75-100
Fungus spores	50-60 (100)	>100
Algae		
Marine algae		
Tropical seas	32-35 (40)	
Temperate seas		
Eulittoral	25-30	
Intertidal	30-35	
Polar seas	(15) 20-28	
Freshwater algae	35-45 (50)	
Aerial algae	40-50	
Eukaryotic thermal algae	45-50	
Lichens		
Temperate regions	33-46	70-100
Mosses		
Temperate zones		
Wet habitats	40-45	
Forest floor	40-50	80-95
Epiphytic mosses		100-110
Poikilohydric ferns	47-50	60-100
Phanerogams		
Ramonda myconi	48	56
Myrothamnus flabellifolia		80
Tropics		
Trees	45-55	
Forest understory	45-48	
Plants of high mountains	ca. 45	
Subtropics		
Evergreen woody plants	50-60	
Subtropical palms	55-60	
Succulents	58-67	
C4 grasses	60-64	

TABLE 6.1 *(continued)*

Temperate zones	
Evergreen woody plants of	
coastal regions with mild winter	46-50 (55)
Dwarf shrubs of Atlantic heaths	45-50
Deciduous trees and shrubs	ca. 50
Herbaceous species of	
sunny habitats	47-52
shady habitats	40-45
Graminoids of steppes	60-65
Succulents	(42) 55-62
Aquatic plants	38-44
Homoiohydric ferns	46-48
Regions with cold winters	
Evergreen conifers	44-50
Boreal deciduous trees	42-45
Arctic-alpine dwarf shrubs	48-54
Herbaceous plants of the high	
mountains and the Arctic	44-54

Source: Simplified from Larcher, 1995.

Note: Numbers in parentheses indicate recorded examples of minimum or maximum temperatures outside the indicated range.

Heat tolerance mechanisms are of major importance for heat resistance in plants. In the following material, we summarize the current views on the molecular mechanisms of plant high-temperature tolerance.

Phenomenon of Acquired Thermotolerance

As pointed out by Howarth and Ougham (1993), the term thermotolerance is used in a number of distinct ways, and it is important to be clear what is meant. High-temperature tolerance in plants has two components: inherent thermotolerance and acquired thermotolerance. Inherent or intrinsic thermotolerance is a constitutive component resulting from the evolutionary thermal adaptation of a species. Acquired thermotolerance is the ability of a plant to survive normally lethal temperatures after an exposure to a mild heat stress. Acquired thermotolerance relies on the induction of specific pathways during the acclimation period and a subsequent development (i.e., acquisition) of thermotolerance. Therefore, it has been postulated that inherent thermotolerance and acquired thermotolerance have different molecular mechanisms. In general, there are significant intraspecies differences in the acquired thermotolerance capacity, whereas variation in inherent thermotolerance within a species is usually small. Thus, the measurement of acquired thermotolerance is more useful in the context of crop breeding and selection.

Acquisition of thermotolerance in plants is a complex phenomenon encompassing an array of physiological, molecular, and biochemical modifications (Levitt, 1980) that are targeted to overcome the major damaging factors of heat stress (as discussed earlier). These may include the changes of isozyme composition of enzymes, modification of the photosynthetic apparatus, alteration of membrane composition, and synthesis and intracellular targeting of protective molecules, for example, HSPs (Nover, 1991), other chaperones, and free-radical quenchers.

In crop plants, cell membrane stability (CMS) and triphenyl tetrazolium chloride (TTC) cell viability assays are routinely used to quantify the level of cellular acquired thermotolerance (Blum, 1988). The CMS test estimates the rate of electrolyte leakage from heat-stressed tissues by conductometric measurements of the solutions in which the stressed tissues are immersed and incubated until equilibrium is reached (Sullivan, Norcio, and Eastin, 1977; Blum, 1988). Therefore, CMS reflects the damage to plant cell membranes caused by heat stress. The TTC cell viability assay is based on the principle of tetrazolium salt reduction to formazan by dehydrogenase respiratory enzymes and, hence, evaluates the residual respiratory activity in the stressed tissue (Towill and Mazur, 1974; Chen, Shen, and Li, 1982; Krishnan, Nguyen, and Burke, 1989). Although these tests measure thermostability of the two important functions of plant cells, other potentially significant components of high-temperature tolerance should be kept in mind.

A long-standing question has been whether the measurements of acquired thermotolerance capacity done in a laboratory setting can be readily translated into a tolerance to high temperatures exhibited in field conditions. Recently, membrane thermostability was shown to be highly correlated with thermotolerance of pearl millet seedlings in field conditions (Howarth, Pollock, and Peacock, 1997). A significant correlation was found between thermotolerance measurements in terms of CMS and TTC at the seedling stage and those at the flowering stage (in flag leaves) in wheat (Fokar, Nguyen, and Blum, 1998). Results of TTC and CMS tests were significantly, but not absolutely, correlated (Chen, Shen, and Li, 1982; Fokar, Nguyen, and Blum, 1998). Moreover, evidence suggests that thermotolerance in terms of CMS correlates with better yield under heat stress in field conditions (Sadalla, Shanahan, and Quick, 1990; Sadalla,

Quick and Shanahan, 1990; Blum, Klueva, and Nguyen, unpublished data). Although some correlations between laboratory-measured TTC reduction or CMS values and crop performance in the field were demonstrated, they were not absolute, and relative standing of genotypes varied depending on a number of other factors. For example, CMS may become a significant factor in stress resistance at extremely high temperatures when membrane damage is likely to occur, while at lower temperatures other components of thermotolerance may be more important (Howarth, Pollock, and Peacock, 1997).

Further progress in the understanding of the molecular mechanisms of acquired thermotolerance is hindered by poor knowledge of the genes contributing to this trait and difficulty obtaining mutants with impaired high-temperature tolerance (Piper, 1993). Although we are yet far from the complete elucidation of the mechanisms of heat tolerance in plants, some progress toward understanding them has been made over the past few years. The following is an effort to review their current status.

Genes Induced by Heat Stress

At the molecular level, the most prominent alterations caused by heat stress in plant cells are modification of protein synthesis, including synthesis of HSPs and antioxidant enzymes, and alteration of membrane composition and structure.

Genes whose expression is induced by heat stress are generally termed "heat shock genes," their products being encompassed by the term "heat shock proteins" (HSPs). A summary of the literature information regarding structure, properties, and putative functions of plant HSPs is presented in Table 6.2.

Plants express, differently from all other organisms, a large set of low molecular weight HSPs (smHSPs) that represents the prominent component of the plant heat shock response (Vierling, 1991; Nguyen, 1994). Heat shock proteins are present in plant cells under actual field conditions and persist under diurnal temperature treatments (Nguyen et al., 1994). After their synthesis in cytoplasm, many HSPs are specifically transported into various intracellular compartments. Both mitochondria and chloroplast-localized HSPs were identified in soybean, pea, maize, and wheat (e.g., Nguyen, Weng, and Joshi, 1993; Lenne et al., 1995; LaFayette et al., 1996). Some organelle-encoded HSPs were also detected (Sinibaldi and Turpen, 1985).

TABLE 6.2. Functions of Heat Shock Proteins and Molecular Chaperones

Class	Molecular Weight, kDa	Subcellular Localization	Putative Functions*	Other Inducers	Homology	Cloned in Crops**
HSP100	100-110	Cytoplasm Nucleus?	ATPase; protein disaggregation, denaturation	Development ABA?	Yeast 104 Prokaryotes ClpB	Soybean
HSP90	80-94	Cytoplasm	Signal transduction	Constitutive Cold stress	Prokaryotes htpG Mammalian HSP90 Yeast HSP82	Maize, rice, tomato
		Endoplasmic reticulum		Pathogens Glycosylation inhibitors	Mammalian GRP94	Barley
HSP70	68-72		ATPase, peptide binding; interacts with DnaJ/GrpE homologues			
		Cytoplasm Nucleus	Movement to nucleus during stress; transport to organelles	Constitutive	Mammalian stress 70 Yeast HSP70	Brassica, carrot, maize, pea, rice, soybean, spinach, tobacco, tomato
		Endoplasmic reticulum	Translocation and folding	Cold stress Heavy metals		Barley, maize, rice, soybean, spinach, tobacco, tomato
				Pathogens Cold stress Glycosylation inhibitors	Yeast KAR2 Mammalian Bip/GRP78	
		Mitochondria	Translocation and folding	Constitutive	Prokaryote DnaK	Bean, pea, potato, spinach, tomato
		Chloroplasts	Translocation and folding	Constitutive	Prokaryote DnaK	Pea, pumpkin, spinach
HSP60 cpn60	a and b 2 × 7 rings	Mitochondria	ATPase Translocation, folding, and assembly with cpn10	Constitutive	Prokaryote GroEL	Brassica, maize, potato, pumpkin
		Chloroplasts (Rubisco large subunit binding protein)	Translocation, folding, and assembly with cpn10	Constitutive	Prokaryote GroEL	Brassica, pea, potato, Ricinus, rye, wheat

TABLE 6.2 (continued)

Class	Molecular Weight, kDa	Subcellular Localization	Putative Functions*	Other Inducers	Homology	Cloned in Crops**
smHSP	16-30	Heat shock granules	Chaperone activity in vitro Prevent aggregation Reactivation			
	200-800	Cytoplasm (I)		Embryogenesis Water stress Heavy metals ABA		Alfalfa, barley, Brassica, carrot, maize, millet, pea, rice, soybean, sunflower, tobacco, tomato, wheat
		Cytoplasm (II)		Embryogenesis Water stress		Alfalfa, maize, parsley, pea, soybean, sunflower, tomato, wheat
		Endoplasmic reticulum				Pea, soybean
		Mitochondria		Cold stress		Pea, soybean
		Chloroplasts	Protection of PSII	Fruit maturation		Barley, maize, pea, soybean, tomato, wheat
		Unknown		Unknown		Soybean GmHSP22.3
HSP10 cpn10	21 7 subunit ring	Mitochondria	Translocation, folding, and assembly with cpn60	Constitutive	Prokaryote GroES	Pea, soybean
		Chloroplasts	Translocation, folding, and assembly with cpn60	Constitutive	prokaryote GroES	Pea, spinach

* Functions of heat shock proteins and molecular chaperones were reviewed in Boston, Viitanen, and Vierling, 1996; Hartl, 1996; Jakob and Buchner, 1994; Miernyk, 1997; Vierling, 1991; Waters, Lee, and Vierling, 1996.

**Compilation of crop plants is based on latest releases of sequence databases: Genbank, European Molecular Biology Laboratory (EMBL), and DNA Data Bank of Japan (DDBJ).

Apart from HSPs, heat stress induces expression of proteins of other classes. In *Arabidopsis thaliana,* heat stress-induced expression of calmodulin, an enzyme modulating intracellular concentration of calcium (Ca) ions (Braam, 1992). In yeast, *Saccharomyces cerevisiae,* heat stress increased the levels of CTT1, a cytoplasmic catalase, and of MnSOD, a mitochondrial superoxide dismutase (Willekens et al., 1995). High temperatures also modified membrane fluidity, acting mainly upon enzymes of the membrane lipid biosynthesis pathway. These examples show that heat stress is a major perturbation causing numerous molecular alterations in plant cells. Which of these alterations may be significant for thermotolerance in plants will be discussed next.

Molecular Factors of Thermotolerance

A causative involvement of a specific molecular component in plant thermotolerance has been conclusively proven in a single case (HSP70, see the following discussion). Meanwhile, a plethora of indirect experimental evidence obtained either through complementation experiments or inferred from experiments on other species points to several molecular traits as significant for plant thermotolerance. According to current views, several mechanisms are involved in a plant's thermotolerance: accumulation of HSPs, antioxidant capacity, components of signaling pathway, composition of membranes, stability of enzymes, and maintenance of RNA function. Among these, the study of induction and accumulation of specific HSPs has attracted most experimental efforts.

Heat Shock Proteins

Since an acquisition of thermotolerance in crops requires exposure to a sublethal high temperature and usually coincides with HSP induction, HSPs have long been hypothesized to be the major factor necessary for the development of the thermotolerant state (Howarth and Ougham, 1993). Temporal and spatial correlation between the expression (and/or accumulation) of HSP and the development of acquired thermotolerance was demonstrated in many studies (O'Connel, 1994). Furthermore, it was postulated that HSP expression plays a role in a plant's capacity to acquire thermotolerance. Quantitative and/or qualitative variation in HSP expression was suggested to underlie the varying capacities of thermotolerant and thermosusceptible strains to acquire thermotolerance.

In soybean seedlings, the development of thermotolerance, achieved following a heat stress or arsenite treatment, paralleled induction and accumulation of HSPs (Lin, Roberts, and Key, 1984). Other correlative observations were made in plants and tissue cultures of barley, wheat, maize, cotton, and other crops (Marmiroli et al., 1996; Nguyen, 1994). Later, function and subcellular localization of HSPs were studied in bacteria, fungi, and animal systems (see Table 6.2). The remarkable evolutionary conservation of HSPs suggested that their functions in higher plants may be similar to those in other systems.

Although some HSPs have a specific mode of action (see Table 6.2), chaperone function is common to a number of HSP families (Hartl, 1996; Miernyk, 1997). Chaperones are defined as proteins that help maintain other proteins in a functional conformation and facilitate their correct processing, including folding, assembly, translocation, and degradation (Ellis, 1990). This function is carried out not only by HSPs per se, but also by the so-called HSPs cognate proteins that are structural homologues of HSPs expressed constitutively at normal temperatures. Under heat stress, chaperones prevent denaturation and aggregation of cellular proteins and allow the maintenance of enzyme activity. In vitro experiments showed that purified HSPs protected enzymes and other plant proteins from denaturation and inactivation (Hartl, 1996). In animal cells, HSPs have been specifically implicated in the protection of cytoskeletal components from the damage caused by heat stress (Liang and MacRae, 1997). In organelles, HSPs and HSP cognate proteins were involved in the import and assembly of other proteins (Hartl, 1996; Miernyk, 1997). Thus, HSP gene families encode numerous HSPs and HSP cognate proteins which are important for plant cell functioning under normal temperatures and which become essential upon exposure to heat stress.

HSP70

The family of HSP70 contains numerous conserved and abundant protein species. In yeast, disruption of HSP70 genes resulted in a lethal phenotype, while mutations in HSP70 genes impaired growth at high temperature (Vogel, Parsell, and Lindquist, 1995). Transgenic *Drosophila* embryos carrying additional copies of the HSP70 gene were able to acquire thermotolerance faster than normal embryos, indicating that HSP70 was a rate-limiting factor for this process (Welte et al., 1993).

Recently, a role of HSP70 in thermotolerance related to cell signaling and buildup of abnormal proteins has been suggested. The depletion of free HSP70 that occurs during heat shock due to the increased demand for its association with other cell proteins could lead to kinase activation, probably through the dissociation of an upstream component of the kinase cascade. In human cells, an activation of this cascade resulted in cell death via apoptosis. However, in acclimated cells with increased concentrations of HSP70, the activation of the kinase cascade and subsequent cell death following a lethal heat stress were prevented (Gabai et al., 1997).

In *Arabidopsis thaliana,* antisense inhibition of HSP70 expression eliminated the plant's ability to acquire thermotolerance, while inherent thermotolerance was not affected (Lee and Schöffl, 1996). In transgenic *Arabidopsis* plants, inherent thermotolerance was not altered up to 48°C. However, after acclimation at 35°C, transgenic *Arabidopsis* plants were less heat tolerant than the wild-type plants (lethal temperature decreased from 56 to 54°C). This is the first direct demonstration of the functional involvement of an HSP in the mechanism of acquisition of thermotolerance in plants. Interestingly, transgenic *Arabidopsis* plants with the over-expressed heat shock transcription factor showed constitutive expression of the full set of HSPs. Under these circumstances, the basal thermotolerance was significantly enhanced (from 48 to 50°C killing temperature). No conclusion on the effect of HSP induction on acquired thermotolerance was made in this study (Lee, Hübel, and Schöffl, 1995). Based on these data, it was hypothesized that the entire set of HSPs was involved in enhancing basal thermotolerance, while the HSP70 family was necessary for acquisition of thermotolerance.

Small HSPs

Small HSPs of plants have several characteristics that distinguish them from HSPs of other classes and, in particular, from smHSPs of animals, fungi, and bacteria. These are (1) diversity of smHSP species belonging to big multigene families, and their abundant expression in most plant tissues under heat stress (Vierling, 1991); (2) targeting to various cellular compartments, including cytoplasm, nuclei, chloroplasts, mitochondria, and endoplasmic reticulum (Loomis and Wheeler, 1982; Nover, 1991; Nguyen, Weng, and Joshi, 1993; Wollgiehn et al., 1994; Waters, Lee, and Vierling, 1996); (3) strict inducibility by elevated temperatures and the absence of expression under normal conditions (Nover, 1991; Vierling, 1991); and (4) developmental regulation during specific stages of plant development, such as seed maturation (Waters, Lee, and Vierling, 1996). Vierling (1991) classified smHSPs of plants into four classes based on their sequence similarity and localization to various intracellular compartments. According to this classification, cytoplasm-localized smHSPs belong to classes I and II, endomembrane-localized smHSPs, to class III, and chloroplast-localized HSPs, to class IV. Later, the fifth class of smHSPs, mitochondria-localized HSPs, was specified (Waters, Lee, and Vierling, 1996).

In the course of high-temperature stress, the induction of smHSPs is very rapid, and their accumulation is positively correlated with both the intensity and the duration of the stress. In addition, stability of smHSPs at high temperatures (Vierling, 1991) suggests that they may play a role in recovery after the termination of heat stress. Curiously, while smHSPs were mostly heat inducible in a vast majority of plant systems studied, immunologically related proteins were found in an unstressed vegetative tissue of the resurrection plant *Craterostigma plantagineum* (Alamillo et al., 1995). Simultaneously, no smHSP accumulation could be detected in desiccation-intolerant callus tissue of the same species. These data suggest that smHSPs may be involved in dessication stress protection in plants.

Small HSPs of different classes may exhibit differences in their accumulation kinetics under heat stress of various intensity. While mRNAs for cytosolic HSPs accumulated to the maximum level after a 2 hour (h) heating at $37°C$ (induction of thermotolerance), mRNAs of certain organelle-localized smHSPs expressed at the higher level when the $37°C$ treatment was followed by a 2 h incubation at $45°C$, a normally lethal temperature (expression of thermotolerance; Lenne et al., 1995; Visioli, Maestri, and Marmiroli, 1997). A possible explanation could be that both groups of smHSPs have a synergistic effect in plant thermotolerance. Cytosolic smHSPs induced immediately upon heat treatment could be involved in the development of the first stages of thermotolerance; organellar HSPs induced at later stages of heat treatment could assist in the establishment of thermotolerance.

Targeted cellular localization to organelles during stress and oligomerization into heat shock granules (HSGs) is considered an indication of the thermoprotective function of smHSPs. The HSGs are specific structures that were first observed in cytoplasm of heat-stressed tomato seedling cells upon electron microscopic examination. The HSGs in tomato *(Lycopersicon esculentum)* cells formed under heat stress were later shown to contain low molecular weight HSPs and RNA. Hence, it was suggested that HSPs play a role in the protection of those mRNAs which are not translated during the heat stress period but which are required during the recovery period (Nover, 1991; Wollgiehn et al., 1994). Recently, it was shown that cytoplasm-localized smHSPs in *Arabidopsis thaliana* formed specific high molecular weight aggregates in vitro (Lee, Pokala, and Vierling, 1995; Helm, Lee, and Vierling, 1997). Chloroplast-localized HSPs also aggregated into high molecular weight complexes both in vivo and in vitro (Chen, Osteryoung, and Vierling, 1994; Osteryoung and Vierling, 1994). Moreover, when incubated together in vitro, smHSPs of classes I and II formed specific complexes in which they did not mix (Lee, Pokala, and Vierling, 1995). Hence, under heat stress, different types of high molecular weight structures consisting of HSPs of different classes could form *in cello;* their function could vary depending on their composition. Protein fractions enriched in smHSPs isolated from stressed plants in vitro prevented heat denaturation of other plant proteins, including membrane proteins, while protein extracts from unstressed plants did not provide such protection (Jinn et al., 1993; Yeh et al., 1994). Furthermore, in the in vitro experiments, smHSPs from *Arabidopsis* bound to heat-denatured substrates and prevented their irreversible aggregation upon heating (Lee et al., 1997). Recently, functional complementation experiments showed that overexpression of a single smHSP from rice (HSP16.9) conferred thermotolerance to *Escherichia coli* cells (Yeh et al., 1997). This is an intriguing observation since most *E. coli* genes are not homologous to the genes of plant smHSPs.

Discernible mutations for major HSP genes are readily available in yeast and *E. coli*. A functional complementation approach, whereby various plant HSP genes are used to substitute for molecular lesions causing temperature-sensitive phenotypes in yeast or *E.coli,* has been successfully employed to demonstrate functions of several plant HSPs. A rice HSP16.9 (cytoplasmic class I smHSP) expression complemented the temperature-sensitive mutation in *E. coli* cells, conferring to them high-temperature resistance (Yeh et al., 1997). Functional complementation was also used to demonstrate a causative involvement of the HSP100 family in thermotolerance (see the following discussion).

HSP100

Targeted disruption of HSP genes in yeast showed that HSP104 was important for thermotolerance in this organism. Mutants for HSP104 grew as well as wild-type cells at both low and high temperatures. In these mutants, basal thermotolerance was not affected; however, when exposed to an acclimation pretreatment, they failed to acquire thermotolerance, and at $50°C$ they died 100 to 1000 times faster than the pretreated wild-type cells (Sanchez and Lindquist, 1990). In transgenic yeast cells that expressed luciferase, heat stress led to inactivation and aggregation of the enzyme. After a return to normal temperatures, thermotolerant cells resolubilized and reactivated luciferase. Mutants lacking HSP104 were thermosusceptible and were unable to resolubilize enzyme aggregates. Thus, HSP104 promoted disaggregation of the denatured proteins both in vivo and in vitro (Vogel, Parsell, and Lindquist, 1995). In the same study, the possible involvement of HSP100 in mRNA splicing under heat stress was studied. At high temperatures, yeast HSP104 mutants were subjected to splicing disruption to the same level as the wild-type cells. After a return to normal temperatures, splicing recovered more rapidly in the wild-type cells than in the mutants. Moreover, in the cells overexpressing HSP104, splicing was not protected, whereas the recovery from stress was accelerated (Vogel, Parsell, and Lindquist, 1995). This indicated that HSP104 was not necessary for the protection of mRNA splicing under the heat stress.

Plants possess structural and functional homologues of yeast HSP104 (HSP100 family), including chloroplast-localized forms (Lee, Nagao, and Key, 1994; Schrimer, Lindquist, and Vierling, 1994). An indication of plant HSP100 gene family potential involvement in thermotolerance was obtained from the complementation experiments. The HSP100 from soybean and *Arabidopsis* functionally complemented HSP104-deficient yeast mutants conferring high-temperature tolerance (Lee, Nagao, and Key, 1994; Schrimer, Lindquist, and Vierling, 1994). Further experiments are needed to study the mechanisms of the HSP100 gene involvement in thermotolerance in plants.

Despite the number of direct and indirect indications of the causal involvement of major HSPs in thermotolerance, in some experiments, HSP synthesis was neither necessary nor sufficient for the acquisition of thermotolerance (e.g., Petko and Lindquist, 1986; Susek and Lindquist, 1989; Smith and Yaffe, 1991; Piper, 1993). Mutations in some HSP70 genes did not significantly affect cell viability in yeast, *Saccharomyces cerevisiae* (Piper, 1993). A mutation in the HSP30 gene of yeast (homologue of plant smHSP) did not result in any discernible phenotype (Smith and Yaffe, 1991). Significant intertaxa differences in HSP expression and HSP function are a possible explanation for the absence of an HSP function in some species compared to others. For example, whereas HSP30 function in yeast could be lost (Piper, 1993), it may be relevant to thermotolerance in other species. Another explanation is that the functions of the mutated HSP genes can be substituted or compensated for by other genes, reflecting the redundancy of high-temperature resistance pathways.

Paclobutrazol, a triazolic growth regulator and fungicide, conferred thermotolerance to wheat seedlings grown in the light without any HSP synthesis (Kraus, Pauls, and Fletcher, 1995). In these experiments, paclobutrazol-induced alteration of constitutive levels of HSP was not ruled out, although it seems to be very likely. Further critical experiments will be needed to elucidate the specific mechanism and mode of HSP involvement in plant thermotolerance. It is clear that not all results obtained from the experiments in other species can be considered true for plants. Moreover, other mechanisms not involving HSP induction may be more important for thermotolerance at certain developmental stages or for certain species.

Antioxidants

Enzymes and nonenzymatic compounds that are able to quench ROS and hence alleviate ROS-induced damage to cells are called antioxidants. Examples of antioxidants present in plant cells are superoxide dismutases, catalases, peroxidases, glutathione, ascorbic acid, and pigments. According to current views, antioxidants play a pivotal role in plant adaptation to abiotic stresses, including heat and light stress.

Studies with tobacco mutants with increased levels of catalase point to an important role for this enzyme in photosynthetic activity during a stress period (Willekens et al., 1995). In yeast, inactivation of CTT1, a heat shock-induced catalase, reduced the capacity to acquire thermotolerance after the mild heat stress (Piper, 1993). Mutants of yeast lacking genes for catalase, superoxide dismutase, and cytochrome peroxidase were more sensitive to heat stress than the isogenic wild-type cells (Davidson et al., 1996). A treatment of *Echinochloa frumentacea* plants with triazole, a synthetic compound known to increase plant stress resistance, stimulated an increase in the levels of antioxidant enzymes in chloroplasts and induced high-temperature tolerance (Sankhla et al., 1992). Another triazolic compound, paclobutrazol, enhanced the synthesis of antioxidant enzymes and conferred thermotolerance in the light-grown but not the dark-grown, wheat seedlings, suggesting chloroplast functions are important in thermotolerance (Kraus and Fletcher, 1994; Kraus, Pauls, and Fletcher, 1995).

Recently, an intriguing connection between a smHSP expression and antioxidant activity was reported. An overexpression of human or *Drosophila* smHSP (HSP27) in murine cells led to a decreased intracellular level of ROS, increased glutathione content, and prevented the tumor necrosis factor-induced oxidative burst and cell death (Mehlen et al., 1996). In this study, it was concluded that the protective effect of HSP27 was realized via an increase in the intracellular glutathione content by an unknown mechanism.

Currently, manipulation of the antioxidant capacity of plant cells seems to be an important component of high-temperature tolerance in plants. More studies of antioxidant activity and its relation to high-temperature tolerance in plants are warranted.

Signaling Cascade

Calcium ions and calmodulin, the protein that regulates Ca levels in eukaryotes, act as general secondary messengers in eukaryotic cells. Fluctuations in cytoplasmic concentrations of Ca are implicated in signaling during heat stress in plant cells and may be involved in the mechanisms of high-temperature tolerance in plants. Heat stress increased calcium concentration in the cytoplasm of tobacco cells (Gong et al., 1998) and induced calmodulin biosynthesis in *Arabidopsis* (Braam, 1992). Involvement of Ca and calmodulin in stress tolerance was demonstrated in maize seedlings and germinating seeds: a decrease in Ca concentration coincided with the reduced thermotolerance, whereas exogenous Ca treatment enhanced thermotolerance (Gong et al., 1997). Pretreatment with Ca or ethylene glycol-bis (β-aminoethyl ether)-N,N,N',N'-tetraacetic acid (EGTA) enchanced or diminished thermotolerance of tobacco seedlings, respectively, compared with thermotolerance of untreated seedlings (Gong et al., 1998). At least one HSP, HSP70, possesses a calmodulin binding site which implies that HSP may be involved in the signaling events (Miernyk, 1997). This data suggests a possible involvement of cytosolic calcium concentration in heat stress signal transduction.

Proteins identified in plants share sequence conservation with elements of Ca-dependent signaling pathways of mammals and yeast (Hare et al., 1996). In yeast, heat stress activates a signaling pathway that regulates cell integrity through a kinase cascade. The following model of the heat stress signaling in yeast has been suggested. Growth of yeast cells at high temperatures causes weakening of cell walls that, in turn, generates alterations in their interaction with the plasma membrane. Membrane stretching activates ion channels (mechanosensitive ion channels), increasing import of Ca ions into the cytoplasm, and the up-regulation of enzymes that modify membrane structure and composition (Kamada et al., 1995). Similarly, tolerance to heat stress in plants could be conferred by the modification of the heat stress-activated signaling chain.

Membrane Structure

Membrane composition and structure is of primary importance for functioning of plant cells during heat stress. Membrane function is especially important for such cell processes as photosynthesis and respiration. Modifications of membrane composition and structure under heat stress were observed in a number of plant species. Whether any of these modifications are significant for the thermotolerance of membrane-dependent functions remains controversial (see the following discussion). Furthermore, HSPs, osmolytes, and isoprene were recently implicated in plant membrane thermoprotection under heat stress.

In a number of studies, saturation of membrane lipids was correlated with heat stability. In chloroplasts, thermotolerance of photosynthetic light reactions in vivo correlated with a decrease in the ratio of monogalactosyl diacylglycerol to digalactosyl diacylglycerol and increased incorporation of saturated digalactosyl diacylglycerol into thylakoid membranes (Suss and Yordanov, 1986). In these experiments, an HSP species with molecular weight of 22 kilodaltons (kDa) accumulated on the outer chloroplast membranes. It was suggested that this HSP enchanced heat stability of photosynthesis and that the formation of digalactosyl diacylglycerol species was beneficial for plant thermotolerance (Suss and Yordanov, 1986).

In a number of studies, no correlation was observed between the level of membrane lipid unsaturation and heat tolerance (see discussion in Gombos et al., 1994), although significant evidence for membrane unsaturation involvement in plant cold tolerance was obtained (Gibson et al., 1994). The first evidence of correlation between lipid unsaturation and heat tolerance was obtained from experiments using *Arabidopsis* mutants with altered membrane lipid compositions (Somerville and Browse, 1991). Recently, Gombos and colleagues (1994), in the mutants of cyanobacterium *Synechocystis* spp. deprived of polyunsaturated lipids, demonstrated that unsaturation of membrane lipids stabilized photosynthesis against heat inactivation to a small, but distinct, extent. Thus, in spite of some controversy in the literature, membrane lipid (un)saturation is considered a potentially important component of the mechanisms of plant high-temperature resistance. Further research is needed to elucidate the impact of this factor in overall plant thermotolerance.

Osmolytes, a class of protective compounds implicated mainly in osmotic stress tolerance in plants, were recently shown to be involved in high-temperature tolerance mechanisms. In yeast, trehalose specifically protected membranes from high-temperature damage, probably substituting for the depleted water (Piper, 1993). Recently, accumulation of an osmolyte, glycinebetaine, was correlated with CMS and activity of PSII under a heat stress in vivo (Yang , Rhodes, and Joly, 1996). Under heat stress, smHSPs were targeted to the plasma membranes (Piper, 1993) and thylakoid membranes (Suss and Yordanov, 1986), suggesting that HSPs can exert a protective effect on membranes.

Sharkey and Singaas (1995) have hypothesized that isoprene could stabilize chloroplast thylakoid membranes at high temperatures. Isoprene, a gaseous biogenic hydrocarbon, is emitted by many vascular plants. Recently, an isoprene treatment was shown to increase thermotolerance of an isoprene-emitting species, kudzu [*Pueraria lobata* (Willd.) Ohwi.] (Singaas et al., 1997). An increase in thermotolerance of 4°C under low light and of 10°C under higher light was demonstrated (Singaas et al., 1997). Isoprene treatment did not affect thermotolerance of a non-isoprene-emitting species, beans (*Phaseolus vulgaris*). In natural environments, isoprene-emission could rapidly fluctuate in a temperature-dependent manner, serving as a rapid heat acclimation agent in the isoprene-emitting species.

Current evidence indicates that stabilization of plant cell membranes under heat stress, through either changes in their composition or interaction with protective compounds, is an important component of high-temperature tolerance in plants.

Stability of Proteins and mRNA

Under heat stress, thermotolerant cells should be able to maintain enzymes and structural proteins in the functional state and be able to recover from any stress-induced inhibition of their functions upon the termination of heating. Thermotolerant plants possess thermostable enzymes and structural proteins. The survey of the TKW of enzymes from various plant species (Burke, Mahan, and Hatfield, 1988) shows that there are indeed such inherent differences in analogous enzymes extracted from plant species adapted to cool and hot environments. The plant species adapted to hot environments possess enzymes with such features of their primary and secondary structures as high dissociation energy and the potential for intramolecular disulfide bond formation.

Moreover, modification of a plant's protein composition induced by high-temperature treatments is the basis of thermotolerance acquisition. According to current views, two groups of proteins may be involved in these processes: chaperones (including HSPs) and antioxidant enzymes. Several specific examples of chaperone- and antioxidant-mediated maintenance of protein function were discussed earlier.

Ensuring integrity, stability, and selective translatability of mRNA under heat stress is another key component of plant high-temperature stress tolerance. The potential involvement of HSPs in the maintenance of mRNA integrity via HSG formation was discussed earlier in this chapter. Another mechanism of heat stress-induced disruption of mRNA function is mRNA processing. Both mRNA stability and splicing were affected by the heat stress in yeast and *Arabidopsis* (Yost and Lindquist, 1988; Osteryoung, Sundberg, and Vierling, 1993). When unprocessed RNA precursors were translated, aberrant proteins were produced in vivo (Yost and Lindquist, 1988). The potential involvement of HSP100 in the maintenance of splicing efficiency under heat stress was studied. It was shown that HSP104 was not necessary for the protection of splicing from heat disruption (Voge, Parsell, and Lindquist, 1995). Other molecular factors that are involved in this process need to be identified.

In vitro splicing was studied in maize protoplasts transformed with plasmids carrying the β-glucuronidase (GUS) reporter gene under the regulation of a heat shock-responsive promoter. These promoters also contained the 5′ untranslated regions of either *hsp18* or *hsp82*. Introns of the genes *adh1* or *hsp82* were utilized for the monitoring of splicing efficiency. Three types of evidence were obtained by measuring the level of GUS formation at normal and high temperatures: (1) splicing efficiency increased from 23°C to around 40°C; (2) higher temperatures, 45 to 50°C, inhibited splicing efficiency; (3) preliminary incubation at 40°C increased the splicing efficiency at higher temperatures (Marmiroli et al., 1994; Sinibaldi and Mettler, 1992).

Polyadenylation of mRNA is another process that is affected by high temperatures. In pea, the length of the poly(A) tail of HSP21 mRNA was negatively correlated with the intensity of the stress (Osteryoung, Sundberg, and Vierling, 1993). The poly(A) tail could have a role in mRNA stability (Osteryoung, Sundberg, and Vierling, 1993). From studies in other organisms, longer poly(A) tails of mRNA were correlated with faster selective degradation of RNA species. It remains unknown whether the same may be true in plants.

A significant body of information has been accumulated over the past decade concerning molecular processes and components that underlie the phenomenon of thermotolerance in plants. Although we are yet far from understanding the complete picture, some significant major components already have been sketched, and our future task will be to bring their details and interactions into full view.

Genetic Basis of Acquired Thermotolerance in Crops

Genetic variation in acquired thermotolerance, as measured by CMS and TTC assays (discussed previously), was observed in various crops (Martineau, Williams, and Specht, 1979; Onwueme, 1979; Blum and Ebercon, 1981; Werner and Watschke, 1981; Chen, Shen, and Li, 1982; Wallner, Bechwar, and Butler, 1982; Marsh, Davis, and Li, 1985; Milliken, 1987; Krishnan, Nguyen, and Burke, 1989). Most genetic studies of thermotolerance have utilized the CMS assay. Martineau and colleagues (1979) evaluated segregating populations of soybean [*Glycine max* (L.) Merr.] for membrane stability and found substantial genetic variability for this trait. No conclusion was made on the number of genes controlling CMS in plants. Marsh and colleagues (1985) reported that a small number of genes controlled the major portion of variability for thermotolerance in common bean. Sadalla, Shanahan, and Quick (1990) found significant variation in CMS among F5 progenies of a winter wheat population and concluded that this trait is heritable but not simply inherited. In maize, heritability of CMS was estimated at 73 percent, and six quantitative trait loci (QTL) accounted for 53 percent of this variation (Ottaviano et al., 1991). Using TTC cell viability assay in a diallel genetic analysis, Porter and colleagues (1995) reported that thermotolerance was controlled mainly by additive genetic effects, suggesting that genetic improvement could be achieved through selection. In a diallel study with six cultivars of winter wheat, chlorophyll fluorescence was used as a measure of thermotolerance, and it was shown that recurrent selection may be an appropriate tool to select for high-temperature tolerance (Moffat et al. 1990). Recently, Fokar and colleagues (1998) reported a broad-sense heritability of 89 percent for the TTC trait in a spring wheat population.

Several studies have attempted to elucidate the genetic involvement of HSPs as a group, or individual HSPs, in acquired thermotolerance. Variation in HSP synthesis within a species has been observed in a number of studies (see Nguyen et al, 1989; O'Connel, 1994, for reviews). A positive correlation between thermotolerance and intensity of synthesis of major HSP or expression of particular HSP isoforms was observed in thermotolerant and thermosensitive cultivars of sorghum, wheat, barley, and maize (Ougham and Stoddart, 1986; Krishnan, Nguyen, and Burke, 1989; Marmiroli et al., 1989; Ristic, Gifford, and Cass, 1991; Weng and Nguyen, 1992; Nguyen et al., 1994). In contrast, Frova and Sari Gorla (1993) reported lack of correlation between thermotolerance, as measured by CMS, and the amount of HSP expression in a maize recombinant inbred line population. As pointed out by Jorgensen and Nguyen (1995), serious drawbacks of this study were the use of one-dimensional electrophoresis for the quantitation of HSP expression and the method of thermotolerance measurement, which allowed the measurement of inherent, but not acquired, thermotolerance.

Recently, using a small subset of recombinant inbred lines of a winter wheat population, Joshi and colleagues (1997) observed a genetic association between an expression of a member of the chloroplast-localized HSP26 family and thermotolerance as measured by TTC cell viability assay. Similarly, Park and colleagues (1996) demonstrated a genetic linkage between thermotolerance and presence of two to three extra isoforms of chloroplast-localized HSP25 in the creeping bent grass (*Agrostis palustris* Huds.). More HSP26 mRNA was associated with polysomes in thermotolerant variants of creeping bent grass than in the thermosusceptible variants during heat stress and recovery, whereas no differences in the abundance of HSP25 mRNA between the two variants were observed (Park et al., 1997). A functional disruption of chloroplast HSP using anti-HSP antibodies or the addition of purified chloroplast HSP to the fraction of lysed chloropasts in vitro indicated that the chloroplast-localized HSP protected the PS II and, consequently, the whole-chain electron transport under heat stress. The presence of this HSP completely accounted for the heat acclimation of electron transport in tomato plants (Heckathorn et al., 1998). These and other observations the point to the possible intimate involvement of the chloroplast-localized HSP family in the phenomenon of acquisition of thermotolerance in plants.

Overall, to date, very limited experimentation has been conducted to elucidate genetic involvement of HSPs in thermotolerance. Part of the difficulty in addressing this issue lies in the need for appropriate genetic materials, reliable methods of scoring acquired thermotolerance, and complications of screening a large number of progenies for HSP patterns. These issues need to be studied further. More evidence on genetic involvement of HSPs or other molecular components in thermotolerance is needed before gene manipulation to improve high-temperature tolerance in crops can be rationally designed.

Thermotolerance and Other Stresses

Plants routinely encounter more than one stress in the fields. Heat stress is commonly associated with drought (water deficits), high light intensity, and ultraviolet (UV) irradiation. Damaging effects of several stresses imposed on a plant simultaneously are usually more than additive. For example, high air temperatures and high irradiance reduced photosynthesis gradually when they were imposed separately and rapidly when they were imposed simultaneously (Al-Khatib and Paulsen, 1990).

Production of ROS caused by both stress conditions and targeting PSII could be the main damaging factor of two stress conditions in this case. Paulsen (1994) noted that heat stress impaired a plant's capacity for osmotic adjustment, an important drought tolerance mechanism in many plant species. Hence, defense mechanisms against induced damage should be simultaneously employed by a plant under adverse field conditions. The experimental evidence highlighted in the following material shows that molecular events observed upon plant cell exposure to various stresses are frequently similar. This knowledge may be useful for elaborating on the strategy of improvement of plant tolerance to several abiotic stresses, or cross-protection.

Current evidence shows that HSPs may play a wider role as common protective agents against other abiotic stresses in plant cells. Many HSPs can be induced by various abiotic stresses (see Table 6.2). Some HSPs were induced by heavy metal (Wollgiehn and Neumann, 1995; Marmiroli et al., 1996), salt stress (Pareek, Singla, and Grover, 1995; Zhu, Hasegawa, and Bressan, 1997), water stress (Almoguera, Coca, and Jordano, 1993), wounding (Heikkila et al., 1984), carbon starvation (Tassi et al., 1992; Zarsky et al., 1995), and low temperatures (Pareek, Singla, and Grover, 1995). Three HSP cognate proteins were induced by drought stress in *Arabidopsis* (Kiyosue, Yamaguchi-Shinozaki, and Shinozaki, 1994). Accumulation of denatured proteins could serve as a common inducer of HSPs under these various stress conditions (Schlesinger and Ryan, 1993; Gabai et al., 1997).

In spite of these common features of various stress responses, there are dissimilarities in the mechanisms of HSP induction under various stress conditions. Not all HSPs can be induced by different stress factors, and no stresses, except heat stress, induce a complete set of HSPs. For example, heavy metal treatment of tobacco culture cells did not affect the expression of the housekeeping proteins while inducing a subset of HSPs (Wollgiehn and Neumann, 1995). Moreover, intracellular localization of HSPs caused by heat and metal stress was different. Expression of HSPs induced by other stresses is usually much less intensive than that under heat stress (Pareek, Singla, and Grover, 1995; Wollgiehn and Neumann, 1995).

At the transcriptional level, induction of HSP genes by various stresses other than heat stress can be mediated through other stress-responsive sequence elements in their promoter regions, different from heat shock factor (HSF). In yeast, *Saccharomyces cerevisiae,* heat shock elements were found in the promoters of metallothionein (MT) genes (Sewell et al., 1995; Liu and Thiele, 1996). Heat stress and superoxide generator menadione, induced MT gene expression; induction of MT genes under oxidative stress was extended compared to that under heat stress. HSF was distinctly differentially phosphorylated under oxidative stress compared to heat stress. HSP genes were not induced by menadione treatment, indicating that HSF was not involved in metallothionein gene transcription under oxidative stress.

Plants become resistant to other abiotic stresses after being exposed to heat stress. Heat-acclimated plants had an increased tolerance to the subsequent osmotic stress (Harrington and Alm, 1988), ethanol treatment (Nover, 1991), and heavy metal exposure (Heenan and Carter, 1977). An exposure to heat stress conferred PSII resistance to photodamage (Havaux, Greppin, and Streasses, 1991), while a preliminary exposure to high light intensity protected PSII against heat stress (Stapel, Kruse, and Kloppstech, 1993). In some cases, such cross-protection coincided with HSP expression. Arsenite induced HSP expression at $28°C$ in soybean and provided thermotolerance to soybean, sorghum, and pearl millet seedlings (Lin, Roberts, and Key, 1984; Howarth, 1989). Heat shock proteins and thermotolerance were induced by incubation of seedlings with amino acid analogs (Nover, 1991), water stress (Almoguera, Coca, and Jordano, 1993), and starvation (Tassi et al., 1992; Zarsky et al., 1995).

Moreover, HSP induction has been spatially and temporally correlated with plant tolerance to other stresses, including oxidative stress, chilling, salt stress, and heavy metal stress. Accumulation of HSP70 and HSP18 was correlated with increased chilling tolerance in tomato *(Lycopersicon esculentum)* fruit (Sabehat, Weiss, and Lurie, 1996). In human cell lines, heat stress protected mitochondria against subsequent oxidative injury (Polla et al., 1996). Conversely, expression of HSP may not always be necessary or sufficient for plant stress protection. In at least one case, tolerance to heavy metal stress did not require HSP expression (Tomsett and Thurman, 1988).

Cross-resistance to various stresses may result from the common damaging factors of different stresses. An exposure to UV light or ionizing radiation can cause heat stress. Water deficits, temperature extremes, and metal toxicity have as a consequence free-radical-induced oxidative damage. Simultaneously, there are response and tolerance pathways specific for various abiotic stress conditions. Overall, the current data indicate that HSPs induced under various abiotic stress conditions may function in a different manner in plant cells. Alternatively, no cause-and-effect relationship may exist between the expression of HSPs and plant tolerance to other abiotic stresses. These issues need to be further investigated.

Overall, the current data indicate that plants' responses to different stresses share some common molecular pathways, including, but probably not limited to, an induction of stress proteins. It is accepted that many abiotic stresses use calcium-dependent signaling pathways (Gong et al., 1997). Calcium signaling is accepted as a general mechanism of regulation of a number of inducible genes under ambient and stress conditions, for example, osmotic stress (Knight, Trewavas, and Knight, 1997). At present, it is unclear how a modification of cytoplasmic Ca concentrations can be translated into an activation of specific sets of genes under various stress conditions. This question needs to be addressed in future studies.

MOLECULAR STRATEGIES FOR ENHANCING THERMOTOLERANCE IN CROPS

Genetic Transformation

To date, genetic manipulation aimed at high-temperature tolerance in crops has been severely hindered by the unavailability of genes with a proven significant impact on plant thermotolerance. The progress in transgenic manipulation of plant thermotolerance has been slow due to our limited understanding of plant high-temperature tolerance mechanisms. The few major efforts in this area are highlighted in the following material.

Modification of Thermal Kinetic Window
of Plant Enzymes

Plants adapted to different thermal environments differ in the in vitro temperature optimums of their enzymes (Burke, 1990; Burke and Oliver, 1993). Recently, in transgenic experiments, it was shown that thermal characteristics of an enzyme could be transferred between species. When a gene of NADH-hydroxypyruvate reductase from cucumber (a species with a higher optimum growth temperature) was introduced into tobacco (a species with a lower optimum growth temperature), its thermal characteristics remained unchanged and the optimum temperature window of the pool of this enzyme in tobacco was broadened (Oliver, Ferguson, and Burke, 1995). Success in exploiting natural plant metabolic differences in such experiments depends on our ability to target rate-limiting steps of a specific enzymatic pathway. Hence, profound knowledge of metabolic pathways and temperature characteristics of plant enzymes is required.

Transformation with HSP Genes
and Acquired Thermotolerance

Although small HSPs in plants seem to be the best candidates for manipulation of this feature, to date, their use in genetic transformation experiments has been mostly unsuccessful. Transgenic tobacco plants with the introduced soybean gene encoding HSP17.6 under the control of ^{35}S promoter expressed this protein constitutively to a level comparable to that of heat induction (Schöffl, Rieping, and Baumann, 1987). Unfortunately, due to the incompetence of the promoter, the expression of the transgene at high temperatures was inhibited, and thus no conclusion was made concerning its effect on plant thermotolerance. In transgenic *Arabidopsis* with sense and antisense constructs of the gene for chloroplast-localized HSP21, no sufficient decrease in HSP21 expression has been achieved (Osteryoung et al., 1994).

Other transgenic experiments on manipulation of heat shock response in general or high molecular weight HSPs gained better results. De-repression of the activity of genetically engineered HSF resulted in the synthesis of heat shock proteins at normal growth temperatures (to about 20 percent of the heat stress-induced level for HSP18) and increased the level of inherent thermotolerance by 4°C in transgenic *Arabidopsis* (Lee, Hübel, and Schöffl, 1995). The transgenic *Arabidopsis* plants with the de-repressed heat shock factor were able to survive a 1 h incubation at 50°C, while the wild-type plants were tolerant only up to 46°C. Recently, an antisense approach efficiently demonstrated involvement of the HSP70 family in thermotolerance in plants. Lee and Schöffl (1996) showed that an introduced antisense gene for a single member of the HSP70 family in *Arabidopsis* prevented the transgenic plants from acquiring thermotoler-ance. This is the first report to conclusively prove the direct function of HSPs in plants. Especially encouraging was that the engineering of a single antisense construct was sufficient to shut down expression of the whole HSP70 multigene family, obviously because of the high degree of sequence similarity among its members. More work needs to be done to dissect the functional involvement of specific HSPs in plant thermotoler-ance.

In conclusion, gene transfer technologies have an enormous potential for crop improvement through introducing novel genes into plants. Al-though many high-temperature stress-induced proteins and genes have been characterized and isolated in plants, our knowledge of their roles and functions remains incomplete. Further research is needed to determine which genes may be useful for genetic engineering of high-temperature stress resistance. When beneficial genes are identified, genetic engineering is a fast and powerful way to incorporate specific resistance traits into existing elite crop cultivars. Field testing will be needed to demonstrate the usefulness of transgenes in an agricultural context. Furthermore, while a single gene may be beneficial in the laboratory, multiple gene transfer would likely be required to provide significant high-temperature stress protection in the fields (e.g., Bohnert and Jensen, 1996).

QTL Mapping and Manipulation

Tanksley and colleagues (1989), Lee (1995), and Staub and Serquen (1996) provided excellent reviews on the theory and application of DNA marker technology in crop improvement. Superior genes can be efficiently selected if molecular markers closely linked to the gene of interest are identi-fied. Selection based on molecular markers, known as marker-assisted selec-tion (MAS), alleviates difficulties associated with low heritability, recessive-ness, and difficult screening assays. This is particularly useful for the genetic improvement of abiotic stress resistance, since most morphological and phys-iological component traits influencing abiotic stress resistance are difficult to manipulate using conventional screening methods. Greaves (1996) provided a useful discussion on the limitations of conventional plant breeding ap-proaches to stress resistance breeding.

Molecular genetic maps are now available for several important crop species, including rice, wheat, barley, maize, sorghum, cotton, soybean, tomato, lettuce, potato, and alfalfa. Genetic markers have been used to map numerous single gene traits and quantitative trait loci (QTLs) in crop species. The power of molecular marker technology in manipulating QTLs for complex agronomic traits such as yield was demonstrated by Eshed and Zamir (1995) and Stuber (1995).

Molecular mapping of the genes controlling plant thermotolerance traits lags behind mapping of other abiotic stress resistance traits. Very limited information is available on QTLs for thermotolerance in plants. In a population of forty-five recombinant inbred lines (RILs) derived from the cross between two contrasting maize inbreds, six QTLs were detected that accounted for 53 percent of the genetic variability for a thermotolerance trait measured by CMS assay (Ottaviano et al., 1991). Shortcomings of this study were that a limited number of RILs were used and inherent thermotolerance was measured instead of acquired thermotolerance. Recently, in an extension of the same analysis, Frova (1996) reported from four to six QTLs for other heat tolerance-related traits, such as radicle growth, pollen germination ability, and pollen tube growth under heat stress. More work needs to be done to map QTLs for thermotolerance traits in crops.

CONCLUSIONS AND PERSPECTIVES

Thermotolerance in plants is a complex multigenic trait. Several components of thermotolerance, such as heat shock proteins, antioxidants, cell membrane stability, protein and mRNA stability, have been identified in recent years. These components control physiological processes that affect whole-plant thermotolerance. A plant's ability to recover from the consequences of a heat stress can be a critical factor for determination of crop yield and stability. Different components may be important for plant thermotolerance at various developmental stages. For example, thermotolerance of reproductive tissues may be determined by a specific set of molecular mechanisms that are not employed in or important for vegetative tissues. Thus, to efficiently manipulate plant thermotolerance, it is crucial to define the target environment and critical stage of development.

In the recent years, some progress has been achieved in the understanding of the molecular basis of thermotolerance in plants. Accumulation of HSPs and antioxidant capacity of the cells are known to be among the major factors in cellular acquired thermotolerance. Although many HSP genes were isolated and characterized in plants, their roles and functions remain largely undetermined, with the exception of the HSP70 family.

Further studies are needed to eluciade the function of HSPs and to identify other genes that play a role in the genetic control of thermotolerance. Recent progress in EST (expressed sequence tags) sequencing and DNA (deoxyribonucleic acid) chip or microarray technology will accelerate gene discovery related to thermotolerance. Once the important genes are identified, gene transfer technologies have enormous potential for crop improvement through introducing novel genes into plants.

Very limited information is available concerning QTLs for thermotolerance traits in plants. Great progress has been made in the development of molecular markers and genetic maps of numerous crop species that will facilitate genetic mapping of thermotolerance. It is mandatory that this work is done in the near future through collaborative efforts among geneticists, physiologists, and plant breeders. Once the major QTLs are identified, marker-assisted selection can be developed for crop improvement, and detailed genetic and physical mapping of those QTLs can be performed to pinpoint the exact genes or clusters of genes controlling various components of thermotolerance. Moreover, comparative genetic mapping analysis will provide better understanding of the evolutionary relationships of stress tolerance genes among crop species.

The potential for genetic improvement of thermotolerance in many modern crops may have been diminished by the limited genetic variation in a few landraces/progenitors captured by a domestication process. Many cereal species originated in the territories located in the Near East Fertile Crescent, and these are the major sites of genetic diversity, where a wealth of gene variability exists for wheat, barley, oat, and rye. Central America, Asia, and Africa are the centers of origin of maize, rice, and pearl millet, respectively. In a recent study, a wild relative of the cultivated rice *Oryza rufipogon* was shown to contain genes that substantially increased the yield of one of the China's highest yielding cultivars, despite its overall inferior appearance and low yield potential (Tanksley and McCouch, 1997). Similar studies are warranted for abiotic stress tolerance. Therefore, conservation of genetic resources and utilization of genomic tools are likely to be critical for our future ability to improve thermotolerance and crop productivity in hot environments.

REFERENCES

Alamillo, J., C. Almoguera, D. Bartels, and J. Jordano (1995). Constitutive expression of small heat shock proteins in vegetative tissues of the resurrection plant *Craterostigma plantagineum*. *Plant Molecular Biology* 29: 1093-1099.

Al-Khatib, K. and G.M. Paulsen (1990). Photosynthesis and productivity during high-temperature stress of wheat genotypes from major world regions. *Crop Science* 30: 1127-1132.

Almoguera, C., M.A. Coca, and J. Jordano (1993). Tissue-specific expression of sunflower heat shock proteins in response to water stress. *Plant Journal* 4: 947-958.

Belanger, F.C., M.R. Brodl, and T.H.D. Ho (1986). Heat shock causes destabilization of specific mRNAs and destruction of endoplasmic reticulum in barley aleurone cells. *Proceedings of the National Academy of Sciences, USA* 83: 1354-1358.

Berry, J.A. and O. Bjorkman (1980). Photosynthetic response and adaptation to temperature in higher plants. *Annual Review of Plant Physiology* 31: 491-543.

Blum, A. (1988). *Plant Breeding for Stress Environments*. Boca Raton, FL: CRC Press, Inc.

Blum, A. and A. Ebercon (1981). Cell membrane stability as a measure of drought and heat tolerance in wheat. *Crop Science* 21: 43-47.

Bohnert, H.J. and R.G. Jensen (1996). Strategies for engineering water-stress tolerance in plants. *Trends in Biotechnology* 14: 89-97.

Boston, R.S., P.V. Viitanen, and E. Vierling (1996). Molecular chaperones and protein folding in plants. *Plant Molecular Biology* 32: 191-222.

Boyer, J.S. (1982). Plant productivity and environment. *Science* 218: 443-448.

Braam, J. (1992). Regulated expression of the calmodulin-related TCH genes in cultured *Arabidopsis* cells: Induction by calcium and heat shock. *Proceedings of the National Academy of Sciences, USA* 89: 3213-3216.

Burke, J.J. (1990). High temperature stress and adaptation in crops. In *Stress Responses in Plants: Adaptation and Acclimation Mechanisms*, eds. R.G. Alscher and J.R. Cumming. New York: Wiley-Liss, Inc., pp. 295-309.

Burke, J.J., J.R. Mahan, and J.L. Hatfield (1988). Crop-specific thermal kinetic windows in relation to wheat and cotton biomass production. *Agronomy Journal* 80: 553-556.

Burke, J.J. and M.J. Oliver (1992). Differential temperature sensitivity of pea superoxide dismutases. *Plant Physiology* 100: 1595-1598.

Burke, J.J. and M.J. Oliver (1993). Optimal thermal environments for plant metabolic processes (*Cucumis sativus* L.). *Plant Physiology* 102: 295-302.

Burke, J.J. and K.A. Orzech (1988). The heat-shock response in higher plants: A biochemical model. *Plant Cell and Environment* 11: 441-444.

Chen H.H., Z.Y. Shen., and P.H. Li (1982). Adaptability of crop plants to high temperature stress. *Crop Science* 22: 719-725.

Chen Q., K. Osteryoung, and E. Vierling (1994). A 21-kDa chloroplast heat shock protein assembles into high molecular weight complexes *in vivo* and *in organelle*. *Journal of Biological Chemistry* 269: 13216-13223.

Davidson, J.F., B. Whyte, P.H. Bissinger, and R.H. Shiestl. (1996). Oxidative stress is involved in heat-induced cell death in *Saccharomyces cerevisiae*. *Proceedings of the National Academy of Sciences, USA* 93: 5116-5121.

Ellis, R.J. (1990). The molecular chaperone concept. *Seminars in Cell Biology* 1: 1-9.

Eshed, Y. and D. Zamir (1995). An introgression line population of *Lycopersicon pennelii* in the cultivated tomato enables the identification and fine mapping of yield-associated QTL. *Genetics* 141: 1147-1162.

Feierabend, J., C. Schaan, and B. Hertwig (1992). Photoinactivation of catalase occurs under both high- and low-temperature stress conditions and accompanies photoinhibition of photosystem II. *Plant Physiology* 100: 1554-1561.

Fokar, M., H.T. Nguyen, and A. Blum (1998). Heat tolerance in spring wheat. I. Estimating cellular thermotolerance and its heritability. *Euphytica* 104: 1-8.

Frova, C. (1996). Genetic dissection of thermotolerance in maize. In *Physical Stresses in Plants,* eds. S. Grillo and A. Leone. Berlin: Springer, pp. 31-38.

Frova, C. and M. Sari Gorla (1993). Quantitative expression of maize HSPs: Genetic dissection and association with thermotolerance. *Theoretical and Applied Genetics* 86: 213-220.

Gabai, V.L., A.B. Meriin, D.D. Mosser, A.W. Caron, S. Rits, V.I. Shifrin, and M.Y. Sherman, (1997). Hsp70 prevents activation of stress kinases. *Journal of Biological Chemistry* 272: 18033-18037.

Gibson, S., D.L. Falcone, J. Browse, and C. Somerville (1994). Use of transgenic plants and mutants to study the regulation and function of lipid composition. *Plant Cell and Environment* 17: 627-637.

Gombos, Z., H. Wada, E. Hideg, and N. Murata (1994). The unsaturation of membrane lipids stabilizes photosynthesis against heat stress. *Plant Physiology* 104: 563-567.

Gong, M., Y.-J. Li, X. Dai, M. Tian, and Z.-G. Li (1997). Involvement of calcium and calmodulin in the acquisition of heat-shock induced thermotolerance in maize seedlings. *Journal of Plant Physiology* 150: 615-621.

Gong, M., A.H. van der Luit, M.R. Knight, and A.J. Trevawas. (1998). Heat-shock induced changes of intracellular Ca^{2+} level in tobacco seedlings in relation to thermotolerance. *Plant Physiology* 116: 429-437.

Greaves, J.A. (1996). Improving suboptimal temperature tolerance in maize—The search for variation. *Journal of Experimental Botany* 47: 307-323.

Grindstaff, K.K., L.A. Fielding, and M.R. Brodl (1996). Effect of gibberellin and heat shock on the lipid composition of endoplasmic reticulum in barley aleurone layers. *Plant Physiology* 110: 571-581.

Hall, A.E. (1992). Breeding for heat tolerance. In *Plant Breeding Reviews,* Volume 10, ed. J. Janic. New York: John Wiley and Sons, pp. 265-293.

Hare, P.D., S. du Plessis, W.A. Cress, and J. van Staden (1996). Stress-induced changes in plant gene expression. *South African Journal of Science* 92: 431-439.

Harrington, H.M. and D.M. Alm (1988). Interaction of heat and salt shock in cultured tobacco cells. *Plant Physiology* 88: 618-625.

Hartl, F.U. (1996). Molecular chaperones in cellular protein folding. *Nature* 381: 571-580.

Havaux, M., H. Greppin, and R.J. Streasses (1991). Functioning of photosystems I and II in pea leaves exposed to heat stress in the presence or absence of light. *Planta* 186: 88-98.

Heckathorn, S.A., C.A. Downs, T.D. Sharkey, and J.S. Coleman (1998). The small, methionine-rich chloroplast heat-shock protein protects photosystem II electron transport during heat stress. *Plant Physiology* 116: 439-444.

Heenan, D.P. and O.G. Carter (1977). Influence of temperature on the expression of manganese toxicity in two soybean varieties. *Plant and Soil* 47: 219-227.

Heikkila, J.J., J.E.T. Papp, G.A. Schultz, and J.D. Bewley (1984). Induction of heat shock protein messenger RNA in maize mesocotyls by water stress, abcsisic acid, and wounding. *Plant Physiology* 76: 270-274.

Helm, K.W., G.J. Lee, and E. Vierling. (1997). Expression and native structure of cytosolic class II small heat shock proteins. *Plant Physiology* 114: 1477-1485.

Howarth, C.J. (1989). Heat shock proteins in *Sorghum bicolor* and *Pennisetum americanum*. *Plant Cell and Environment* 12: 471-477.

Howarth, C.J. and H.J. Ougham (1993). Gene expression under temperature stress. *New Phytologist* 125: 1-26.

Howarth, C.J., C.J. Pollock, and J.M. Peacock (1997). Development of laboratory-based methods for assessing seedling thermotolerance in pearl millet. *New Phytologist* 137: 129-139.

Ivanovici, A.M. and R.J. Wiebe (1981). Towards a working "definition" of "stress": A review and critique. In *Stress Effects on Natural Ecosystems,* eds. G.W. Barrett and R. Rosenberg. New York: John Wiley, pp. 13-27.

Jakob, U. and J. Buchner (1994). Assisting spontaneity: The role of Hsp90 and small Hsps as molecular chaperones. *Trends in Biochemical Sciences* 19: 205-211.

Jinn, T.-L., S.-H. Wu, C.-H. Yeh, M.-H. Hsieh, Y.-C. Yeh, Y.-M. Chen, and C.-Y. Lin (1993). Immunological kinship of class I low molecular weight heat shock proteins and thermostabilization of soluble proteins in vitro among plants. *Plant Cell and Cell Physiology* 34: 1055-1062.

Jorgensen, J.A. and H.T. Nguyen (1995). Genetic analysis of heat shock proteins in maize. *Theoretical and Applied Genetics* 91: 38-46.

Joshi, C. P., N.Y. Klueva, K.J. Morrow, and H.T. Nguyen (1997). Expression of a unique plastid-localized heat shock protein is genetically linked to acquired thermotolerance in wheat. *Theoretical and Applied Genetics* 95: 834-841.

Kamada, Y., U.S. Jung, J. Piotrowski, and D.E. Levin (1995). The protein kinase C-activated MAP kinase pathway of *Saccharomyces cerevisiae* mediates a novel aspect of the heat shock response. *Genes and Development* 9: 1559-1571.

Kiyosue, T., K. Yamaguchi-Shinozaki, and K. Shinozaki (1994). Cloning of cDNAs for genes that are early-responsive to dehydration stress (ERDs) in *Arabidopsis thaliana* L. Identification of three ERDs as HSP cognate genes. *Plant Molecular Biology* 25: 791-798.

Klueva, N.Y., J. Zhang, and H.T. Nguyen (1998). Molecular strategies for managing environmental stress. In *Crop Productivity and Substantiality and Shaping the Future. Proceedings of Second International Crop Science congress, 1996,* eds. V.L. Chopra, R.B. Singh, and A. Varma. New Delhi, India: Oxford and IBH Publishing Company, pp. 501-524.

Knight, H., A.J. Trewavas, and M.R. Knight (1997). Calcium signalling in *Arabidopsis thaliana* responding to drought and salinity. *Plant Journal* 12: 1067-1078.

Kraus, T.E. and R.A. Fletcher (1994). Paclobutrazol protects wheat seedlings from heat and paraquat injury. Is detoxification of active oxygen involved? *Plant and Cell Physiology* 35: 45-52.

Kraus, T.E., K.P. Pauls, and R.A. Fletcher, R.A. (1995). Paclobutrazol- and hardening-induced thermotolerance of wheat: Are heat shock proteins involved? *Plant and Cell Physiology* 36: 59-67.

Krishnan M., H.T. Nguyen and J.J. Burke (1989). Heat shock protein synthesis and thermal tolerance in wheat. *Plant Physiology* 90: 140-145.

LaFayette, P.R., R.T. Nagao, K. O'Grady, E.Vierling, and J.L Key (1996). Molecular characterization of cDNAs encoding low-molecular-weight heat shock proteins of soybean. *Plant Molecular Biology* 30: 159-169.

Larcher, W. (1995). *Physiological Plant Ecology*. Berlin and Heidelberg: Springer-Verlag.

Lee, G.J., N. Pokala, and E. Vierling (1995). Structure and in vitro molecular chaperone activity of cytosolic small heat shock propteins from pea. *Journal of Biological Chemistry* 270: 10432-10438.

Lee, G.J., A.M. Roseman, H.R. Saibil, and E. Vierling (1997). A small heat shock protein stably binds heat denatured model substrates and can maintain a substrate in a folding-competent state. *EMBO Journal* 16: 659-671.

Lee, J.H., A. Hübel, and F. Schöffl (1995). Depression of the activity of genetically engineered heat shock factor causes constitutive synthesis of heat shock proteins and increased thermotolerance in transgenic *Arabidopsis*. *Plant Journal* 8: 603-612.

Lee, J.H. and F. Schöffl (1996). An Hsp70 antisense gene affects the expression of HSP70/HSC70, the regulation of HSF, and the acquisition of thermotolerance in transgenic *Arabidopsis thaliana*. *Molecular and General Genetics* 252: 11-19.

Lee, M. (1995). DNA markers and plant breeding programs. *Advances in Agronomy* 55: 265-344.

Lee, Y.-R.J., R.T. Nagao, and J.L. Key. (1994). A soybean 101-kDa heat shock protein complements a yeast HSP104 deletion mutant in acquiring thermotolerance. *Plant Cell* 6: 1889-1897.

Lenne, C., M.A. Block, J. Garin, and R. Douce (1995). Sequence and expression of the mRNA encoding HSP22, the mitochondrial small heat-shock protein in pea leaves. *Biochemical Journal* 311: 805-813.

Levitt, J. (1980). *Responses of Plants to Environmental Stresses*. New York: Academic Press.

Liang, P. and T.H. MacRae (1997). Molecular chaperones and the cytoskeleton. *Journal of Cellular Science* 110: 1431-1440.

Lin, C.-Y., J.K. Roberts, and J.L. Key (1984). Acquisition of thermotolerance in soybean seedlings. *Plant Physiology* 74: 152-160.

Liu, X.-D. and D.J. Thiele (1996). Oxidative stress induces heat shock factor phosphorylation and HSF-dependent activation of yeast metallothionein gene transcription. *Genes and Development* 10: 592-603.

Loomis, W.F. and S.A. Wheeler (1982). Chromatin-associated heat shock proteins of *Dictyostelium*. *Developmental Biology* 90: 412-418.

Mahan, J.R., J.J. Burke, and K.A. Orzech (1987). The "thermal kinetic window" as an indicator of optimum plant temperature. *Plant Physiology* 83: S-87.

Mahan, J.R., B.L. McMichael and D.F. Wanjura (1997). Reduction of high temperature stress in plants. In *Mechanisms of Environmental Stress in Plants*, eds. A.S. Basra, and R.K. Bastra. Amsterdam: The Netherlands: Harwood Academic Publishers, pp. 137-150.

Marmiroli, N., M. Gulli, E. Maestri, C. Calestani, A. Malcevschi, C. Perrotta, S.A. Quarrie, K.M. Devos, G. Raho, H. Hartings, and E. Lupotto (1996). Specific and general gene induction in limiting environmental conditions. In *Physical Stresses in Plants*, eds. S. Grillo and A. Leone. Berlin: Springer, pp. 171-185.

Marmiroli, N., C. Lorenzoni, L. Cattivelli, A.M. Stanca, and V. Terzi (1989). Induction of heat shock proteins and acquisition of thermotolerance in barley (*Hordeum vulgare* L.). Variations associated with growth habit and plant development. *Journal of Plant Physiology* 135: 267-273.

Marmiroli, N., E. Maestri, V. Terzi, M. Gulli, A. Pavesi, G. Raho, E. Lupotto, G. Di Cola, R. Sinibaldi, and C. Perrotta (1994). Genetic and molecular evidences of the regulation of gene expression during heat shock in plants. In *Biochemical and Cellular Mechanisms of Stress Tolerance in Plants. NATO ASI Series H: Cell Biology,* Volume 86, ed. J.H. Cherry. Berlin: Springer-Verlag, pp. 157-190.

Marsh, L.E., D.W. Davis, P.H., and Li (1985). Selection and inheritance of heat tolerance in the common bean by the use of conductivity. *Horticultural Science* 110: 630-683.

Martineau, J.R., J.H. Williams, and J.E. Specht (1979). Temperature tolerance in soybeans. II. Evaluation of segregating populations for membrane thermostability. *Crop Science* 19: 79-81.

Matters, G.L. and J.G. Scandalios (1986). Effect of elevated temperature on catalase and superoxide dismutase during maize development. *Differentiation* 30: 190-196.

Mehlen, P., C. Kretz-Remy, X. Preville, and A.-P. Arrigo (1996). Human hsp27, Drosophila hsp27 and human αB-crystallin expression-mediated increase in glutathione is essential for the protective activity of these proteins against TNFα-induced cell death. *EMBO Journal* 15: 2695-2706.

Miernyk, J.A. (1997). The 70 kDa stress-related proteins as molecular chaperones. *Trends in Plant Science* 2: 180-187.

Milliken, G.A. (1987). Comparison of TTC and electrical conductivity heat tolerance screening techniques in *Phaseolus*. *Horticultural Science* 22: 642-645.

Moffat, J.M., G. Sears, T.S. Cox, and G.M. Paulsen (1990). Wheat high temperature tolerance during reproductive growth. II. Genetic analysis of chlorophyll fluorescence. *Crop Science* 30: 886-889.

Monteith, J.L. and J. Elston (1993). Climatic constraints on crop production. In *Plant Adaptation to Environmental Stress,* eds. L. Fowden, T. Mansfield, and J. Stoddart. London: Chapman and Hall, pp. 3-18.

Nguyen, H.T. (1994). Genetic and molecular aspects of high temperature stress responses. In *Physiology and Determination of Crop Yield,* eds. K. Boote, J.M. Bennette, T.R. Sinclair, and G.M. Paulsen. Madison, WI: American Society of Agronomy, pp. 391-394.

Nguyen, H.T., C.P. Joshi, N. Klueva, J. Weng, K.L. Hendershot, and A. Blum (1994). The heat-shock response and expression of heat-shock proteins in wheat under diurnal heat stress and field conditions. *Australian Journal Plant Physiology* 21: 857-867.

Nguyen, H.T., M. Krishnan, J.J. Burke, D.R. Porter, and R.A. Vierling (1989). Genetic diversity of heat shock protein synthesis in cereal plants. In *Biochemical and Physiological Mechanisms Associated with Environmental Stress Tolerance in Plants,* eds. J.H. Cherry and D.D. Davies. Heidelberg: Springer-Verlag, pp. 319-330.

Nguyen, H.T., J. Weng and C.P. Joshi. (1993). A wheat *(Triticum aestivum)* cDNA clone encoding a plastid-localized heat-shock protein. *Plant Physiology* 103: 675-676.

Nover, L. (1991). Induced thermotolerance. In *Heat Shock Response,* ed. L. Nover. Boca Raton, FL: CRC Press, pp. 409-452.

O'Connel, M.A. (1994). Heat shock proteins and thermotolerance. In *Stress-Induced Gene Expression in Plants,* ed. A.S. Basra. Chur, Switzerland: Harwood Academic Publishers, pp. 163-184.

Oliver, M.J., D.L. Ferguson, and J.J. Burke (1995). Interspecific gene transfer. Implications for broadening temperature characteristics of plant metabolic processes. *Plant Physiology* 107: 429-434.

Onwueme, I.C. (1979). Rapid plant conserving estimation of heat tolerance in plants. *Journal of Agricultural Sciences of Cambridge* 92: 527-531.

Osteryoung, K.W., B. Pipes, N. Wehmeyer, and E. Vierling (1994). Studies of a chloroplast-localized small heat shock protein in *Arabidopsis.* In *Biochemical and Cellular Mechanisms of Stress Tolerance in Plants,* ed. J.H. Cherry. Berlin: Springer-Verlag, pp. 97-113.

Osteryoung, K.W., H. Sundberg, and E. Vierling (1993). Poly(A) tail length of a heat shock protein RNA is increased by severe heat stress, but intron splicing is unaffected. *Molecular and General Genetics* 239: 323-333.

Osteryoung, K.W., and E. Vierling (1994). Dynamics of small heat shock protein distribution within the chloroplasts of higher plants. *Journal of Biological Chemistry* 269: 28676-28682.

Ottaviano, E., M. Sari Gorla, E. Pe, and C. Frova. (1991). Molecular markers (RFLPs and HSPs) or the genetic dissection of thermotolerance in maize. *Theoretical and Applied Genetics* 81: 713-719.

Ougham, H.J. and J.L. Stoddart (1986). Synthesis of heat shock proteins and acquisition of thermotolerance in high temperature tolerant and high temperature susceptible lines of sorghum. *Plant Science* 44: 163-167.

Pareek, A., S.L. Singla, and A. Grover (1995). Immunological evidence for accumulation of two high-molecular-weight (104 and 90 kDa) HSPs in response to different stresses in rice and in response to high temperature stress in diverse plant genera. *Plant Molecular Biology* 29: 293-301.

Park, S.-Y., K.-C. Chang, R. Shivaji, and D.S. Luthe (1997). Recovery from heat shock in heat-tolerant and nontolerant variants of creeping bentgrass. *Plant Physiology* 115: 229-240.

Park, S.-Y., R. Shivaji, J.V. Krans, and D.S. Luthe (1996). Heat-shock response in heat-tolerant and nontolerant variants of *Agrostis palustris* Huds. *Plant Physiology* 111: 515-524

Paulsen, G.M. (1994). High temperature responses of crop plants. In *Physiology and Determination of Crop Yield,* eds. K.J. Boote, J.M. Bennette, T.R. Sinclair, and G.M. Paulsen. Madison, WI: American Society of Agronomy, pp. 365-389.

Petko, L., and S.L. Lindquist (1986). Hsp26 is not required for growth at high temperatures, nor for thermotolerance, spore development, or germination. *Cell* 45: 885-894.

Piper, P.W. (1993). Molecular events associated with acquisition of heat tolerance by the yeast *Saccharomyces cerevisiae. FEMS Microbiological Review* 11: 339-356.

Polla, B.S., S. Kantengwa, D. Francois, S. Salvioli, C. Franceschi, C. Marsac, and A. Cossariza (1996). Mitochondria are selective targets for the protective effects of heat shock against oxidative injury. *Proceedings of the National Academy of Sciences, USA* 93: 6458-6463.

Pollock, C.J., C.F. Eagles, C.J. Howarth, P.H.D. Schünmann, and J.L. Stoddart (1993). Temperature stress. In *Plant Adaptation to Environmental Stress,* eds. L. Fowden, T. Mansfield, and J. Stoddart. London: Chapman and Hall, pp. 109-132.

Porter, D.R., H.T. Nguyen, and J.J. Burke (1995). Genetic control of acquired high temperature tolerance in winter wheat. *Euphytica* 83: 153-157.

Rijven, A.H.G.C. (1986). Heat inactivation of starch synthase in wheat endosperm tissue. *Plant Physiology* 81: 448-453.

Ristic, Z., D.J. Gifford, and D.D. Cass (1991). Heat shock proteins in two lines of *Zea mays* L. that differ in drought and heat resistance. *Plant Physiology* 97: 1430-1434.

Sabehat, A., D. Weiss, and S. Lurie (1996). The correlation between heat-shock protein accumulation and persistence of chilling tolerance in tomato fruit. *Plant Physiology* 110: 531-537.

Sadalla, M.M., J.S. Quick, and J.F. Shanahan (1990) Heat tolerance in winter wheat: II. Membrane thermostability and field performance. *Crop Science* 30: 1248-1251.

Sadalla, M.M., J.F. Shanahan, and J.S. Quick (1990). Heat tolerance in winter wheat. I. Hardening and genetic effects on membrane thermostability. *Crop Science* 30: 1243-1247.

Sanchez, Y., and S.L. Lindquist (1990). HSP104 required for induced thermotolerance. *Crop Science* 248: 1112-1115.

Sankhla, N., A. Upadhyaya, T.D. Davis, and D. Sankhla (1992). Hydrogen perox-ide-scavenging enzymes and antioxidants in *Echinochloa frumentacea* as af-fected by triazole growth regulators. *Plant Growth Regulation* 11: 441-442.

Schlesinger, M.J. and C. Ryan (1993). An ATP- and hsc70-dependent oligomer-ization of nascent heat-shock factor (HSF) polypeptide suggests that HSF itself could be a "sensor" for the cellular stress response. *Protein Science* 2: 1356-1360.

Schöffl, F., M. Rieping and G. Baumann (1987). Constitutive transcription of a soybean heat-shock gene by a cauliflower mosaic virus promoter in transgenic tobacco. *Developmental Genetics* 8: 365-374.

Schrimer, E.C., S. Lindquist, and E. Vierling (1994). An *Arabidopsis* heat shock protein complements a thermotolerance defect in yeast. *Plant Cell* 6: 1899-1909.

Schuster, G., D. Even, K. Kloppstech, and I. Ohad (1988). Evidence for protection by heat-shock proteins against photoinhibition during heat-shock. *EMBO Journal* 7: 1-6.

Sewell, A.K., F. Yokoya, W. Yu, T. Miyagawa, T. Murayama, and D.R. Winge (1995). Mutated yeast heat shock transcription factor exhibits elevated basal transcriptional activation and confers metal resistance. *Journal of Biological Chemistry* 270: 25079-25086.

Sharkey, T.D. and E.L. Singaas. (1995). Why plants emit isoprene. *Nature* 374: 769.

Singaas, E.L., M. Lerdau, K. Winter, and T.D. Sharkey (1997). Isoprene increases thermotolerance of isoprene-emitting species. *Plant Physiology* 115: 1413-1420.

Sinibaldi, R.M. and I.J. Mettler (1992). Intron splicing and intron-mediated en-hanced expression in monocots. *Progress in Nucleic Acids Research* 42: 229-257.

Sinibaldi, R.M. and T. Turpen (1985). A heat shock protein is encoded within mito-chondria of higher plants. *Journal of Biological Chemistry* 260: 15382-15385.

Smith, B.J. and M.P. Yaffe (1991). Uncoupling thermotolerance from the induc-tion of heat shock proteins. *Proceedings of the National Academy of Sciences, USA* 88: 11091-11094.

Somerville, C. and J. Browse (1991). Plant lipids: Metabolism, mutants, and membrane. *Science* 252: 80-87.

Stapel, D., E. Kruse, and K. Kloppstech (1993). The protective effect of heat shock proteins against photoinhibition under heat shock in barley (*Hordeum vulgare*). *Journal of Photochemistry and Photobiology* 21: 211-218.

Staub, J.E. and F.C. Serquen (1996). Genetic markers, map construction, and their application in plant breeding. *Horticultural Science* 31: 729-741.

Stuber, C.W. (1995). Mapping and manipulating quantitative traits in maize. *Trends in Genetics* 11: 477-481.

Sullivan, C.Y., N.V. Norcio, and J.D. Eastin (1977). Plant responses to high temperatures. In *Genetic Diversity in Plants,* eds. A. Muhammed, R. Aksel, and R.C. von Borstel. New York: Plenum Press, pp. 301-312.

Susek, R.E. and S.L. Lindquist (1989). Hsp26 of *Saccharomyces cerevisiae* is related to the superfamily of small heat shock proteins but is without a demon-strable function. *Molecular and Cellular Biology* 9: 5265-5271.

Suss, K.-H. and I.T. Yordanov (1986). Biosynthetic causes of in vivo acquired thermotolerance of photosynthetic light reactions and metabolic responses of chloroplast to heat stress. *Plant Physiology* 81: 192-199.

Tanksley, S.D. and S.R. McCouch (1997). Seed banks and molecular maps: Unlocking genetic potential from the wild. *Science* 277: 1063-1066.

Tanksley, S.D., N.D. Young, A.H. Paterson, and M.W. Bonierbale (1989). RFLP mapping in plant breeding: New tools for an old science. *Biotechnology* 7: 257-264.

Tassi, F., E. Maestri, F.M. Restivo, and N. Marmiroli (1992). The effects of carbon starvation on cellular metabolism and protein and RNA synthesis in *Gerbera* callus cultures. *Plant Sciences* 83: 127-136.

Tomsett, A.B. and D.A. Thurman (1988). Molecular biology of metal tolerances in plants. *Plant Cell and Environment* 11: 383-394.

Towill, L.E. and P. Mazur (1974). Studies on the reduction of 2,3,5-triphenyl tetrazolium chloride as a viability assay for plant tissue culture. *Canadian Journal of Botany* 53: 1097-1102.

Trent, J.D. (1996). A review of acquired thermotolerance, heat-shock proteins, and molecular chaperones in archaea. *FEMS Microbiology Review* 18: 249-258.

Vierling, E. (1991). The roles of heat shock proteins in plants. *Annual Review of Plant Physiology and Plant Molecular Biology* 42: 579-620.

Visioli, G., E. Maestri, and N. Marmiroli (1997). Differential display-mediated isolation of a genomic sequence for a putative mitochondrial LMW HSP specifically expressed in condition of induced thermotolerance in *Arabidopsis thaliana* (L.) Heynh. *Plant Molecular Biology* 34: 517-527.

Vogel, J.L., D.A. Parsell, and S. Lindquist (1995). Heat-shock proteins Hsp104 and Hsp70 reactivate mRNA splicing after heat inactivation. *Current Biology* 5: 306-317.

Wallner, S.J., M.R. Bechwar, and J.D. Butler (1982). Measurement of turfgrass heat tolerance in vitro. *Journal of American Society of Horticultural Sciences* 107: 608-611.

Waters, E.R., G.J. Lee, and E. Vierling (1996). Evolution, structure and function of the small heat shock proteins in plants. *Journal of Experimental Botany* 47: 325-338.

Weis, E. (1981). The temperature-sensitivity of dark-inactivation of the ribulose-1,5-biphosphate carboxylase in spinach chloroplasts. *FEBS Letters* 129: 197-200.

Welte, M.A., J.M. Tetrault, R.P. Dellavalle, and S.L. Lindquist (1993). A new method for manipulating transgenes: Engineering heat tolerance in a complex, multicellular organism. *Current Biology* 3: 842-853.

Weng, J. and H.T. Nguyen (1992). Differences in the heat-shock response between thermotolerant and thermosusceptible cultivars of hexaploid wheat. *Theoretical and Applied Genetics* 84: 941-946.

Werner, D.J. and T.L. Watschke (1981). Heat tolerance of Kentucky bluegrasses, perennial ryegrasses, and annual bluegrasses. *Agronomy Journal* 73: 79-83.

Willekens, H., D. Inzé, M. van Montagu, and W. van Camp (1995). Catalases in plants. *Molecular Breeding* 1: 207-228.

Wollgiehn, R. and D. Neumann (1995). Stress response of tomato cell cultures to toxic metals and heat shock: Differences and similarities. *Journal of Plant Physiology* 146: 736-742.

Wollgiehn, R., D. Neumann, U. zur Nieden, A. Müsch, K.-D. Scharf, and L. Nover (1994). Intracellular distribution of small heat stress proteins in cultured cells of *Lycopersicon peruvianum. Journal of Plant Physiology* 144: 491-499.

Yang, G., D. Rhodes and R.I. Joly (1996). Effects of high temperature on membrane stability and chlorophyll fluorescence in glycinebetaine-containing maize lines. *Australian Journal of Plant Physiology* 23: 437-443.

Yeh, C.-H., P.F.L. Chang, K.-W. Yeh, W.-C. Lin, Y.-M. Chen, and C.-Y. Lin (1997). Expression of a gene encoding a 16.9 kDa heat shock protein, Oshsp16.9, in *Escherichia coli* enhances thermotolerance. *Proceedings of the National Academy of Sciences, USA* 94: 10967-10972.

Yeh, K.-W., T.-L. Jinn, C.-H. Yeh, Y.-M. Chen, and C.-Y. Lin (1994). Plant low-molecular-mass heat-shock proteins: Their relationship to the acquisition of thermotolerance in plants. *Biotechnology and Applied Biochemistry* 19: 41-49.

Yost, H.J. and S. Lindquist (1988). Translation of unspliced transcripts after heat shock. *Science* 242: 1544-1548.

Zarsky, V., D. Garrido, N. Eller, J. Tupy, O. Vicente, F. Schöffl, and E. Heberle-Bors (1995). The expression of a small heat shock gene is activated during induction of tobacco pollen embryogenesis by starvation. *Plant Cell and Environment* 18: 139-147.

Zhang, J., N. Klueva, and H.T. Nguyen (1999). Plant adaptation and crop improvement for arid and semi-arid environments. In *Desert Development: The Endless Frontier,* Volume 2, eds. I.R. Taylor Jr., H. Dregne, and K. Mathis. Proceedings of the Fifth International Conference on Desert Development. Lubbock, TX: Texas Tech University, pp. 604-620.

Zhu, J.-K., P.M. Hasegawa, and R.A. Bressan (1997). Molecular aspects of osmotic stress in plants. *Critical Reviews in Plant Sciences* 16: 253-277.

Chapter 7

Control of the Heat Shock Response in Crop Plants

Daniel R. Gallie

INTRODUCTION

The heat shock (HS) response is one of the earliest cellular mechanisms to have evolved, as several aspects of the response are conserved in bacteria, yeast, plants, and animals, underscoring the threat that hyperthermia presents to cellular functions. Nevertheless, organisms have adapted to environments of widely different temperatures, and, consequently, the temperature that constitutes heat shock is a relative one. For instance, an alga *(Plocamium cartilagineum)* found in antarctic waters, whose normal growth conditions vary in temperature from 0.3 to $-1.8°C$, experiences heat shock at only $5°C$ (Vayda and Yuan, 1994), whereas thermophiles must require a substantially higher temperature before mounting an HS response. Regardless of the growth temperature of an organism, its heat shock response is typically mounted at approximately $10°C$ above the normal ambient temperature. Although crop plants grow within a much more narrow temperature range, the temperature at which they mount a heat shock response shows some variation. A number of reviews have focused on several aspects of the heat shock response, including the structure and function of heat shock proteins (HSPs) (Vierling, 1991; Winter and Sinibaldi, 1991; Becker and Craig, 1994; Yeh et al., 1994; Waters, 1995), the developmental response to heat shock and the expression of HSPs during development (Neumann, Scharf, and Nover 1984; Winter

The work in the author's laboratory is supported by the U.S. Department of Agriculture (NRICGP 95-37100-1618, NRICGP 96-35301-3144, and NRICGP 97-35301-4404).

and Sinibaldi, 1991; Nover and Scharf, 1997), and the transcriptional control of HSPs (Czarnecka-Verner, Dulce-Barros, and Gurley, 1994; Nover and Scharf, 1997). This chapter will focus mainly on the recent advances in our understanding of the changes in gene expression as part of the HS response in crop plants, and also how the HS response relates to developmental control of HSP expression and HSP induction by other stresses.

EXPRESSION OF HSPs FOLLOWING STRESS AND DURING DEVELOPMENT

A hallmark of the heat shock response in all organisms examined to date is the induction of a specific set of proteins, several of which are structurally or functionally conserved throughout evolution. The designation for these proteins, i.e., heat shock proteins (HSPs), is somewhat misleading, as the expression of several is developmentally regulated or induced by other stresses. A second category, the heat shock cognate proteins, is used to indicate those proteins which are related to "true" heat shock proteins but are either constitutively expressed or otherwise regulated.

Expression of HSPs was first demonstrated in plants using controlled laboratory conditions, and only later was their expression demonstrated under field conditions. Synthesis of proteins with similar molecular weights as several well-characterized HSPs was detected in field-grown cotton when daily canopy temperatures reached 40°C (Burke et al., 1985). HSP levels were higher in nonirrigated cotton than in irrigated. Similar results were observed in field-grown soybean (Kimpel and Key, 1985). Several low molecular weight (LMW) HSP mRNAs (messenger ribonucleic acid) were detected in leaves of soybean subject to a midday temperature of 40°C, and, as with cotton, higher levels were observed in nonirrigated plants (Kimpel and Key, 1985). However, the water stress induced by the lack of irrigation was not, in itself, sufficient to induce these HSP mRNAs.

Subsequent studies in *Medicago sativa* demonstrated the effect that transpiration has on leaf temperature (Hernandez and Vierling, 1993). Leaf temperature increased approximately 0.64°C for every 1.0°C increase in air temperature (Hernandez and Vierling, 1993). Therefore, during the hottest part of the day, leaf temperature can be more than 8°C lower than ambient temperature. The temperature of a leaf exposed directly to the sun will nevertheless be higher than a neighboring leaf that is shielded from direct exposure to the sun even though the ambient temperature is the same for both. The ability of plant organs to maintain a temperature differential will depend on their ability to transpire. As nonirrigated plants experiencing water stress close their stomata in an attempt to conserve water, their ability to reduce the leaf temperature through transpiration will be compromised. Consequently, the water status of a crop plant can have an indirect effect on the severity of the heat shock. Organs within a crop plant can differ in their ability to regulate their temperature. The temperature of flowers and pods of *Medicago sativa* was shown to exceed that of leaves by 2 to 3°C (Hernandez and Vierling, 1993). Consequently, these organs were observed to express HSPs at an air temperature that was too low to elicit HSP expression in leaves (Hernandez and Vierling, 1993). The difference in the ability of organs to reduce their temperature through transpiration may well account for many of the conflicting reports concerning the ability of different tissues to mount a response to a heat shock.

The extensive evidence documenting HSP expression during select stages of development strongly suggests that HSPs play a more extensive role in the biology of plants than was originally thought. Expression of HSPs, measured by the presence of protein, transcript, or promoter activity, has been demonstrated during embryo development in *Arabidopsis* (Wehmeyer et al., 1996); pea (DeRocher and Vierling, 1994, 1995); bean (Zur Nieden et al., 1995); sunflower (Coca, Almoguera, and Jordano, 1994); sorghum (Howarth, 1990); and maize (Marrs et al., 1993; Zur Nieden et al., 1995). HSP expression has also been observed during somatic embryogenesis in carrot (Darwish et al., 1991; Apuya and Zimmerman, 1992), alfalfa (Gyorgyey et al., 1991), white spruce (Dong and Dunstan, 1996), and during induction in tobacco pollen (Zarsky et al., 1995). Translation of HSP mRNAs present in mature embryos appears to extend into the earliest stages of germination in several species, including wheat (Helm and Abernathy, 1990), *Arabidopsis* (Wehmeyer et al., 1996), pea (DeRocher and Vierling, 1994, 1995), and sunflower (Coca, Almoguera, and Jordano, 1994). The persistence of HSPs and their expression typically declines rapidly during subsequent germination, although some studies report persistence in two-day-old seedlings (Kruse, Liu, and Kloppstech, 1993; D. Gallie, unpublished data). HSP gene expression or HSPs have been detected in non-seed tissues such as during the meiotic stages of pollen development in maize (Frova, Taramino, and Binelli, 1989; Atkinson et al., 1993; Marrs et al., 1993; Magnard, Vergne, and Dumas, 1996), tobacco (Zur Nieden et al., 1995), and lily (Kobayashi et al., 1994); in specific floral organs in *Arabidopsis* (Tsukaya et al., 1993); in meristematic and provascular tissues in transgenic *Arabidopsis* (Prandl, Kloske, and Schöffl, 1995); and in vegetative tissues of the resurrection plant *Craterostigma plantagineum* (Alamillo et al., 1995).

Whether the developmental expression of HSPs occurs from the same genes induced by heat shock or whether the developmentally regulated genes are separate from those which are heat induced has been a question for some years. Recent evidence suggests that at least some HSP genes induced by heat shock are also under developmental control. Analysis of the promoter region for *Gmhsp17.3* of soybean demonstrated that the regulatory elements required for heat induction and developmental expression colocalized (Prandl and Schöffl, 1996). The expression of HSPs during embryogenesis is not limited to plants, as HSPs are expressed during early embryogenesis in *Drosophila* (Michaud, Marin, and Tanguauy, 1997), *Xenopus* (Heikkila et al., 1997), and zebrafish (Krone, Sass, and Lele, 1997). The role of the HSPs during heat shock and during development is thought to be the same: to prevent aggregation or misfolding of proteins to promote either the adoption of the proper conformation or, failing this, to be presented to the proteolytic machinery (Gottesman, Wickner, and Maurizi, 1997). The assembly of HSPs into high molecular weight complexes that has been observed in several plant species, including pea, soybean, rice and *Brassica napus* (Chen, Osteryoung, and Vierling, 1994; Jinn, Chen, and Lin, 1995; Lee, Pokala, and Vierling, 1995; Helm, Lee, and Vierling, 1997; Krishna et al., 1997), agrees well with results in other species and is consistent with the multimeric nature associated with molecular chaperone complexes.

Expression of HSPs is not limited to conditions of elevated temperatures. Induction of an LMW HSP was observed following oxidative stress in parsley caused by exposure to ozone (Eckey-Kaltenbach et al., 1997), and the induction of high molecular weight HSPs was observed following either cold treatment (Neven et al., 1992; Pareek, Singla, and Grover, 1995; Didierjean et al., 1996), salt stress (Pareek, Singla, and Grover, 1995), or wounding (Henry-Vian et al., 1995; Didierjean et al., 1996). Some heavy metals, such as arsenite, cadium, or Cu^{2+} (copper), can elicit the synthesis of some HSPs (Nagao et al., 1986; Czarnecka et al., 1988; Feussner et al., 1997), as can application of methomyl (Rees, Gullons, and Walden, 1989) or calcium (Braam, 1992). In addition to abiotic stress, biotic stress can also induce HSP expression: induction of HSP70 expression was observed in pea following infection by pea seed-borne mosaic virus (Aranda et al., 1996). These observations suggest that HSPs may play critical roles in meeting the challenge of biological or environmental insults that share aspects with thermal stress.

THE IMPACT OF HEAT SHOCK ON TRANSCRIPTION

Heat shock affects transcription in plants in two ways: the transcription of non-HSP genes is repressed, whereas the transcription of HSP genes is induced. Although the means by which transcription of non-HSP genes is repressed in plants has not been well studied, the heat-induced activation of transcription of HSP genes in plants as well as in yeast and animal systems has received considerable attention. The transcriptional regulation of HSP genes is one of the most conserved aspects in eukaryotes and requires the presence of heat shock elements (HSEs) within or adjacent to the HSP promoter and the transcriptional activator, the heat shock factor (HSF), which specifically binds to the HSE (Gurley and Key, 1991; Gurley, Czarnecka, and Barros, 1993; Czarnecka-Verner, Dulce-Barros, and Gurley, 1994; Nover et al., 1996; Nover and Scharf, 1997). HSF is functionally conserved between plants and animals in that HSFs from *Drosophila* or humans could function correctly in tobacco to direct the heat-mediated activation of transcription (Treuter et al., 1993). Surprisingly, the yeast HSF failed to function in this same plant (Treuter et al., 1993). An additional adenine-thymine (AT)-rich element upstream of some HSP genes examined also contributes to HSP expression (Baumann et al., 1987; Czarnecka, Key and Gurley, 1989; Czarnecka, Fox, and Gurley, 1990; Schöffl et al., 1989; Gurley and Key, 1991; Rieping and Schöffl, 1992; Gurley, Czarnecka, and Barros, 1993; Schöffl et al., 1993). In one case, this AT-rich region functions as a matrix attachment region (MAR) or scaffold attachment region (SAR). MARs/SARs are AT rich (>70 percent) chromosomal DNA regions that attach to the nuclear matrix and often flank actively expressed genes (Schöffl et al., 1993; van der Geest et al., 1994; Chinn and Comai, 1996) to affect their expression through influencing chromatin structure. For the *Gmhsp17.6* gene of soybean, a 395 base pair region was identified as an SAR (Schöffl et al., 1993). An MAR/SAR flanking a transgene can increase expression fourfold to 140-fold in plants (Spiker and Thompson, 1996), perhaps by maintaining the chromosomal region containing the transgene in an open configuration to facilitate communication between an enhancer and a promoter (van der Geest and Hall, 1996).

As shown for the *Gmhsp17.3* gene in soybean, the regulatory sequences that direct its developmental expression colocalized with the regulatory elements required for its heat induction (Prandl and Schöffl, 1996). The HSEs are elegantly simple in that they are a trinucleotide core sequence, 5'-nGAAn-3' or 5'-nTTCn-3', present in alternating orientations, separated by two nucleotides, in all species examined, including plants (Baumann et al., 1987; Schöffl et al., 1989; Dulce-Barros, Czarnecka, and Gurley, 1992). There, typically, multiple copies of HSEs are found in HSP genes (Gurley, Czarnecka, and Barros, 1993; Czarnecka-Verne, Dulce-

Barros, and Gurley, 1994), and their head-to-head or tail-to-tail orientation is thought important in binding HSF in its trimer conformation.

HSF is expressed from small multigene families in plants, of which at least one member is constitutively expressed, whereas the remaining members are heat inducible (Nover and Scharf, 1997). The constitutive expression of HSF permits the plant to be continually poised to respond to a heat shock with the speed necessary to limit heat-induced damage to proteins. The purpose of the heat-induced expression of HSF is less well understood. Under normal conditions, the HSF exists in the cytoplasm in a monomer state and, following the onset of a heat shock, forms homotrimers, enters the nucleus, and binds to HSEs, resulting in the activation of HSP transcription. The purpose of the trimerization has not been established, but it is this form that binds to the HSEs where two homotrimers bind the tail-to-tail or head-to-head core domains of the HSEs (Nover and Scharf, 1997).

The trimerization may serve to stabilize complex formation between the HSF and the HSEs. Not surprisingly, multiple domains within HSFs are required for cellular localization, trimerization, DNA binding, and transcription activation. Under normal conditions, the plant HSF does not exhibit DNA-binding activity but requires exposure to a heat shock to activate this function (Hubel et al., 1995). The regulation of the DNA-binding activity appears to involve several hydrophobic repeats present throughout the HSF that can form a leucine zipper domain required for protein-protein interactions. If expressed in other species, such as in *Drosophila* or human cells, the *Arabidopsis* HSF (ATHSF1) fails to exhibit proper regulation and binds in a constitutive manner to HSP promoter sequences containing HSEs, suggesting that a specific factor may be required to inhibit DNA-binding and transcriptional regulatory activities under non-heat shock conditions (Hubel et al., 1995). N- or C-terminus fusions to ATHSF1 resulted in a constitutive DNA-binding phenotype and partial activation of HSP gene expression, leading to a higher basic thermotolerance in transgenic *Arabidopsis* (Lee, Pokala, and Vierling, 1995).

EFFECTS OF HEAT SHOCK ON PROTEIN YIELD

Although many of the molecular events involved in the induction of heat shock genes are remarkably conserved in eukaryotes, the control of translation following heat shock varies considerably among species. Yeast respond to heat shock mostly through transcriptional changes, whereas in *Xenopus* oocytes, the HS response is mediated entirely at the translational level (Bienz and Gurdon, 1982; Bienz, 1984). In plants, the HS response lies between these extremes, with gene regulation involving both transcriptional and translational mechanisms. The degree to which synthesis of non-HSPs is inhibited by heat shock is proportional to the severity of the stress. Moreover, the degree to which translational repression occurs is amplified in the absence of transcription during the heat shock (Gallie and Pitto, 1996). In spite of the heat-mediated translational repression, a low level of non-HSP mRNA translation continues in plants following many heat shock treatments (Key, Lin, and Chen, 1981; Callis, Fromm, and Walbot, 1988; Hwang and Zimmerman, 1989; Krishnan, Nguyen, and Burke, 1989). For those species which experience translational repression following a heat shock, a rapid disassembly of polyribosomes in many species results in a reduction in the translation of non-HSP mRNAs (Storti et al., 1980; Lin, Roberts, and Key, 1984; Gallie, Caldwell, and Pitto, 1995). Non-HSP mRNAs are not destroyed but are maintained in heat shock granules (HSGs) that include two major HSPs, HSP70 and HSP17 (Nover, Scharf, and Neumann, 1983, 1989), and the mRNAs are subsequently recruited for translation upon recovery (Storti et al., 1980). The HSGs associate with the cytoskeleton, forming perinuclear complexes in plants (Nover et al., 1989; Apuya and Zimmerman, 1992), in invertebrates (Arrigo, 1987; Leicht et al., 1986), and in vertebrates (Collier and Schlesinger, 1986).

Heat shock not only causes a reduction in translational efficiency but increases the mRNA half-life of non-HSP mRNAs in plants, and both effects are proportional to the severity of the stress (Gallie, Caldwell, and Pitto, 1995). The inverse correlation between the effect on translation and mRNA stability following heat shocks of increasing severity results in a complicated combinatorial effect on protein synthesis. The degree to which each of these two processes is affected determines whether the final protein yield will be higher or lower than that achieved at the control temperature. For example, a mild heat shock can cause a small increase in both the translational efficiency and message stability in carrot, resulting in increased protein synthesis (Gallie, Caldwell, and Pitto, 1995). Alternatively, a moderate heat shock can result in a severalfold increase in mRNA stability, but only a moderate loss in translational efficiency, such that the final protein yield from a given species of mRNA can be greater than that same pool of mRNA in nonstressed cells. In contrast, even with nearly an order of magnitude increase in the stability of an mRNA following a more severe heat shock, the reduction in translational efficiency is so great that the final yield of protein from an mRNA is far less than that from the same mRNA in nonstressed cells (Gallie, Caldwell, and Pitto, 1995). This increase in mRNA stability may result from either an increase in protection from the mRNA degradatory apparatus and/or a reduction in the expression or activity of those RNases responsible for mRNA turnover. Protein yield resulting from these heat-mediated effects on translation and RNA turnover are further complicated by the repression of transcription of non-HSP mRNAs that occurs following a heat shock.

HEAT-INDUCED CHANGES IN THE TRANSLATIONAL MACHINERY OF CROP PLANTS

The cap and poly(A) tail function in a cooperative manner to establish an efficient level of translation (Gallie, 1991). This means that neither element functions well in the absence of the other and suggests communication between these two regulatory elements. Following a heat shock, the translational machinery loses its ability to discriminate between capped and uncapped mRNAs. As a consequence, translation becomes less cap dependent, and the functional codependency between the cap and poly(A) tail is reduced (Gallie, 1991; Gallie, Caldwell, and Pitto, 1995). Heat shock causes not only a loss in cap function but also a loss in the codependency between the cap and poly(A) tail (Gallie, Caldwell, and Pitto, 1995), which is a consequence of a reduction in the communication between the 5' cap and the poly(A)-binding protein (PABP) at the poly(A) tail. PABP, therefore, can be considered as a participant in the translation initiation process. Multiple PABP isoforms were observed in wheat leaves; however, no change in the distribution of the PABP isoforms was observed in leaves or following a heat shock (Gallie et al., 1997). This also suggests that the loss in codependency between the cap and the poly(A) tail following a heat shock may be a consequence of the heat-induced modifications of the initiation factors that associate with the 5' cap.

In animals, the reprogramming of translation following thermal stress correlates with changes in phosphorylation for several initiation factors (Duncan and Hershey, 1989). Two of the best-studied examples are (1) the dephosphorylation of the cap-binding protein subunit of the eukaryotic initiation factor eIF4E and (2) the phosphorylation of eIF2α. eIF4F (of which eIF4E is a subunit) binds to the cap structure at the 5' terminus of the mRNA and stimulates the binding of eIF4A and eIF4B. eIF4A is an RNA-dependent RNA helicase that, together with eIF4B and eIF4F, unwinds any secondary structure present in the 5'-untranslated leader, thereby preparing the mRNA for binding to the 40S ribosomal subunit. eIF4E is the smallest of the three subunits that comprise eIF4F, whereas eIF4G is the largest.

The loss in cap-dependent translation that follows a heat shock in plants is proportional to the severity of the stress (Gallie, Caldwell, and Pitto, 1995) and suggests that the activity of one or more of the cap-associated initiation factors, for example, eIF4B or eIF4F, may be altered following thermal stress. The translational machinery of plants differs significantly from that in animals and yeast in that plants contain not only eIF4F but also an isoform, designated eIFiso4F. eIF4E and eIFiso4E (the small sub-unit of eIFiso4F) share homology and are functionally analogous, as are eIF4G and eIFiso4G (the large subunit of eIFiso4F). Dephosphorylation of mammalian eIF4E occurs following heat shock (Duncan and Hershey, 1989) and correlates with reduced eIF4F binding to the cap and protein synthesis activity (Panniers et al., 1985; Lamphear and Panniers, 1990, 1991; Zapata, Maroto, and Sierra, 1991). eIF4E that is present in wheat leaves is present as a single pair of isoforms, whereas only a single isoform of the eIFiso4E is observed. However, no change in the number or dis-tribution of the isoforms was observed in leaves subject to a heat shock (Gallie et al., 1997), suggesting that, unlike mammalian eIF4E, wheat eIF4E and eIFiso4E do not undergo dephosphorylation following heat shock.

Wheat eIF4A, a single 47 kilodalton (kDa) polypeptide, is an adenosine triphosphate (ATP)-dependent RNA helicase that, in conjunction with eIF4B and eIF4F, is thought to remove secondary structure present within a $5'$ leader (Browning et al., 1989; Jaramillo et al., 1990) and is also an RNA-dependent ATPase in plants, animals, and yeast (Lax et al., 1986; Schmid and Linder, 1992). Phosphorylation of maize eIF4A occurs in roots following oxygen deprivation; consequently, phosphorylation of eIF4A is part of the hypoxic stress response (Webster et al., 1991). Two eIF4A species that represent dephosphorylated and monophosphorylated isoforms present in an approximately 9:1 ratio, respectively (Gallie et al., 1997), are observed in wheat leaves. No change in the distribution of these isoforms was observed in leaves subject to a 15 minute (min) treatment at $45°C$. A substantially longer heat shock increased the amount of phospho-rylated eIF4A such that both isoforms were present in equal amounts (Gallie et al., 1997). Therefore, as in hypoxically treated maize roots, a prolonged heat shock can result in the phosphorylation of wheat eIF4A, suggesting that phosphorylation of this initiation factor may be a charac-teristic response to stress in plants.

Mammalian eIF4B is a single subunit of approximately 70 kDa present in eight to ten isoforms (Duncan and Hershey, 1989) and is dephosphory-lated following heat shock (Duncan and Hershey, 1984) that correlates with a reduction in translation, whereas its phosphorylation following insulin treatment (Manzella et al., 1991) correlates with an increase in translation. eIF4B, a 59 kDa polypeptide in wheat, is present in leaves in several isoelectric states similar to those observed in mammalian cells (Gallie et al., 1997). An acidic cluster (approximately four to six isoforms) are the phosphorylated isoforms of the basic cluster (approximately four isoforms). The acidic cluster is completely absent when eIF4B from wheat embryos is examined, which suggests that eIF4B is developmentally regulated (Gallie et al., 1997). In leaves subject to even a brief exposure to heat shock (15 min at 45°C), eIF4B undergoes dephosphorylation (Gallie et al., 1997). The presence of phosphorylated eIF4B correlates with the resumption of active translation during wheat seed germination, and the absence of phosphorylated eIF4B correlates with the repression of translation following heat shock (Gallie et al., 1997).

eIF2 is a three subunit complex that is responsible for bringing the Met-tRNA$_i^{met}$ (methionine-tRNA) to the 40S subunit in both plants and animals. The α subunit is subject to phosphorylation in animal cells by a number of physiological events, including heat shock (Duncan, Milburn, and Hershey, 1987). Phosphorylation of the α subunit prevents GDP/GTP (guanine diphosphate/guanine triphosphate) exchange by eIF2B and consequently inhibits eIF2 activity and, therefore, translation (Murtha-Riel et al., 1993). Overexpression of an eIF2α mutant that is resistant to phosphorylation partially protected Chinese hamster ovary cells from the inhibition of protein synthesis following a heat shock (Murtha-Riel et al., 1993), suggesting that the control of this initiation factor constitutes an important regulatory point following heat shock in animal cells. Similar to its animal counterpart, wheat eIF2 is also a three subunit complex composed of α (42 kDa), ß (38 kDa), and γ (50 kDa) subunits (Metz and Browning, 1997). Multiple isoforms for the ß subunit of eIF2 and an acidic doublet representing the α subunit were observed in wheat embryos (Gallie et al., 1997). eIF2α exists in a phosphorylated state in wheat embryos and is converted to a hypophosphorylated or dephosphorylated state in leaves. A heat shock, whether of short or long duration, has little detectable impact on the phosphorylation state of the α subunit (Gallie et al., 1997). If phosphorylation of wheat eIF2α results in its inactivation, as it does in mammalian cells and yeast, then wheat eIF2α may be maintained in an inactive (phosphorylated) state in the seed and shifted to an active (dephosphorylated) state upon germination. The dephosphorylation of eIF2α in leaves does correlate with the activation of translation that occurs during germination. Following a heat shock, however, the lack of phosphorylation of the α subunit and only minor changes in the ß subunit suggest that the response to heat stress by the translation apparatus in plants may differ substantially from that in yeast and animal cells.

The changes in phosphorylation of both transcription and translation factors following thermal stress can be summarized as follows: whereas eIF4E, eIF4B, and eIF2α are modified in mammals following a heat shock, in wheat, only eIF4A and eIF4B are subject to heat-induced modifications. Although phosphorylation of eIF4A following a prolonged stress may not account for the rapid changes in translation observed in plants following a short heat shock, eIF4B does undergo rapid dephosphorylation following thermal stress. Thus, in their response to thermal stress, plants may differ from animals in two key points: in the regulation of eIF4E/eIFiso4E and of eIF2α. As heat shock causes a loss in the functional interaction between the cap and the poly(A) tail (Gallie, Caldwell, and Pitto, 1995), it is possible that the heat shock-induced dephosphorylation of eIF4B (Gallie et al., 1997) is responsible for the loss in its interaction with PABP that is observed under non-heat shock conditions (Le et al., 1997).

HSP mRNAs ESCAPE HEAT SHOCK-INDUCED TRANSLATIONAL REPRESSION

Concurrent with the translation repression of non-heat shock mRNAs is the active translation from HSP mRNAs. Analysis of *Hsp70* and *Hsp22* mRNAs in *Drosophila* (McGarry and Lindquist, 1985; Klemenz, Hultmark, and Gehring, 1985; Hultmark, Klemenz, and Gehring, 1986) and the maize *HSP70* (Pitto, Gallie, and Walbot, 1992), revealed that the 5′ leader of each was sufficient and necessary to confer translational competence, on a reporter mRNA that was otherwise subject to repression in the absence of the heat shock leader. A subsequence close to the 5′ terminus of the *Hsp22* 5′ leader was required for the translational competence although no subsequence could be identified for the *Drosophila Hsp70* mRNA. One hypothesis to explain the effect of HSP 5′ leaders on translation following a heat shock is that they could be adenine-uridine (AU)-rich and therefore lack significant secondary structure. Consequently, scanning of the HSP leader by the 40S ribosomal subunit would be unimpeded. Analysis of a large number of HSP 5′ leaders, including those from higher and lower plants, demonstrated that they do not differ significantly in AU content or length from those present in non-HSP mRNAs (Joshi and Nguyen, 1995). Moreover, the sequence context surrounding the initiator AUG codon, which is important in identifying the start codon, differed little between HSP and non-HSP mRNAs (Joshi and Nguyen, 1995). An alternative explanation for the preferential translation of HSP mRNAs following a heat shock may be that they are recognized by a *trans*-acting factor that mediates their translational competence. This may involve heat-induced modifications to the ribosomal subunits that have been observed to occur in tomato and wheat (Scharf and Nover, 1982; Fehling and Weidner, 1988). In addition to the role that HSP leaders play in translation regulation, changes in the length of the poly(A) tail may also influence the continued translational competence of these mRNAs. An increase in the length of the poly(A) tail of the *HSP21* transcript from *Arabidopsis* was observed following a heat shock (Osteryoung, Sundberg, and Vierling, 1993). As the poly(A) tail is required for efficient translational initiation (Gallie, 1991; Le et al., 1997), a heat-induced increase in the length of the poly(A) tail would be a means by which the translational efficiency of the HSP mRNAs could be selectively increased. Such a mechanism would also require the specific recognition of this class of mRNA.

The ability of a 5′ leader to confer translational competence on an mRNA is not limited to those from a heat shock mRNA. The 5′ leader from tobacco mosaic virus was shown to be functionally equivalent to a heat shock leader in heat-shocked carrot, tobacco, and maize (Pitto, Gallie, and Walbot, 1992). Translation from reporter mRNA containing this viral leader actually increased following a heat shock, in contrast to the same reporter mRNA with a control leader that was translationally repressed under the same stress conditions (Pitto, Gallie, and Walbot, 1992). The fact that a 5′ leader of an mRNA is required to escape the heat shock-mediated repression of translation suggests that the initiation step of translation functions as the regulatory point targeted by heat shock as the means to control protein synthesis in plants.

REGULATION OF RNA DEGRADATORY MACHINERY FOLLOWING HEAT SHOCK

As mentioned earlier, heat shock causes an increase in mRNA stability that is a function of the severity of the stress, resulting in an increase in the time over which a given mRNA is translationally active (Gallie, Caldwell, and Pitto, 1995). For example, following a 15 min exposure to 37°C, the functional half-life of a reporter mRNA in carrot increased only 50 percent, but it increased fivefold following a 42°C heat shock, and nearly ninefold following a 45°C heat shock (Gallie, Caldwell, and Pitto, 1995). In contrast, no change in message stability was detected in heat-shocked mammalian cells (Gallie, Caldwell, and Pitto, 1995). The thermally induced increase in mRNA half-life in plants could be a result of two non-mutually exclusive possibilities: a reduction in the amount or activity of those RNases responsible for mRNA degradation, or a sequestration of mRNAs from RNase attack, perhaps through their incorporation into HSGs.

The response to heat shock is complex and varies with the temperature of the stress, its duration, and the rate of heating. However, the activities of all detectable RNases decrease in wheat leaves following a heat shock that occurs as a function of the length and severity of the heat treatment (Chang and Gallie, 1997). The decrease in RNase activity is small following a 37°C heat treatment but is more apparent as the temperature of the heat shock increases. The changes in RNase activity following a heat shock (Chang and Gallie, 1997) correlate well with the increase in in vivo message stability discussed previously (Gallie, Caldwell, and Pitto, 1995). In contrast to the heat-induced effects on RNase activities, nucleases present in wheat leaves whose expression is regulated by senescence and light (Blank and McKeon, 1989) are not significantly affected by heat shock (Chang and Gallie, 1997). The loss in RNase activity observed following a heat shock is not explained by the thermal instability of the RNases, as they remain active in vitro following heat treatments up to 100°C, whereas nuclease activities are thermal labile in vitro. Moreover, there is no change in the amount of RNase protein in wheat leaves before or after a heat shock (Chang and Gallie, 1997). The observation that wheat leaf RNases retain full activity in vitro following heat treatments as high as 100°C suggests that their in vivo loss of activity following heat treatments ranging from only 37°C to 45°C results from a reduction in activity of the RNases and not from thermal lability or protein turnover.

Recovery of RNase activity in wheat leaves begins between 20 and 35 hours (h) after the cessation of 41°C heat shock and is complete by approximately 35 h (Chang and Gallie, 1997). Recovery following a more severe heat shock (45°C) is detectable only by 35 h post-heat shock, suggesting that the length of the time required for recovery of RNase activity increases with the severity of the thermal stress (Chang and Gallie, 1997). Likewise, the time required for the recovery from a heat shock-induced translational repression is a function of the severity of the stress (D. Gallie, unpublished observations). The time required for the recovery of the pre-heat shock level of RNase activities following a heat shock might be needed for either renewed RNase synthesis or for the reactivation of the preexisting RNase activities that were inactivated by the heat shock. These observations correlate with the suggestion that the increase in mRNA stability following a heat shock is also a function of the severity of the thermal stress (Gallie, Caldwell, and Pitto, 1995).

The heat-mediated regulation of RNase activity may play a role in controlling the stability of HSP as well as non-HSP mRNAs. The half-life of HSP mRNA was substantially shorter at control temperatures than during heat shock (Kimpel et al., 1990), suggesting that expression from HSP mRNAs may be controlled not only by a translational mechanism but also through heat-mediated alteration in its stability.

Although the decrease in RNase activity following a heat shock might be responsible for the increase in mRNA stability, the heat-mediated regulation of RNase activity does not exclude the possibility that mRNAs are sequestered into HSGs following a heat shock as an additional means by which non-heat shock mRNAs might be protected (Nover, Scharf, and Neumann, 1983, 1989) until their subsequent recruitment for translation during recovery. Moreover, although these observations concerning the effects of heat shock on mRNA stability may apply to many mRNAs, there are important exceptions. 22 kDa zein mRNAs were rapidly destabilized in maize endosperm subject to a moderate heat shock (Plotnikov, Bakaldina, and Efimov, 1992). mRNAs encoding secreted proteins such as α-amylase in barley aleurone cells, for instance, are selectively destabilized following a heat shock (Brodl and Ho, 1991; Lanciloti, Cwik, and Brodl, 1996), perhaps as a consequence of the heat-mediated disruption of some aspects of endoplasmic reticulum function that are normally required for the translation of these mRNAs (Belanger, Brodl, and Ho, 1986; Sticher et al., 1990). Similar results were observed in heat-shocked carrot roots for secretory protein mRNAs that were stable under normal conditions (Brodl and Ho, 1992). For such mRNAs, specific degradatory mechanisms may have evolved as a means by which some tissues can effectively respond to a heat shock.

CONCLUSIONS

The HSP response in crop plants can now be viewed in a wider context of those conditions in which the requirement for chaperone activity is particularly acute. Therefore, expression of HSP during certain stages in development and in response to biotic and abiotic stresses constitutes conditions under which protein misfolding or aggregation represents a challenge that must be met through induction of HSP expression. Heat shock may represent only an extreme example of the need to limit the injury inflicted on cellular proteins. Controlling gene expression plays an important role in limiting cellular damage. In addition to the induction of HSP expression to increase the presence of molecular chaperone activity, the expression of non-HSP genes is repressed in an attempt to reduce the production of protein that might only contribute further to the pool of misfolded or aggregated protein. Such changes in gene expression have been observed at virtually every level, including the repression of transcription and translation. Moreover, the activity of the RNA degradatory machinery, itself, is reduced as a means to maintain the existing non-HSP mRNAs that will be recruited for translation once again during recovery from heat shock. Our increasing knowledge of these regulatory controls provides the means to manipulate the responses of crop plants at the level of gene expression to assist their adaptation to new growing areas or to adapt crops to climate changes in existing areas of cultivation.

REFERENCES

Alamillo, J., C. Almoguera, D. Bartels, and J. Jordano (1995). Constitutive expression of small heat shock proteins in vegetative tissues of the resurrection plant *Craterostigma plantagineum. Plant Molecular Biology* 29: 1093-1099.

Apuya, N.R. and J.L. Zimmerman (1992). Heat shock gene expression is controlled primarily at the translational level in carrot cells and somatic embryos. *The Plant Cell* 4: 657-665.

Aranda, M.A., M. Escaler, D. Wang, and A.J. Maule (1996). Induction of HSP70 and polyubiquitin expression associated with plant virus replication. *Proceedings of the National Academy of Sciences, USA* 93: 15289-15293.

Arrigo, A.-P. (1987). Cellular localization of HSP23 during *Drosophila* development and following subsequent heat shock. *Developmental Biology* 122: 39-48.

Atkinson, B.G., M. Raizada, R.A. Bouchard, J.R.H. Frappier, and D.B. Walden (1993). The independent stage-specific expression of the 18-kDa heat shock protein genes during microsporogenesis in *Zea mays* L. *Developmental Genetics* 14: 15-26.

Baumann, G., E. Raschke, M. Bevan, and F. Schöffl (1987). Functional analysis of sequences required for transcriptional activation of a soybean heat shock gene in transgenic tobacco plants. *EMBO Journal* 6: 1161-1166.

Becker, J. and E.A. Craig (1994). Heat-shock proteins as molecular chaperones. *European Journal of Biochemistry* 219: 11-23.

Belanger, F.C., M.R. Brodl, and T.H.D. Ho (1986). Heat shock causes destabilization of specific mRNAs and destruction of endoplasmic reticulum in barley aleurone cells. *Proceedings of the National Academy of Sciences, USA* 83: 1354-1358.

Bienz, M. (1984). Developmental control of the heat shock response in *Xenopus. Proceedings of the National Academy of Sciences, USA* 81: 3138-3142.

Bienz, M. and J.B. Gurdon (1982). The heat shock response in *Xenopus* oocytes is controlled at the translational level. *Cell* 29: 811-819.

Blank, A. and T.A. McKeon (1989). Single-strand-preferring nuclease activity in wheat leaves is increased in senescence and is negatively photoregulated. *Proceedings of the National Academy of Sciences, USA* 86: 3169-3173.

Braam, J. (1992). Regulated expression of the calmodulin-related *TCH* genes in cultured *Arabidopsis* cells: Induction by calcium and heat shock. *Proceedings of the National Academy of Sciences, USA* 83: 3213-3216.

Brodl, M.R. and T.H.D. Ho (1991). Heat shock causes selective destabilization of secretory protein mRNAs in barley aleurone cells. *Plant Physiology* 96: 1048-1052.

Brodl, M.R. and T.H.D. Ho (1992). Heat shock in mechanically wounded carrot root disks causes destabilization of stable secretory protein mRNA and dissociation of endoplasmic reticulum lamellae. *Physiologia Plantarum* 86: 253-262.

Browning, K.S., L. Fletcher, S.R. Lax, and J.M. Ravel (1989). Evidence that the 59-kDa protein synthesis initiation factor from wheat germ is functionally similiar to the 80-kDa initiation factor 4B from mammalian cells. *Journal of Biological Chemistry* 264: 8491-8494.

Burke, J.J., J.L. Hatfield, R.R. Klein, and J.E. Mullet (1985). Accumulation of heat shock proteins in field-grown cotton. *Plant Physiology* 78: 394-398.

Callis, J., M. Fromm, and V. Walbot (1988). Heat-inducible expression of a chimeric maize hsp70CAT gene in maize protoplasts. *Plant Physiology* 88: 965-968.

Chang, S.-C. and D.R. Gallie (1997). RNase activity decreases following a heat shock in wheat leaves and correlates with its posttranslational modification. *Plant Physiology* 113: 1253-1263.

Chen, Q., K. Osteryoung, and E. Vierling (1994). A 21-kDa chloroplast heat shock protein assembles into high molecular weight complexes in in vivo and *in organelle. Journal of Biological Chemistry* 269: 13216-13223.

Chinn, A.M. and L. Comai (1996). The heat shock cognate 80 gene of tomato is flanked by matrix attachment regions. *Plant Molecular Biology* 32: 959-968.

Coca, M.A., C. Almoguera, and J. Jordano (1994). Expression of sunflower low-molecular-weight heat-shock proteins during embryogenesis and persistence after germination: Localization and possible functional implications. *Plant Molecular Biology* 25: 479-492.

Collier, N.C. and M.J. Schlesinger (1986). The dynamic state of heat shock proteins in chicken embryo fibroblasts. *Journal of Cell Biology* 103: 1495-1507.

Czarnecka, E., P.C. Fox, and W.B. Gurley (1990). In vitro interaction of nuclear proteins with the promoter of soybean heat shock gene *Gmhsp17.5E. Plant Physiology* 94: 935-943.

Czarnecka, E., J.L. Key, and W.B. Gurley (1989). Regulatory domains of the *Gmhsp17.5E* heat shock promoter of soybean. *Molecular and Cellular Biology* 9: 3457-3463.

Czarnecka, E., R.T. Nagao, J.L. Key, and W.B. Gurley (1988). Characterization of *Gmhsp26-A*, a stress gene encoding a divergent heat shock protein of soybean: Heavy-metal-induced inhibition of intron processing. *Molecular and Cellular Biology* 8: 1113-1122.

Czarnecka-Verner, E., M. Dulce-Barros, and W.B. Gurley (1994). Regulation of heat shock gene expression. In *Stress-Induced Gene Expression in Plants,* ed. A.S. Basra. Amsterdam: Harwood Academic Publishers, pp. 131-161.

Darwish, K., L.Q. Wang, C.H. Hwang, N. Apuya, and J.L. Zimmerman (1991). Cloning and characterization of genes encoding low molecular weight heat shock proteins from carrot. *Plant Molecular Biology* 16: 729-731.

DeRocher, A.E. and E. Vierling (1994). Developmental control of small heat shock protein expression during pea seed maturation. *The Plant Journal* 5: 93-102.

DeRocher, A.E. and E. Vierling (1995). Cytoplasmic HSP70 homologues of pea: Differential expression in vegetative and embryonic organs. *Plant Molecular Biology* 27: 441-456.

Didierjean, L., P. Frendo, W. Nasser, G. Genot, J. Marivet, and G. Burkard (1996). Heavy-metal-responsive genes in maize: Identification and comparison of their expression upon various forms of abiotic stress. *Planta* 200: 85-91.

Dong, J.-Z. and D.I. Dunstan (1996). Characterization of three heat-shock-protein genes and their developmental regulation during somatic embryogenesis in white spruce [*Picea glauca* (Moench) Voss]. *Planta* 200: 85-91.

Dulce-Barros, M., E. Czarnecka, and W.B. Gurley (1992). Mutational analysis of a plant heat shock element. *Plant Molecular Biology* 19: 665-675.

Duncan, R. and J.W.B. Hershey (1984). Heat shock-induced translational alterations in HeLa cells. *Journal of Biological Chemistry* 259: 11882-11889.

Duncan, R. and J.W.B. Hershey (1989). Protein synthesis and protein phosphorylation during heat stress, recovery, and adaptation. *Journal of Cell Biology* 109: 1467-1481.

Duncan, R., S.C. Milburn, and J.W.B. Hershey (1987). Regulated phosphorylation and low abundance of HeLa cell initiation factor eIF-4F suggest a role in translational control. *Journal of Biological Chemistry* 262: 380-388.

Eckey-Kaltenbach, H., E. Kiefer, E. Grosskopf, D. Ernst, and H. Sandermann Jr. (1997). Differential transcript induction of parsley pathogenesis-related proteins and of a small heat shock protein by ozone and heat shock. *Plant Molecular Biology* 33: 343-350.

Fehling, E. and M. Weidner (1988). Adaptive potential of wheat ribosomes toward heat depends on the large ribosomal subunit and ribosomal protein phosphorylation. *Plant Physiology* 87: 562-565.

Feussner, K., I. Feussner, I. Leopold, and C. Wasternack (1997). Isolation of a cDNA coding for an ubiquitin-conjugating enzyme UBC1 of tomato—The first stress-induced UBC of higher plants. *FEBS Letters* 409: 211-215.

Frova, C., G. Taramino, and G. Binelli (1989). Heat-shock proteins during pollen development in maize. *Developmental Genetics* 10: 324-332.

Gallie, D.R (1991). The cap and poly(A) tail function synergistically to regulate mRNA translational efficiency. *Genes and Development* 5: 2108-2116.

Gallie, D.R., C. Caldwell, and L. Pitto (1995). Heat shock disrupts cap and poly(A) tail function during translation and increases mRNA stability of introduced reporter mRNA. *Plant Physiology* 108: 1703-1713.

Gallie, D.R., H. Le, C. Caldwell, R.L. Tanguay, N.X. Hoang, and K.S. Browning (1997). The phosphorylation state of translation initiation factors is regulated developmentally and following heat shock in wheat. *Journal of Biological Chemistry* 272: 1046-1053.

Gallie, D.R. and L. Pitto (1996). Translational control during recovery from heat shock in the absence of heat shock proteins. *Biochemical and Biophysical Research Communications* 227: 462-467.

Gottesman, S., S. Wickner, and M.R. Maurizi (1997). Protein quality control: Triage by chaperones and proteases. *Genes and Development* 11: 815-823.

Gurley, W.B., E. Czarnecka, and M.D.C. Barros (1993). Anatomy of a soybean heat shock promoter. In *Control of Plant Gene Expression,* ed. D.P.S. Verma. Boca Raton, FL: CRC Press, pp. 103-123.

Gurley, W.B. and J.L. Key (1991). Transcriptional regulation of the heat-shock response: A plant perspective. *Biochemistry* 30: 1-12.

Gyorgyey, J., A. Gartner, K. Nemeth, Z. Magyar, H. Hirt, E. Heberle-Bors, and D. Dudits (1991). Alfalfa heat shock genes are differentially expressed during somatic embryogenesis. *Plant Molecular Biology* 16: 999-1007.

Heikkila, J.J., N. Ohan, Y. Tam, and A. Ali (1997). Heat shock protein gene expression during *Xenopus* development. *Cellular and Molecular Life Science* 53: 114-121.

Helm, K.W. and R.H. Abernathy (1990). Heat shock proteins and their mRNAs in dry and early imbibing embryos of wheat. *Plant Physiology* 93: 1626-1633.

Helm, K.W., G.J. Lee, and E. Vierling (1997). Expression and native structure of cytosolic class II small heat-shock proteins. *Plant Physiology* 114: 1477-1485.

Henry-Vian, C., A. Vian, E. Davies, G. Ledoigt, and M.-O. Desbiez (1995). Wounding regulates polysomal incorporation of *hsp70* and *tch1* transcripts during signal storage and retrieval. *Physiologia Plantarum* 95: 387-392.

Hernandez, L.D. and E. Vierling (1993). Expression of low molecular weight heat-shock proteins under field conditions. *Plant Physiology* 101: 1209-1216.

Howarth, C (1990). Heat shock proteins in *Sorghum bicolor* and *Pennisetum americanum.* II. Stored RNA in sorghum seed and its relationship to heat shock protein synthesis during germination. *Plant, Cell, and Environment* 18: 57-64.

Hubel, A., J.H. Lee, C. Wu, and F. Schöffl (1995). *Arabidopsis* heat shock factor is constitutively active in *Drosophila* and human cells. *Molecular and General Genetics* 248: 136-141.

Hultmark, D., R. Klemenz, and W.J. Gehring (1986). Translation and transcriptional control elements in the untranslated leader of the heat-shock gene *hsp22.* *Cell* 44: 429-438.

Hwang, C.H. and J.L. Zimmerman (1989). Heat shock response of carrot: Protein variations between cultured cell lines. *Plant Physiology* 91: 552-558.

Jaramillo, M., K. Browning, T.E. Dever, S. Blum, H. Trachsel, W.C. Merrick, J.M. Ravel, and N. Sonenberg (1990). Translation initiation factors that function as RNA helicases from mammals, plants and yeast. *Biochimica et Biophysica Acta* 1050: 134-139.

Jinn, T.-L., Y.-M. Chen, and C.-Y. Lin (1995). Characterization and physiological function of class I low-molecular-mass, heat-shock protein complex in soybean. *Plant Physiology* 108: 693-701.

Joshi, C.P. and H.T. Nguyen (1995). 5′ untranslated leader sequences of eukaryotic mRNAs encoding heat shock-induced proteins. *Nucleic Acids Research* 23: 541-549.

Key, J.L., C.-Y. Lin, and Y.-M. Chen (1981). Heat shock proteins of higher plants. *Proceedings of the National Academy of Sciences, USA* 78: 3526-3530.

Kimpel, J.A. and J.L. Key (1985). Presence of heat shock mRNAs in field grown soybeans. *Plant Physiology* 79: 672-678.

Kimpel, J.A., R.T. Nagao, V. Goekjian, and J.L. Key (1990). Regulation of the heat shock response soybean seedlings. *Plant Physiology* 94: 988-995.

Klemenz, R., D. Hultmark, and W.J. Gehring (1985). Selective translation of the heat shock mRNA in *Drosophila melanogaster* depends on sequence information in the leader. *EMBO Journal* 4: 2053-2060.

Kobayashi, T., E. Kobayashi, S. Sato, Y. Hotta, N. Miyajima, A. Tanaka, and S. Tabata (1994). Characterization of cDNAs induced in meiotic prophase in lily microsporocytes. *DNA Research* 1: 15-26.

Krishna, P., R.K. Reddy, M. Sacco, J.R.H. Frappier, and R.F. Felsheim (1997). Analysis of the native forms of the 90 kDa heat shock protein (hsp90) in plant cytosolic extracts. *Plant Molecular Biology* 33: 457-466.

Krishnan, M., H.T. Nguyen, and J.J. Burke (1989). Heat shock protein synthesis and thermal tolerance in wheat. *Plant Physiology* 90: 140-145.

Krone, P.H., J.B. Sass, and Z. Lele (1997). Heat shock protein gene expression during embryonic development of the zebrafish. *Cellular and Molecular Life Science* 53: 122-129.

Kruse, E., Z. Liu, and K. Kloppstech (1993). Expression of heat shock proteins during development of barley. *Plant Molecular Biology* 23: 111-122.

Lamphear, B.J. and R. Panniers (1990). Cap-binding protein complex that restores protein synthesis in heat-shocked Ehrlich cell lysates contains highly phosphorylated eIF-4E. *Journal of Biological Chemistry* 265: 5333-5336.

Lamphear, B.J. and R. Panniers (1991). Heat shock impairs the interaction of cap-binding protein complex with 5′ mRNA cap. *Journal of Biological Chemistry* 266: 2789-2794.

Lanciloti, D.F., C. Cwik, and M.R. Brodl (1996). Heat shock proteins do not provide thermoprotection to normal cellular protein synthesis, α-amylase RNA and endoplasmic reticulum lamellae in barley aleurone layers. *Physiologia Plantarum* 97: 513-523.

Lax, S.R., K.S. Browning, D.M. Maia, and J.M. Ravel (1986). ATPase activities of wheat germ initiation factors 4A, 4B, and 4F. *Journal of Biological Chemistry* 261: 15632-15636.

Le, H., R.L. Tanguay, L. Balasta, C.-C. Wei, K.S. Browning, A.M. Metz, D.J. Goss, and D.R. Gallie (1997). Translation initiation factors eIF-iso4G and eIF-4B interact with the poly(A)-binding protein and increase its RNA binding activity. *Journal of Biological Chemistry* 272: 16247-16255.

Lee, G.J., N. Pokala, and E. Vierling (1995). Structure and in vitro molecular chaperone activity of cytosolic small heat shock proteins from pea. *Journal of Biological Chemistry* 270: 10432-10438.

Leicht, B.G., H. Biessmann, K.B. Palter, and J.J. Bonner (1986). Small heat shock proteins of *Drosophila* associate with the cytoskeleton. *Proceedings of the National Academy of Sciences, USA* 83: 90-94.

Lin, C.-Y., J.K. Roberts, and J.L. Key (1984). Acquisition of thermotolerance in soybean seedlings. *Plant Physiology* 74: 152-160.

Magnard, J.U.-L., P. Vergne, and C. Dumas (1996). Complexity and genetic variability of heat-shock protein expression in isolated maize microspores. *Plant Physiology* 111: 1085-1096.

Manzella, J.M., W. Rychlik, R.E. Rhoads, J.W. Hershey, and P.J. Blackshear (1991). Insulin induction of ornithine decarboxylase. Importance of mRNA secondary structure and phosphorylation of eukaryotic initiation factors eIF-4B and eIF-4E. *Journal of Biological Chemistry* 266: 2383-2389.

Marrs, K.A., E.S. Casey, S.A. Capitant, R.A. Bouchard, P.S. Dietrich, I.J. Mettler, and R.M. Sinibaldi (1993). Characterization of two maize HSP90 heat shock protein genes: Expression during heat shock, embryogenesis, and pollen development. *Developmental Genetics* 14: 27-41.

McGarry, T.J. and S. Lindquist (1985). The preferential translation of *Drosophila hsp70* mRNA requires sequences in the untranslated leader. *Cell* 42: 903-911.

Metz, A.M. and K.S. Browning (1997). Assignment of the ß subunit of wheat eIF2 by protein and DNA sequence analysis and immunoanalysis. *Archives of Biochemistry and Biophysics* 342: 187-189.

Michaud, S., R. Marin, and R.M. Tanguay (1997). Regulation of heat shock gene induction and expression during *Drosophila* development. *Cellular and Molecular Life Science* 53: 104-113.

Murtha-Riel, P., M.V. Davies, B.J. Scherer, S.-Y. Choi, J.W.B. Hershey, and R.J. Kaufman (1993). Expression of a phosphorylation-resistant eukaryotic initiation factor 2 α-subunit mitigates heat shock inhibition of protein synthesis. *Journal of Biological Chemistry* 268: 12946-12951.

Nagao, R.T., J.A. Kimpel, E. Vierling, and J.L. Key (1986). The heat shock response: A comparative analysis. In *Oxford Surveys of Plant Molecular and Cell Biology,* ed. B.J. Miflin. Oxford: Oxford University Press, pp. 384-438.

Neumann, D., K.D. Scharf, and L. Nover (1984). Heat shock-induced changes of plant cell ultrastructure and autoradiographic localization of heat shock proteins. *European Journal of Cell Biology* 34: 254-264.

Neven, L.G., D.W. Haskell, C.L. Guy, N. Denslow, P.A. Klein, L.G. Green, and A. Silverman (1992). Association of 70-kilodalton heat-shock cognate proteins with acclimation to cold. *Plant Physiology* 99: 1362-1369.

Nover, L. and K.-D. Scharf (1997). Heat stress proteins and transcription factors. *Cellular and Molecular Life Science* 53: 80-103.

Nover, L., K.-D. Scharf, D. Gagliardi, P. Vergne, E. Czarnecka-Verner, and W.B. Gurley (1996). The Hsf world: Classification and properties of plant heat stress transcription factors. *Cell Stress & Chaperones* 1: 215-223.

Nover, L., K.-D. Scharf, and D. Neumann (1983). Formation of cytoplasmic heat shock granules in tomato cell cultures and leaves. *Molecular and Cellular Biology* 3: 1648-1655.

Nover, L., K.-D. Scharf, and D. Neumann (1989). Cytoplasmic heat shock granules are formed from precursor particles and are associated with a specific set of mRNAs. *Molecular and Cellular Biology* 9: 1298-1308.

Osteryoung, K.W., H. Sundberg, and E. Vierling (1993). Poly(A) tail length of a heat shock protein RNA is increased by severe heat stress, but intron splicing is unaffected. *Molecular and General Genetics* 239: 323-333.

Panniers, R., E.B. Stewart, W.C. Merrick, and E.C. Henshaw (1985). Mechanism of inhibition of polypeptide chain initiation in heat-shocked Ehrlich cells involves reduction of eukaryotic initiation factor 4F activity. *Journal of Biological Chemistry* 260: 9648-9653.

Pareek, A., S.L. Singla, and A. Grover (1995). Immunological evidence for accumulation of two high-molecular-weight (104 and 90 kDa) HSPs in response to different stresses in rice and in response to high temperature stress in diverse plant genera. *Plant Molecular Biology* 29: 293-301.

Pitto, L., D.R. Gallie, and V. Walbot (1992). Role of the leader sequence during thermal repression of translation in maize, tobacco, and carrot protoplasts. *Plant Physiology* 100: 1827-1833.

Plotnikov, V.K., N.B. Bakaldina, and V.A. Efimov (1992). Stability of corn zein mRNAs under conditions of normal and high temperatures. *Fiziologiya Rastenii* 38: 981-990.

Prandl, R., E. Kloske, and F. Schöffl (1995). Developmental regulation and tissue-specific differences of heat shock gene expression in transgenic tobacco and *Arabidopsis* plants. *Plant Molecular Biology* 28: 73-82.

Prandl, R. and F. Schöffl (1996). Heat shock elements are involved in heat shock promoter activation during tobacco seed maturation. *Plant Molecular Biology* 31: 157-162.

Rees, C.A.B., A.M. Gullons, and D.B. Walden (1989). Heat shock protein synthesis induced by methomyl in maize (*Zea mays* L.) seedlings. *Plant Physiology* 90: 1256-1261.

Rieping, M. and F. Schöffl (1992). Synergistic effect of upstream sequences, CCAAT box elements, and HSE sequences for enhanced expression of chimaeric heat shock genes in transgenic tobacco. *Molecular and General Genetics* 231: 226-232.

Scharf, K.-D. and L. Nover (1982). Heat-shock-induced alterations of ribosomal protein phosphorylation in plant cell cultures. *Cell* 30: 427-437.

Schmid, S.R. and P. Linder (1992). D-E-A-D protein family of putative helicases. *Molecular Microbiology* 6: 283-292.

Schöffl, F., M. Rieping, G. Baumann, M. Bevan, and S. Angermuller (1989). The function of plant heat shock promoter elements in the regulated expression of chimaeric genes in transgenic tobacco. *Molecular and General Genetics* 217: 246-253.

Schöffl, F., G. Schroder, M. Kliem, and M. Rieping (1993). An SAR sequence containing 395 bp DNA fragment mediates enhanced, gene-dosage-correlated expression of a chimaeric heat shock gene in transgenic tobacco plants. *Transgenic Research* 2: 93-100.

Spiker, S. and W.F. Thompson (1996). Nuclear matrix attachment regions and transgene expression in plants. *Plant Physiology* 110: 15-21.

Sticher, L., A.K. Biswas, D.S. Bush, and R.L. Jones (1990). Heat shock inhibits α-amylase synthesis in barley aleurone without inhibiting the activity of endoplasmic reticulum marker enzymes. *Plant Physiology* 92: 506-513.

Storti, R.V., M.P. Scott, A. Rich, and M.L. Pardue (1980). Translational control of protein synthesis in response to heat shock in *D. melanogaster* cells. *Cell* 22: 825-834.

Treuter, E., L. Nover, K. Ohme, and K.-D. Scharf (1993). Promoter specificity and deletion analysis of three heat stress transcription factors of tomato. *Molecular and General Genetics* 240: 113-125.

Tsukaya, H., T. Takahashi, S. Naito, and Y. Komeda (1993). Floral organ-specific and constitutive expression of an *Arabidopsis thaliana* heat-shock HSP18.2: GUS fusion gene is retained even after homeotic conversion of flowers by mutation. *Molecular and General Genetics* 237: 26-32.

van der Geest, A.H.M. and T.C. Hall (1996). The ß-phaseolin 5′ matrix attachment region acts as an enhancer facilitator. *Plant Molecular Biology* 33: 553-557.

van der Geest, A.H.M., G.E. Hall, S. Spiker, and T.C. Hall (1994). The ß-phaseolin gene is flanked by matrix attachment regions. *Plant Journal* 6: 413-423.

Vayda, M.E. and M.-L. Yuan (1994). The heat shock response of an antarctic alga is evident at 5°C. *Plant Molecular Biology* 24: 229-233.

Vierling, E (1991). The roles of heat shock proteins in plants. *Annual Reviews of Plant Physiology and Plant Molecular Biology* 42: 579-620.

Waters, E.R (1995). The molecular evolution of the small heat-shock proteins in plants. *Genetics* 141: 785-795.

Webster, C., R.L. Gaut, K.S. Browning, J.M. Ravel, and J.K.M. Roberts (1991). Hypoxia enhances phosphorylation of eukaryotic initiation factor 4A in maize root tips. *Journal of Biological Chemistry* 266: 23341-23346.

Wehmeyer, N., L.D. Hernandez, R.R. Finkelstein, and E. Vierling (1996). Synthesis of small heat-shock proteins is part of the developmental program of late seed maturation. *Plant Physiology* 112: 747-757.

Winter, J. and R. Sinibaldi (1991). The expression of heat shock protein and cognate genes during plant development. In *Heat Shock and Development,* eds. L. Hightower and L. Nover. Berlin: Springer, pp. 85-105.

Yeh, K.-W., T.-L. Jinn, C.-H. Yeh, Y.-M. Chen, and C.-Y. Lin (1994). Plant low-molecular-mass heat-shock proteins: Their relationship to the acquisition of thermotolerance in plants. *Biotechnological Applications in Biochemistry* 19: 41-49.

Zapata, J.M., F.G. Maroto, and J.M. Sierra (1991). Inactivation of mRNA cap-binding protein complex in *Drosophila melanogaster* embryos under heat shock. *Journal of Biological Chemistry* 266: 16007-16014.

Zarsky, V., D. Garrido, N. Eller, J. Tupy, O. Vicente, F. Schöffl, and E. Heberle-Bors (1995). The expression of a small heat shock gene is activated during induction of tobacco pollen embryogenesis by starvation. *Plant, Cell and Environment* 18: 139-147.

Zur Nieden, U., D. Neumann, A. Bucka, and L. Nover (1995). Tissue-specific localization of heat-stress proteins during embryo development. *Planta* 196: 530-538.

Chapter 8

The Effects of Heat Stress on Cereal Yield and Quality

Peter Stone

INTRODUCTION

What Is Heat Stress?

Myriad definitions have been used for the term heat stress, but here it is defined as a deleterious response to elevated temperature. For cereal crops, heat stress reduces yield or changes quality. Elevated temperatures that do not result in one or both of these responses may be of considerable academic interest, but in the practical world of food production, it is ultimately yield and quality that count.

Heat stress can be moderate or extreme in intensity, acute or chronic in degree, and can occur from before a crop has emerged to during maturation. It is little wonder then that the effects of heat stress on crop yield and quality are many and varied. Nor should it be surprising that a vast number of plant processes are involved in the response to heat stress. The aim here is not to catalog the responses of crops to heat stress but to identify those processes which are most likely to limit yield or quality under heat stress conditions.

What Is High Temperature?

While admitting the risk of advancing a circular argument, high temperatures may be considered to be those which lead to heat stress. The level and duration of temperature events that result in heat stress vary with crop type, growth stage, and the plant property in question. Different crops vary in their response to temperature at a given stage of crop growth, and high temperatures that affect yield may not affect quality, and vice versa. Whether a temperature is "high" is therefore something of a dynamic consideration.

Notwithstanding that it is relative, we must try to make some firm definitions about heat stress, and the high temperatures that cause it. A general consensus appears to have been reached, whereby stress-inducing temperature has been divided into three general classes: (1) moderately high temperature (ca 15 to 32°C); (2) very high temperature (ca 32 to 50°C); and (3) lethal temperature (ca > 50°C). The divisions between these classes are not merely ones of linguistic convention, but rather a reflection of the differing effects of temperature on the physiology of plant function in each temperature range.

Plant responses to moderately high temperature result largely from changes in the rates and durations of existing processes, whereas under very high temperatures, some processes are severely retarded and new physiological processes may be induced. This chapter delineates these two temperature ranges and responses, where appropriate. Scant attention is given to lethal temperatures, which most cropping culture successfully avoids.

How Prevalent Are High Temperatures?

Both moderately high and very high temperatures are common in the world's major cropping regions. Chronic high temperature is a fact of life within the vast region bounded by the tropics. Mean annual temperatures of 21 to 27°C predominate, although mean annual temperatures of > 27°C occur in large tracts of arable land, particularly in West Africa, India, and Southeast Asia (Russell and Coupe, 1994). In these regions, unabated moderately high temperatures are the norm, although short periods of very high temperatures are not uncommon.

In the regions outside the tropics, lower mean annual temperatures do not necessarily provide relief from heat stress. Despite the fact that the mean annual temperature in vast cereal-producing regions such as North America and Kazakhstan is only 4 to 10°C, moderately high temperatures are chronic in the latter parts of the cropping season, and acute, very high temperatures are frequent during grain filling (see Figure 8.1).

Figure 8.2 shows the mean monthly maximum temperatures for some of the world's most important cropping regions. Evidently, most of the world's crops are exposed to heat stress during some stage of their life cycle. In fact, as Paulsen (1994) has noted, yield losses from heat stress are so common that high temperature is frequently not recognized as a yield-limiting factor, and changes in quality are perhaps even less obvious.

FIGURE 8.1. Percentage Time Above a Given Temperature During Grain Filling of Wheat and Maize in North America

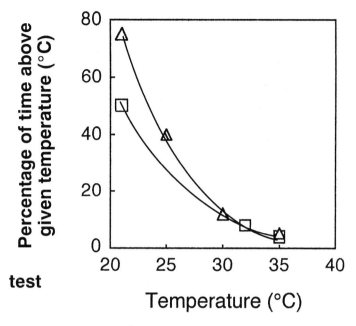

test

Temperature (°C)

Note: □ wheat (Nebraska); Δ maize (Iowa).

Source: Adapted from Keeling et al., 1994, p. 814. Wheat data calculated from Graybosch et al., 1995, p. 48.

EFFECTS OF HIGH TEMPERATURE ON YIELD

Vegetative Growth

Seedling Emergence, Vigor, and Survival

Many crops are exposed to severe heat stress even before they have germinated. Bare soil surfaces may be exposed to high radiation loads, and unless this energy can be dissipated, the soil may warm considerably. In moist soils, much of the incident energy is lost through latent heat of vaporization as water evaporates from the soil surface, but as dry soils do not have this avenue of heat loss available to them, they may attain tem-

FIGURE 8.2. Mean Monthly Maximum Temperature for Selected Cropping Regions

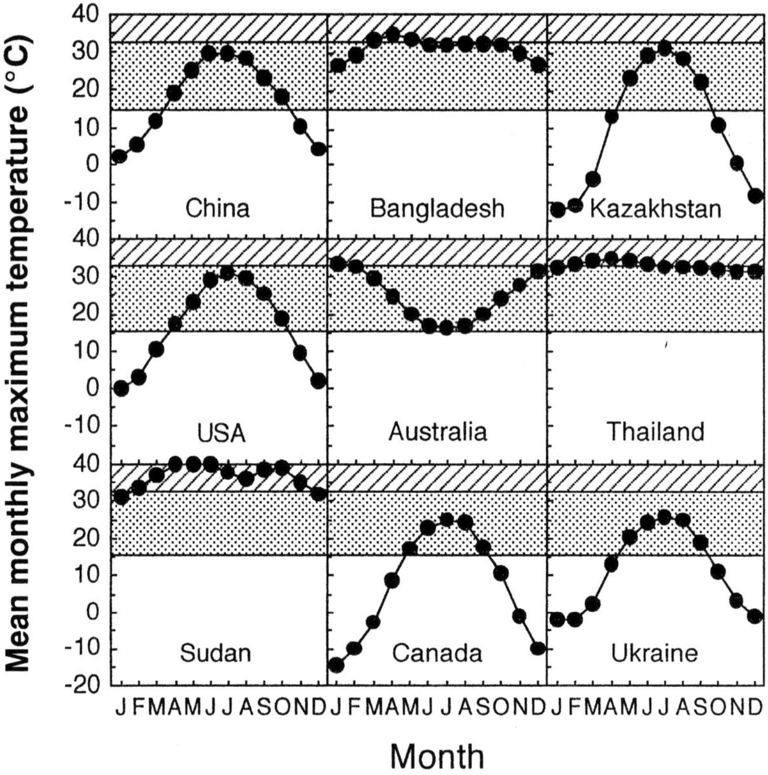

Note: Shaded region shows moderately high temperature range; hatched region shows very high temperature range. Sites chosen from central regions of important cropping areas. Site details: China, Yancheng, 35.5N, 112.4E; Bangladesh, Dhaka, 23.8N, 90.4E; Kazakhstan, Turgaj, 49.6N, 65.3E; USA, Omaha, 41.4N, 96.0W; Australia, Merredin, 31.5S, 118.3E; Thailand, 13.7N, 100.6E; Sudan, Khartoum, 15.6N, 32.6E; Canada, Winnepeg, 50.0N, 97.2W; Ukraine, Kharkiv, 50.0N, 36.1E.

Source: Data from Vose et al., 1992.

peratures much higher than the ambient ones. Soil temperatures at sowing as high as 45°C have been reported for wheat (Morris, Belaid, and Byerlee, 1991), and for summer crops in the semiarid tropics, seed is frequently sown into soil with temperatures around 55°C at sowing depth (Abrecht and Bristow, 1996).

While the plant is autotrophic, its source of energy is directly affected by soil temperature. Even though the rate of wheat seed imbibition does not reach a maximum until temperature has exceeded 35°C, imbibition at these temperatures significantly reduces germination. Furthermore, the final mass of wheat seedlings declines with temperatures above 20°C, and this appears to be related to increased rates of seed respiration and, consequently, lower energy available for growth (Blum and Sinmena, 1994).

Seedling growth is particularly sensitive to soil temperature, as both roots and shoot apex are located in the soil. For some crops, such as maize, this can mean that soil temperature exerts a greater influence on crop development than air temperature until as late as the eight-leaf stage, when the apical meristem rises above the soil surface (Wilson, Muchow, and Murgatroyd, 1995). As a result, crop development can be advanced significantly by elevated soil temperatures, and this tends to reduce yields, as will be discussed later.

Evidence suggests that some crops may have adapted to high soil surface temperatures, in that the optimum temperature for root growth is higher early in the season than when the crop is more mature (Nielsen and Humphries, 1966). Despite this, soil temperatures are frequently well above the optimum for root or shoot growth of seedlings. For maize, seedling growth is maximized at a soil temperature of 26°C, and above this temperature, root and shoot mass both decline by about 10 percent per °C, until at 35°C growth is severely retarded (Walker, 1969). Seedlings stressed by high soil temperatures are more likely to suffer further damage by this means, as their diminished growth delays canopy closure and, consequently, reduces soil shading.

For most crops, seedling growth continues at a diminished rate until temperatures reach about 45°C, at which point growth stops but the seedling survives (Ougham et al., 1988). It seems incredible that in many crops high levels of seedling mortality do not occur until the temperature exceeds 52°C (Weaich, Bristow, and Cass, 1996).

Even though many crops show genotypic variation in the ability of seeds and seedlings to withstand high temperature (Ougham et al., 1988; Blum and Sinmena, 1994), breeding may not be the most effective means of reducing the effects of heat stress on crop establishment. Plant residues at the soil surface can significantly reduce soil temperature by reducing radiation load and, when present as mulches, by insulating the soil (Felton et al., 1987). These benefits have been shown to increase crop yields in a heat-stressed environment (Lal, 1974), although it should be noted that by reducing soil temperature during spring, mulches may delay planting and, consequently, reduce yield.

The effects of heat stress on seeds and seedlings are not always visible, and this may be in part because the damage occurs underground or is small on an absolute scale. This should not be allowed to disguise the fact that, in many parts of the world, heat stress of seeds and seedlings imposes severe and insidious limitations on crop yield.

Photosynthesis, Transpiration, and Respiration

Prima facie, it would be logical to assume that the response of crop biomass and yield to temperature is very largely determined by the temperature response of photosynthesis: after all, photosynthesis is highly temperature sensitive, and biomass and yield are frequently proportional to the potential for crop photosynthesis (Loomis and Williams, 1963; Monteith, 1972; Tollenaar and Bruulsema, 1988a). This discussion shall therefore examine the effects of temperature on the processes involved in photoassimilation.

Stomatal conductance. This processs is directly responsible for the ingress of carbon dioxide (CO_2) to the leaf and may therefore be seen as the first point of "control" of photosynthesis. It is widely reported that high temperatures usually cause stomatal closure, although it appears that this is actually an indirect response to the effect of temperature on vapor pressure deficit (Lösch, 1979) and leaf respiration, which induces stomatal closure by raising the internal CO_2 concentration of the leaf. So, although it is possible to absolve high temperature of the crime of directly restricting gas exchange, it is nevertheless an accessory before the fact. In effect, however, this is a crime of little consequence. Temperature-induced stomatal closure does not appear to reduce photosynthesis, as shown by the fact that changes in stomatal conductance are not consistent with the response of photosynthesis to temperature (Slatyer, 1977); that is, stomatal conductance increases at temperatures that are high enough to irreversibly damage photosynthesis (Björkman, Mooney, and Ehleringer, 1975).

Although stomatal conductance does not appear to directly affect rate of photosynthesis, it does help regulate transpiration, which is most likely to affect the response of photosynthesis to heat stress by modifying leaf temperature. In crops with an adequate supply of moisture, canopy temperatures may be as much as $8°C$ below the ambient as a result of transpirational cooling (Reynolds et al., 1994). This contrasts with droughted crops or those exposed to a rapid increase in temperature, where leaf temperatures may be up to $15°C$ above the ambient (Altschuler and Mascarenhas, 1982). Clearly, a given high temperature will result in greater stress if accompanied by drought-induced stomatal closure.

A number of workers have examined the possibility that the propensity of a crop to cool itself during heat stress events may be an index of heat tolerance (or avoidance) that can be used to aid selection (Hall, 1992; Reynolds et al., 1994). It is not yet clear whether heat tolerance confers the ability to transpire during heat stress or whether transpirational cooling is the cause of heat avoidance (by reducing tissue temperature).

Reversible effects of heat stress on photosynthesis (up to ca 40 °C). For wheat, and most C_3 crops, net photosynthesis is stable in the range 15 to 30°C, and above or below this plateau region, the rate of net photosynthesis declines by 5 to 10 percent per °C (see Figure 8.3). The relatively minor response of net photosynthesis to temperature in the range of 15 to 30°C should not be allowed to disguise the fact that the gross rate of CO_2 fixation actually increases with temperature. Very little change in net photosynthesis is evident from 15 to 30°C, because increased rates of CO_2 fixation are counterbalanced by increased rates of whole-plant respiration

FIGURE 8.3. Effect of Leaf Temperature on Rate of Net Photosynthesis in C_3 and C_4 Crop Species

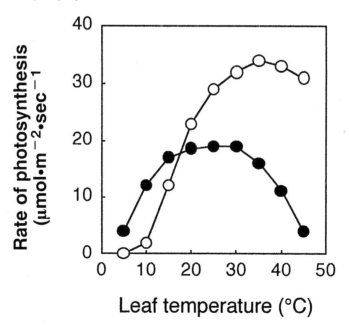

Note: ● C_3 (wheat); ○ C_4 (maize).

and photorespiration, in particular. As temperature rises from 30 to 40°C, the increase in rate of respiration becomes greater than the increase in rate of CO_2 fixation, so that photorespiraton consumes an increasing proportion of gross photosynthetic production. In short, elevated temperatures reduce the net rate of C_3 photosynthesis because the temperature optimum for respiration is higher than that for CO_2 fixation.

This differs markedly from the C_4 crops, for which optimum rates of net photosynthesis occur at relatively high (30 to 40°C) temperatures (Christy, Williamson, and Wideman, 1985). Photorespiration is negligible in C_4 plants, and as a result, the stimulation of CO_2 fixation by temperature is not as severely eroded by respiratory losses. C_4 plants therefore exhibit higher temperature optima for net photosynthesis than C_3 plants, not so much because they more effectively fix CO_2, but because they respire less wastefully at elevated temperatures.

The response of net photosynthesis to temperatures in the range of 10 to 40°C is therefore governed largely by changes in the rates of normal metabolic processes. It is not until leaf temperatures exceed about 40°C that specific or irreversible damage to photosynthetic apparatus occurs.

Irreversible effects of high temperature (HT) on photosynthesis (> 40 °C). For both C_3 and C_4 crops, photosynthetic activity declines markedly as leaf temperatures exceed approximately 40°C. In these extreme temperatures, chloroplast function is disrupted almost entirely by deactivation of photosystem II (PSII) (Burke, 1990), which may occur within a matter of minutes. Damage to PSII at these temperatures is generally considered to be "irreversible," although partial or full recovery of activity may occur with time. Clearly, leaf temperatures greater than 40°C have the potential to limit both the current and future photosynthetic capacity of a crop.

Thus far, the discussion of the effects of heat stress on CO_2 assimilation has concentrated on photosynthesis, largely as a rate-important biochemical phenomenon. Although the rate at which photosynthesis occurs is significant in determining crop productivity, it is ultimately much less important than the total amount of photosynthesis undertaken during the life of a crop. Because of this, the simple ability to intercept light usually exerts a greater influence on crop yield than any biochemical limitations to photosynthesis itself. As light interception is very largely governed by the pattern and duration of crop development, the discussion will now examine the effects of temperature on crop phenology.

Development

Temperature exerts a fundamental influence on crop development, as evidenced by the widespread use of thermal time as a predictor of the duration of developmental phases. Basically, rate of development increases curvilinearly with temperature (Slafer and Rawson, 1994), and as a result, the duration of each developmental phase declines as temperature rises.

Through this basic effect on development, elevated temperatures may reduce crop yields via three main mechanisms: (1) if the increase in rate of organ production at elevated temperature is not as great as the increase in rate of development, then elevated temperatures may result in fewer organs or phytomers being produced in each developmental phase; (2) if at elevated temperature the increase in rate of growth of a given phytomer or yield component is not commensurate with the increase in rate of development, then the organ or yield component will be reduced in size; and (3) biomass and yield are frequently related to the total amount of radiation intercepted by a crop (Loomis and Williams, 1963; Hunt and Pararajasingham, 1995), and by reducing the duration of crop growth, elevated temperatures reduce the time over which radiation can be absorbed and, hence, cumulative radiation interception. Each mechanism is now examined in turn, but to avoid an excessively complicated discussion, effects of photoperiod and vernalization will be ignored by assuming that (1) crops are being grown in their "normal" photoperiod environment, and (2) that any vernalization requirements have been satisfied.

Heat Stress May Result in Fewer Organs

In wheat, leaf number is relatively insensitive to temperature (Rawson and Zajac, 1993), so that the response of total leaf area to temperature is generally mediated through effects on leaf size and duration. By contrast, most determinants of potential grain number are highly sensitive to temperature and are reduced in proportion to the duration of preanthesis development. The number of tillers, and consequently ears per plant, is reduced by elevated temperature (Rawson, 1986), and both spikelets per ear and the number of florets per spikelet tend to decrease as temperature rises above about $15°C$ in the preanthesis period (Fischer and Maurer, 1976; Shpiler and Blum, 1986). Clearly, the potential number of grains/ m^2 (square meter) is significantly reduced by elevated temperatures, and this generally results in severe yield losses, as the reduction in grain number does not appear to be compensated for by an increase in individual kernel mass (Wardlaw et al., 1989). It is clear from the strong positive relationship between spikelet and floret number and duration of the preanthesis period (Shpiler and Blum, 1986) that under elevated temperature, increased rates of organ production are generally not sufficient to compensate for the reduced length of the preanthesis period. It should be noted, however, that there is significant genotypic variation in the response of spikelet and floret number to temperature, with a number of cultivars showing a remarkable lack of temperature sensitivity for these traits (Bagga and Rawson, 1977; Rawson and Zajac, 1993). This has clear implications for selection for heat tolerance.

For wheat and other winter cereals, heat stress in the preanthesis period can manifest itself by changing the rate of development, and this may occur above a threshold as low as 15°C. For summer crops such as maize, reductions in leaf number are not apparent until a maximum of 30°C occurs during the vegetative growth stage (Warrington and Kanemasu, 1983), and although effects of temperature on potential grain number are similar to wheat, the threshold tends to be higher in maize than in wheat.

Heat Stress May Result in Smaller Organs

Heat stress frequently reduces the size of organs because rate of organ growth is not as responsive to temperature as rate of organ development. In wheat, mature leaf size declined by 50 percent when temperature was increased from an average of about 15 to 27°C, and leaf area at a given growth stage was strongly positively correlated with time taken to reach that growth stage (Rawson, 1986). Similarly, maize leaf length declined by about 2 percent per °C above 20°C (Ritchie and NeSmith, 1991). Plant height of wheat is reduced by elevated temperatures (Fischer and Maurer, 1976), presumably because the duration of stem elongation decreases more than the rate of stem elongation increases. Clearly, the yield-bearing organs are also smaller as a result of enhanced rates of preanthesis development, but as discussed previously, this may be considered to be a consequence of fewer organs being produced. The plant part in which we are most interested is generally the grain, and we shall discuss the effects of temperature on grain development in a later section.

Heat Stress May Result in Reduced Total Light Interception

It has long been known that the biomass and yield of crops is frequently related to the total amount of radiation intercepted by the crop from emergence to maturity (Loomis and Williams, 1963; Monteith and Scott, 1982). This largely explains why crop yield and leaf area duration are so closely related, under both "normal" (Spiertz, 1974) and heat stress conditions (Bagga and Rawson, 1977; Kuroyanagi and Paulsen, 1985).

As we have seen, increased temperature reduces the duration of all stages of crop growth and, consequently, the duration of the crop itself.

Muchow and colleagues (1990) assimilated these two ideas and showed that high temperatures limit maize growth and yield primarily by reducing the total amount of light intercepted by the crop. That is, by increasing the rate of development, high temperatures reduce the length of time over which a crop intercepts light and, hence, the integral of light intercepted. Following this logic, it would seem that cool temperatures and high irradiance would tend to result in high yields, and for maize (Muchow, Sinclair, and Bennett, 1990) and many other crops (Monteith and Scott, 1982), this is the case.

Plants with enhanced rates of development frequently look "normal" because no symptoms of heat stress are readily apparent. Yet, by accelerating crop development, elevated temperatures may significantly reduce yield by reducing the size and number of plant structures and the total amount of assimilate available for their growth.

Reproductive Growth

The effects of preanthesis temperature on reproductive growth have been outlined previously, and as such, this discussion shall examine only fertilization and the postanthesis phase.

Fertilization

Fertilization and early grain growth is a period in which sensitivity to temperature makes grain number highly responsive to supraoptimal temperatures (Langer and Olugbemi, 1970; Satake and Yoshida, 1978). Furthermore, in many crops, the temperature threshold for damage in this period is considerably lower than for other stages of crop growth. This has clear implications for yield, as the extent to which potential grain number is realized is largely determined by the successful coordination of the events leading to fertilization and embryo growth: viable pollen must be shed onto receptive stigma, and ovules must, in turn, receive functional pollen tubes. The successful sequence of these processes is reduced by high temperature.

Pollen. Pollen is one of only two plant organs that is unable to synthesize heat shock proteins (Cooper, Ho, and Hauptmann, 1984; Xiao and Mascarenhas, 1985), a class of peptide that (as detailed later) protects cells against certain types of damage arising from heat and other stresses. This may help to explain the particular sensitivity of pollen viability to high temperatures in a wide range of crop species, including wheat (Sainim Sedgley, and Aspinall, 1984), maize (Herrero and Johnson, 1980), and rice

(Matsui et al., 1997). Pollen viability is particularly sensitive to heat stress in that it has both a relatively low damage threshold and a significant loss in viability once that threshold is surpassed. The critical temperature for pollen viability in wheat is only 30°C (Saini and Aspinall, 1982), and in a tropical species such as rice, it is no higher than 34°C (Matsui et al., 1997). This makes grain number susceptible to drastic reductions as a result of short periods of heat stress, especially during pollen mother cell meiosis, when the ear is still within the leaf sheath. As little as three days exposure to 30°C during this period can reduce grain set in wheat by almost 70 percent (Saini and Aspinall, 1982), and floret fertility of rice by up to 20 percent when temperature rises above 34°C (Matsui et al., 1997).

Stigma and style. Receptivity of the stigma and style does not appear to limit floret fertility under heat stress (Dawson and Wardlaw, 1989), and this is probably because of the higher temperature threshold for damage to these organs. The critical temperature for stigma function in rice is at least 6°C higher than that for pollen (Satake and Yoshida, 1978; Matsui et al., 1997).

Ovules. The preservation of functional stigma and style during heat stress does not guarantee efficient fertilization, even in the presence of viable pollen. Under normal conditions, the embryo sac guides the tubes of germinated pollen toward ovules via hormonal signal (Smyth, 1997), but it appears that this signal is disrupted by temperatures similar to those which render pollen unviable. In wheat, it has been shown that unheated pollen which germinates successfully on heat-stressed stigma tends to produce truncated pollen tubes that grow haphazardly rather than in the direction of the ovules (Saini, Sedgley, and Aspinall, 1983). This ineffective guidance of pollen tubes appears to result from ovule dysfunction caused by proliferation of the integument or nucellus, either of which may "crowd out" an otherwise functional embryo sac in heat-stressed plants (Saini, Sedgley, and Aspinall, 1983).

Of the three main processes required for successful fertilization, pollen viability appears to be the most limiting under heat stress. This was elegantly shown by Saini and Aspinall (1982) who increased floret fertility in heat-damaged wheat from 30 to 80 percent by pollinating heat-stressed pistils with unstressed pollen.

Postanthesis

In many crops, the frequency and severity of exposure to heat stress increases in the postanthesis period. This is particularly true for winter or spring crops such as wheat and barley but is also common for summer crops such as maize, sorghum, and rice. Studies concerned with the effects of high postanthesis temperature on grain yield and quality of crops generally fall into one of two categories: (1) responses to sustained periods of moderately high temperature (ca 25 to 32°C) (Asaoka et al., 1985; Wardlaw, Dawson, and Munibi, 1989); and (2) responses to shorter periods of very high temperature (ca 33 to 40°C) (Singletary, Banisadr, and Keeling, 1994; Savin et al., 1997). As noted in the introduction, the distinction between the moderately high and very high temperatures is not merely one of convenience but reflects the fact that plants appear to respond differently to temperatures in each range. Consequently, even though moderately and very high temperatures are frequently interspersed in the field, it is logical to consider their effects on crop yield and quality separately.

The majority of the discussion that follows focuses on wheat as a model crop, mainly because it has been so extensively studied. The implications arising from the discussion are, however, applicable to all of the major crops, for which the mechanisms of response of grain yield to postanthesis temperature are remarkably similar. The main difference between the major crops arises from differences in the extent rather than the nature of response to temperature. These differences will be alluded to where necessary.

Moderately High Temperature

The effects of moderately high temperature on cereal yields have been quantified using both field and controlled-environment studies. Together, the results suggest that, in many parts of the world, wheat yield losses due to moderately high temperature average approximately 10 to 15 percent per annum (Wardlaw and Wrigley, 1994), and it would appear from Figure 8.4 that losses of at least this magnitude could be expected in summer crops such as rice and maize.

Summer crops such as rice and maize are generally less sensitive to temperature than winter crops such as wheat in that they have both a higher temperature threshold above which kernel mass is reduced and a lower sensitivity once that threshold is passed (see Figure 8.4). This is especially the case in the moderately high temperature range, where wheat (4 percent) loses a greater proportion of its maximum kernel mass per °C than maize (3 percent) and, especially, rice (1 percent).

The effect of moderately high temperature on individual kernel mass has been shown to result from the differing sensitivities to temperature of the rate and duration of grain filling. The effect is the same for all crops: as temperature rises, the increase in rate of grain growth does not adequately compensate for the reduced duration of grain filling, in wheat (Sofield, Evans et al., 1977), in maize (Tollenaar and Bruulsema, 1988b), in rice

FIGURE 8.4. The Effect of Temperature on Individual Kernel Mass of Cereals

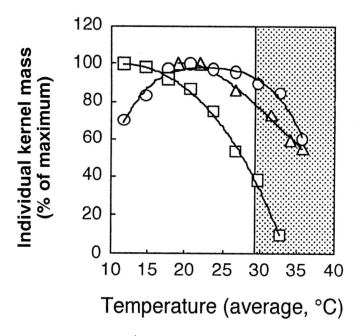

Note: □ wheat; ○ rice; and △ maize. Shading shows very high temperature range.

Source: Data for wheat and rice calculated from Chowdhury and Wardlaw, 1978, p. 216, and Tashiro and Wardlaw, 1989, p. 61; and for maize from Singletary, Banisadr, and Keeling, 1994, p. 833.

(Tashiro and Wardlaw, 1989; 1991a), and in sorghum (Chowdhury and Wardlaw, 1978).

The example of wheat will more fully illustrate the mechanism by which moderately high postanthesis temperature reduces kernel mass and, hence, yield. In wheat, the duration of grain filling declines markedly as temperature increases. Linear reductions of approximately three days for every 1°C increase in the range of 16 to 26°C have been reported (Wiegand and Cuellar, 1981), although above this range the response becomes asymptotic (Marcellos and Single, 1972; Sofield, Evans, et al., 1977). This shows that the effect of temperature on wheat grain development declines with each incremental increase in temperature above an average of about 26°C, and that the duration of grain growth is almost halved by an increase in temperature from 21/16 to 30/25°C (day/night) (see Figure 8.5a). The rate of grain growth, by contrast, increases by less than 10 percent (see Figure 8.5b), with the net result that mature kernel mass declines (linearly) by almost 50 percent over the range 21/16 to 30/25°C (see Figure 8.5c) (Tashiro and Wardlaw, 1989). The response of duration of grain growth to moderately high temperature is therefore the primary determinant of the response of individual kernel mass to moderate heat stress (Wiegand and Cuellar, 1981).

It is interesting to note that in a tropically adapted crop such as rice, duration of grain filling is most sensitive to temperature under relatively cool conditions and is comparatively insensitive as temperature exceeds about 20°C (see Figure 8.6). This contrasts with tropically adapted maize, for which duration of grain filling is strongly negatively related to temperatures above 20°C. In maize, the response of duration of grain filling to temperature is of a similar magnitude to that of wheat but occurs at a 10°C higher threshold (see Figure 8.6). The general pattern of response of duration of grain filling to temperature is therefore similar in each of the major crops, although the species are distinguished by differences in temperature threshold and rate of response.

Similarly, the effects of temperature on rate of grain filling follow the same general pattern in the three major crops, although the specifics of response differ. Cool-temperate-adapted wheat exhibits a maximum rate of grain filling over a broad range of cool temperatures, whereas warm-adapted rice and maize have a much more narrow range at which the maximum rate of grain filling occurs (see Figure 8.7). All species share a major trait in that the rate of grain filling declines rapidly once the "very high temperature" threshold has been crossed. This phenomenon will be discussed further in subsequent material.

The temperature-induced reductions in mature kernel mass discussed earlier do not appear to be consistently related to the effects of temperature on either potential grain size or assimilate supply. In wheat, moderately high temperatures have no significant effect on the foundation elements of grain size. Endosperm cell number and size are maintained, as increases in the rates of endosperm cell division (Wardlaw, 1970) and enlargement (Nicolas, Gleadow, and Dalling 1984) compensate for reduced durations of these cellular processes. In maize, evidence suggests that endosperm cell size and number are reduced by temperature, although this has not been shown in the moderately high range (Jones, Quattar, and Crookston, 1984; Jones, Roessler, and Quattar, 1985).

Similarly, moderately high temperature does not affect the initiation (and consequently number) of A-type starch granules of wheat, but it does

FIGURE 8.5. Effect of Temperature on Components of Grain Growth of Wheat

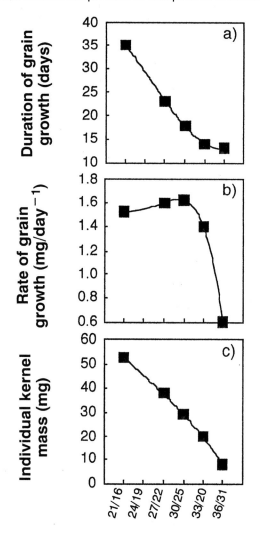

Note: Figure 8.5a, duration of grain growth; Figure 8.5b, rate of grain growth; Figure 8.5c, mature individual kernel mass.

Source: Adapted from Tashiro and Wardlaw, 1989, p. 61.

FIGURE 8.6. The Effect of Temperature on Duration of Grain Filling of Cereals

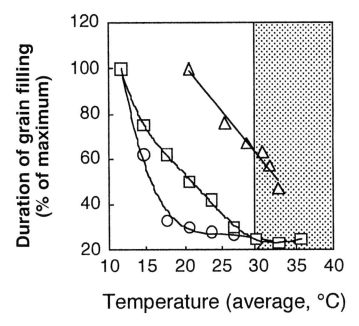

Note: □ wheat; ○ rice; and △ maize. Shading shows very high temperature range.

Source: Data for wheat and rice calculated from Chowdhury and Wardlaw, 1978, p. 218, and Tashiro and Wardlaw, 1989, p. 61; and for maize from Muchow, 1990, p. 153, and Singletary, Banisadr, and Keeling, 1994, p. 833.

reduce their size (Bhullar and Jenner, 1985). This is consistent with the lack of effect of moderately high temperature on *potential* individual kernel mass in wheat. By contrast, the number of B-type starch granules is significantly reduced by moderately high temperature (Bhullar and Jenner, 1985), although this is likely to be a symptom, rather than a cause, of reduced duration of grain filling.

Grain sucrose levels are not significantly reduced by moderately high temperature in any of the major crops (Ford, Pearman, and Thorne, 1976; Bhullar and Jenner, 1986; Singletary, Banisadr, and Keeling, 1994), despite the fact that heat stress increases the rates of grain respiration (Chowdhury and Wardlaw, 1978) and leaf senescence (Kuroyanagi and

FIGURE 8.7. The Effect of Temperature on Rate of Grain Filling of Cereals

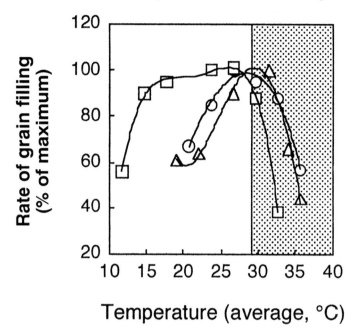

Note: □ wheat; ○ rice; and △ maize. Shading shows very high temperature range.

Source: Data for wheat and rice calculated from Chowdhury and Wardlaw, 1978, pp. 216, 218, and Tashiro and Wardlaw, 1989, p. 61; and for maize from Muchow, 1990, p. 153, and Singletary, Banisadr, and Keeling, 1994, p. 833.

Paulsen, 1985). Indeed, significant reductions in maize grain mass occur at temperatures above 22°C, despite increases in photoassimilate production with temperatures up to 32°C (Christy, Williamson, and Wideman, 1985). Nor is the reduced duration of grain growth related to premature desiccation of the grain, for in wheat, grain water and osmotic potential are not responsive to temperature in the 21/16 to 30/25°C range (Bhullar and Jenner, 1983). Considering that grain growth is reduced almost as much by heating only the ear of wheat or maize as it is by heating the entire plant (Ford, Pearman, and Thorne, 1976; Bhullar, 1984; Lu et al., 1996), it seems likely that reductions in grain mass due to postanthesis heat stress are related to sink, rather than source, limitation.

Studies referred to thus far have tended to use heat stresses applied continuously during grain filling, whereas temperature data suggest that, particularly in subtropical and temperate regions, heat stress is unlikely to occur throughout the entire postanthesis period. The stage during grain filling at which heat stress occurs is therefore likely to influence the response of grain yield to temperature.

Early grain development is particularly sensitive to the effects of high temperature, which can significantly reduce the proportion of potential grains that bear fruit. Because pollination is not synchronous even within the ear of determinate crops, grain number remains sensitive to heat stress even after many ovules have been fertilized. Within the wheat ear, for instance, fertilization of florets occurs over a period of about seven days, with central and apical spikelets being fertilized before basal spikelets, and basal florets having priority over distal florets (Rawson and Evans, 1970). Similarly, maize silks first appear from the base of the cob and work their way toward the tip over a period of three to five days (Aldrich, Scott, and Hoeft, 1986). A heat stress occurring two days after first anthesis/silk will therefore have a greater impact on fertility of basal spikelets and on distal florets in wheat, and on distal grains of maize. Clearly, the time at which heat stress occurs affects both grain number and the distribution of grains within reproductive structures.

Generally, earlier heat stress has a more drastic impact on the development of a grain than later heat stress, but later stress events tend to affect a greater proportion of grains. Grain set is most sensitive to temperature in the three days after first anthesis (Tashiro and Wardlaw, 1990a), and significant reductions in grain number can occur in response to high temperatures occurring up to the time of grain formation (Langer and Olugbemi, 1970), which may be up to seven days after first anthesis (Tashiro and Wardlaw, 1990b).

Because grain number of wheat is generally set by this point, high temperatures imposed from about seven days after anthesis reduce yield mainly through effects on individual kernel mass (Kolderup, 1979a; Bhullar and Jenner, 1985). In most crops, the responsiveness of individual kernel mass to temperature generally declines with time, so that early heat exposure reduces individual kernel mass more than later heat treatment (Randall and Moss, 1990; Tashiro and Wardlaw, 1990b). This contrasts with the effects of heat stress on grain water content, which become more pronounced toward maturity (Tashiro and Wardlaw, 1990b), probably because of a hastening of the loss of grain water that occurs toward harvest ripeness.

Although it is clear that yield would tend to decrease with increased duration of exposure to heat stress, there does not appear to have been a systematic study of this phenomenon in the moderately high temperature range. Such studies are difficult to interpret in that they are naturally confounded by the effects of timing of heat stress.

It is interesting to note that a prevalence of moderately high temperatures can reduce the exposure of crops to very high temperatures. By enhancing rate of development, moderately high temperatures can ensure that crop growth has ceased by the time that very high temperature events become frequent. This phenomenon is common in the Australian wheat belt, where crops grown in subtropical northern regions mature before the end of spring, whereas crops grown in the temperate south develop slowly during winter and spring and do not mature until summer, when exposure to very high temperature events is frequent.

Very High Temperature

The effects of very high temperature on grain yield have been studied less extensively than those of moderately high temperature, despite the fact that the former can result in greater yield reductions than their frequency or duration would suggest. A number of workers have shown that exposure of wheat to temperatures of 35/30 or 35/25°C for as little as four days during grain filling can reduce individual kernel mass by over 20 percent (Hawker and Jenner, 1993; Stone and Nicolas, 1994). Given the prevalence of such temperatures, it is clear that short periods of very high temperature may seriously reduce grain yields in both tropical and temperate environments.

Grains are particularly sensitive to very high temperature in their early stages of development, and the systematic study of Tashiro and Wardlaw, (1990a) showed that only two to three days difference in the timing of heat stress (two days of 36/31°C, day/night) had a marked effect on grain morphology. High temperatures occurring within a day of fertilization induced parthenocarpy or abortion, yet shrunken grains formed when stress was applied three days after fertilization. A further delay in stress to about eight days after fertilization did not damage whole grains as much as the packing of endosperm, resulting in notched, split, and opaque grains. All of these responses, by reducing the size or density of the grain will obviously affect yield, but quality is also affected. As will be shown later, the timing of heat stress also exerts a significant influence on grain quality.

As in the moderately high temperature range, the duration of grain growth is reduced by exposure to very high temperature (Bagga and Rawson, 1977; Stone and Nicolas, 1995a), although the asymptotic relationship between temperature and duration of grain filling for wheat and rice (referred to earlier) suggests that the effects of temperature on duration of grain growth becomes less important as temperature increases above about $33/28°C$ (see Figure 8.6). Even if the relationship between temperature and the duration of grain filling was linear (as it appears to be for maize), the influence of duration of grain filling on individual kernel mass would decline as temperature increased above approximately $30/25°C$, primarily because the rate of grain filling declines very sharply above this threshold, for all the major crop species (see Figure 8.7). As in the moderately high temperature range, the effects of very high temperature on individual kernel mass appear to be independent of the response of assimilate supply to heat stress. A number of workers have shown that yield reductions under very high temperature are largely caused by an impaired ability of grains to convert assimilate to starch, in wheat (Rijven, 1986; Jenner, 1991a, b), in maize (Singletary, Banisadr, and Keeling, 1994; Cheikh and Jones, 1995), and in rice (Inaba and Sato, 1976; Sato and Inaba, 1976). In particular, the activity of soluble-starch synthase (SSS) (which, with branching enzyme, converts adenosine diphosphate [ADP]-glucose to amylopectin within the amyloplast) has been shown to be greatly reduced by temperatures greater than around $30°C$ (Hawker, Jenner, and Niemietz, 1991; Keeling et al., 1994). As amylopectin accounts for 75 to 80 percent of the mass of wheat, maize, and rice starch (which in turn comprise ca 70 percent of grain mass), a reduction in the activity of SSS can significantly reduce grain yield. For wheat, the effects of heat stress on the activity of SSS (and consequently grain growth) are more apparent after the cessation of heat stress (Jenner, 1991b; Hawker and Jenner, 1993), than during the application of stress itself (Jenner, 1991a). In addition, it has been shown that continual exposure to very high temperature reduces SSS activity to a greater extent than intermittent heat stress (Hawker and Jenner, 1993), which suggests that cool night temperatures following a hot day may improve the speed and extent of recovery from heat stress. Although this may be the case over short time frames, the available evidence suggests that cool days following heat stress events do not help to ameliorate the deleterious effects of exposure to very high temperature on crop yield (Stone et al., 1995; Savin, Stone, and Nicolas, 1996).

Recent work with wheat has shown that although the mechanisms by which moderately high and very high temperatures reduce grain yield may differ, the overall pattern of response is similar. The effects of very high temperature on grain yield of wheat vary widely among cultivars (Stone and Nicolas, 1995c), and varietal differences in heat tolerance appear to be stable for stresses occurring at various stages of grain growth (Stone and

Nicolas, 1995a) (see Figure 8.8) and for varying lengths of time (Stone and Nicolas, 1998) (see Figure 8.9). Together, this suggests that selection and possibly breeding may help to improve tolerance to very high temperatures during grain filling.

As with moderately high temperature, the effects of very high temperature on kernel mass decrease linearly with time from anthesis (Stone and Nicolas, 1995a) (see Figure 8.8). This would help to reduce the impact of heat stress on winter and spring crops, for which the prevalence of high postanthesis temperatures tends to increase with time from anthesis.

FIGURE 8.8. The Effect of Timing of Very High Temperature Stress on Individual Kernel Mass of Wheat

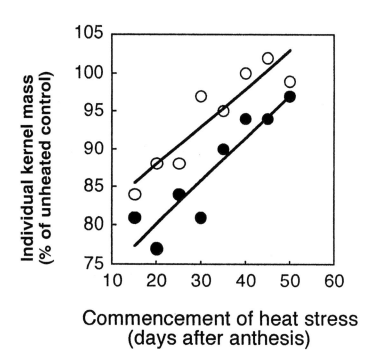

Commencement of heat stress (days after anthesis)

Note: ○ heat-tolerant cultivar; ● heat-sensitive cultivar. Heat stress (40°C, for six hours per day) was applied for five days, and the response compared with controls maintained at 21/16°C.

Source: Adapted from Stone and Nicolas, 1995a, p. 928.

FIGURE 8.9. The Effect of Duration of Very High Temperature Stress on Individual Kernel Mass of Wheat

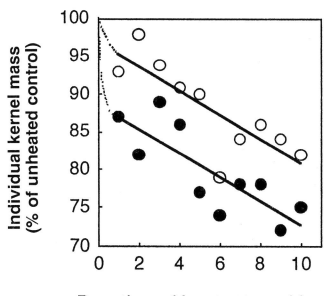

Duration of heat stress (days)

Note: ○ heat-tolerant cultivar; ● heat-sensitive cultivar. Heat stress treatment (40°C, for six hours per day) was centered around twenty days after anthesis, to reduce the confounding effects of timing of heat stress. Response is compared with controls maintained at 21/16°C.

Source: Adapted from Stone and Nicolas, 1998, p. 14.

Grain mass decreases linearly with increased duration of heat stress, and it appears that the first day of stress reduces grain mass more than subsequent stress events, particularly in a heat-sensitive cultivar (Stone and Nicolas, 1998) (see Figure 8.9). This may be evidence of either adaptation to very high temperature or a reduction in yield potential caused by earlier stress. Adaptation to elevated temperatures through heat shock protein-mediated acquired thermotolerance is reportedly an important factor in reducing the effects of heat stress on wheat (Krishnan, Nguyen, and Burke, 1989; Porter, Nguyen, and Burke 1994), maize (Smillie and Gibbons, 1981; Cooper and Ho, 1984), and potentially all living organisms (Petersen and Mitchell, 1981; Subjeck, Sciandra, and Johnson,

1982). Although the benefits of acquired thermotolerance have been widely reported for a variety of plant organs, such as leaves (Ristic, Gifford, and Cass, 1991), roots (Skogqvist and Fries, 1970), and germinating seeds (Abernethy et al., 1989), direct evidence of acquired thermotolerance having protected against stress-induced yield loss is less common and more equivocal (Stone and Nicolas, 1995b; Savin et al., 1997). Alternatively, a reduced sensitivity to heat stress following an initial stress event may be evidence of reduced yield potential. That is, the first heat stress event may reduce the capacity of grains to accumulate mass such that the response to subsequent stress events is lessened, as a consequence of occurring against a lower potential for grain mass (Stone et al., 1995). Whatever the mechanism of response, it is clear that a short period (one to two days) of very high temperature causes proportionally more damage to yield than an extended period of very high temperature (five to ten days) (Stone and Nicolas, 1998).

Elevated postanthesis temperatures, therefore, reduce yield in two-stages: Moderately high temperatures reduce yield by reducing the duration of grain growth more than they increase the rate of grain growth—the response is therefore primarily developmental. Very high postanthesis temperatures reduce grain mass by reducing the duration of grain growth *and* the rate of grain growth—the response is therefore both developmental and directly physiological.

EFFECTS OF HIGH TEMPERATURE ON QUALITY

What Is Quality?

Grain quality is a simple term that describes the complex balance of grain constituents. We may think of a grain as being made up of a vast array of ingredients, each of which has different properties. If yield is the sum of these ingredients, quality is the result of their relative proportion. Because the grain constituents differ in their physical, chemical, and nutritional properties, the balance between them will determine which properties predominate and how they will interact to determine the physical, chemical, and nutritional properties that define a grain's quality characteristics.

Because grains have such a wide variety of potential end uses (see Table 8.1), it would be difficult to describe a given batch as simply "good" or "bad" in quality; a grain is really best suited to just one or more of these uses. Grain quality is perhaps less an index of the inherent properties of a grain than the specific requirements of those who wish to process, feed, or

TABLE 8.1. Common End Uses for Wheat, Rice, and Maize Grains

End Use	Wheat	Rice	Maize
Bread	✓		✓
Cookies	✓		
Crackers	✓	✓	
Cakes	✓	✓	✓
Noodles	✓	✓	
Polenta			✓
Puddings	✓	✓	✓
Breakfast cereals	✓	✓	✓
Snack foods	✓	✓	✓
Brewing	✓	✓	✓
Distilling	✓	✓	✓
Stock feed	✓		✓
Vegetable oil			✓
Emulsifiers	✓		✓
Paper and textiles	✓	✓	✓
Glucose	✓		✓
Synthetic polymers	✓		✓
Industrial solvents	✓		✓

eat grain. Consequently, it would appear that definition of grain quality is something of a moving target; hence, a discussion of multiple uses of multiple grains could be very complex; and, consequently, unenlightening. Here such difficulties are avoided by discussing the general principles by which temperature alters grain quality. Each of the three major grains (wheat, rice, and maize) will be discussed separately, with allusions to the effects of temperature on major grain products.

Why Are High Temperatures a Problem for Grain Quality?

Probably more important than the absolute effect on quality is the fact that high temperature induces a *change* in grain quality. Sellers, buyers, and processors of grain often have decades of experience in matching grain of a given variety and region of origin with a given end use. Their marketing and processing decisions are based upon an expectation of certain quality characteristics being derived from a given combination of genotype, environment, and crop management. Problems arise when subtle variations in environment, such as slightly higher daily temperatures or a few days of hot weather, cause unpredicted changes in quality. Grain that normally would have been ideal for bread making may suddenly be best suited for stock feed, and, conversely, grain usually relegated to low-value uses may acquire properties that make it a valuable input. It is not hard to imagine the loss in time and money caused by trying to rectify problems caused by mismatches of grain with intended end use.

On the other hand, with such a wide variety of potential end uses, it is not usually difficult to find a suitable market for a given batch of grain—all but the most contaminated batches have some commercial value. The material problems caused by high temperature, therefore, are related more to distribution and marketing than to science or technology. Nevertheless, a scientific understanding of the mechanisms by which temperature changes grain quality is essential if we are to increase the average quality of grain. This will involve either (1) selection for grain that is less sensitive to changes in temperature, (2) matching less sensitive genotypes with less stable growth regions, or (3) being able to predict the effects of temperature on grain quality. Advances made using any of these three approaches will improve the likelihood of grain being matched with its optimum end use, which would increase both the average quality and value of a given grain crop.

Moderately High Temperature

The distinction between the effects of moderately high and very high temperature on quality for rice and maize appears to be less clear-cut than that for wheat. This is almost certainly related to the fact that quality in rice and maize is dictated very largely by starch properties, for which the response to temperature appears to be more or less continuous throughout the physiological temperature range. For wheat, however, quality is very largely determined by protein content and composition. The latter of these, in particular, shows distinct responses to temperature in the moderately high and very high temperature ranges. For this reason, the discussion focuses on the effects of temperature on rice and maize only in the moderately high temperature section of this chapter, although readers are reminded that quality of these crops appears to be similarly affected by very high temperatures.

Preanthesis Temperature

Despite the fact that the events of the postanthesis period are largely a product of the physiological events that precede them, very few reports have documented the way in which preanthesis temperatures affect grain quality. Clearly, shortening or lengthening the duration of the preanthesis period by exposure to warmer or cooler conditions, respectively, will result in different environmental conditions during the postanthesis growth of the crop. Although this is obviously likely to have an important impact on grain quality, it is not discussed here, for the vast proportion of the work to date concerns itself with the effects of postanthesis temperature on grain quality. Although unfortunate, this is logical, given that the effects of preanthesis temperature on grain quality would be very difficult to interpret without a thorough understanding of the effects of postanthesis temperature on quality.

Postanthesis Temperature

As with the effects on yield, the effects of moderately high postanthesis temperature on grain quality are primarily caused by changes in the rates and durations of existing processes. Most important with respect to quality is the fact that the processes contributing to yield are frequently differentially affected by temperature. As a consequence, changes in postanthesis temperature alter the balance between the different constituents of the grain and, hence, grain quality.

Wheat. Perhaps the best-characterized response of quality to temperature is that of grain protein percentage in wheat. Grain protein percentage is fundamental to the quality of wheat grain, as it is the overriding determinant of dough strength, the property for which wheat grain is most valued (MacRitchie, 1984; Wrigley, 1996).

Grain protein percentage frequently increases as a result of moderately high temperature during grain filling (Sosulski, Paul, and Hutcheon, 1963; Schipper, Jahn-Deesbach, and Weipert, 1986), and this has been ascribed to the greater temperature sensitivity of starch protein, compared to accumulation (Jenner, Ugalde, and Aspinall, 1991; Stone and Nicolas, 1996a). This may manifest itself in a number of ways: (1) a greater reduction in starch, than in protein, accumulation with elevated temperature (Chowdhury and Wardlaw, 1978; Stone, Nicolas, and Wardlaw, 1996), or (2) a reduction in starch, but not protein, accumulation with increased temperature (Kolderup, 1975, 1979b). Conversely, (3) it has also been shown that elevated temperature can increase grain protein accumulation in the range 15/10 to 21/16°C (with only a minor impact on starch) and decrease protein accumulation as temperature rises from 21/16 to 30/25°C (with a greater reduction in starch) (Sofield, Wardlaw, et al., 1977). All three mechanisms of response result in a linear to slightly (concave) curvilinear increase in grain protein percentage throughout the moderately high temperature range (Figure 8.10).

For bread dough quality, flour protein percentage usually correlates strongly with dough strength and hence desirable traits such as sodium dodecylsulfate (SDS)-sedimentation volume (Moonen, Scheepstra, and Graveland, 1982), loaf volume (Doekes and Wennekes, 1982), loaf score (Sutton, Hay, and Griffin, 1989), and tolerance to overmixing (Preston, Lukow, and Morgan, 1992). Very few studies, however, have shown a direct effect of moderately high temperature on dough quality for bread making (Schipper, Jahn-Deesbach, and Weipert, 1986). The work of Randall and Moss (1990) suggests that bread-making quality increases with temperature to the upper

FIGURE 8.10. The Effect of Moderately High Temperature on Grain Protein Percentage of Wheat

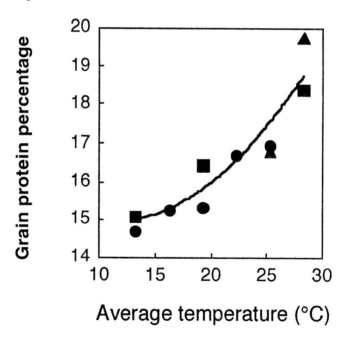

Source: Calculated from ● Chowdhury and Wardlaw, 1978, p. 212; ■ Sofield, Wardlaw, et al., 1977, p. 802; ▲Stone, Nicolas, and Wardlaw, 1996, p. 608.

limit of the moderately high temperature range (ca 30°C), but that further increases in the daily temperature maximum may reduce quality. More recent work suggests that this response may be genotype dependent: in heat-sensitive cultivars the threshold temperature for reductions in dough strength may be as low as 25°C (Stone, Gras, and Nicolas, 1997), whereas in other cultivars it appears to be as high as 35°C (Borghi et al., 1995).

This breakdown in the usually positive relationship between protein grain percentage and dough strength is evidence of a temperature-induced change in protein composition. A number of workers have shown that the proportion of gluten in flour protein increases as average daily temperature rises across the range 12 to 24°C (Sosulski, Paul, and Hutcheon, 1963; Kolderup, 1975), and this would tend to increase the contribution of protein percentage to dough strength. In heat-sensitive varieties of wheat, however, increases in the proportion of dough-weakening monomer (gliadin) in flour protein have been noted at relatively low temperatures (Stone, Gras, and Nicolas, 1997). Evidently, the sensitivity of grain quality to moderately high temperature differs among wheat genotypes.

Why does protein composition vary with temperature? Essentially, the mechanism is the same as that for protein percentage described previously—different components of grain protein respond differentially to elevated temperature. It was recently reported that as average daily temperature increased from 18 to 28°C, the amount of all proteins accumulating in the grain declined, but that the decline for strength-giving polymer (glutenin) was greater than that for dough-weakening monomer (gliadin) (Stone, Nicolas, and Wardlaw, 1996). Changes of this nature have been demonstrated for a wide variety of protein classes and subclasses (Stone, Nicolas, and Wardlaw, 1996; Stone, Gras, and Nicolas, 1997), and we are consequently gaining a greater understanding of the way in which moderately high temperature alters protein composition and, hence, wheat quality.

The discussion of the effects of temperature on grain quality has, thus far, centered on the effects on protein, but exposure to elevated temperature during grain development can also alter the composition of starch and its associated lipids, which may influence the quality of products such as bread, food additives, and noodles. Both moderately high (Tester et al., 1995) and very high (Shi, Seib, and Bernardin, 1994) temperatures during grain filling tend to increase amylose content and gelatinization temperature of wheat starch, while reducing starch swelling power. Together, these factors suggest that noodle quality would decline in response to elevated temperatures during grain growth (Oda et al., 1980; Toyokawa et al., 1989), although there has been no direct study reported to date. Starch and nonstarch lipid content of wheat has been shown to decrease in response to a 4°C rise in the moderately high temperature range, and it is predicted that this would have a negative impact on loaf volume (Williams, Shewry, and Harwood, 1994). By contrast, both brief (Blumenthal, Bekes, et al., 1995) and extended (Shi, Seib, and Bernardin, 1994) periods of very high temperature increase starch and flour lipid content.

Rice. Rice differs significantly from wheat in that its quality is determined primarily by carbohydrate, rather than protein, composition. In addition, because rice is frequently consumed as a near whole-grain product (as opposed to a highly processed material), the physical appearance of the grain is also important in determining consumer preference. Moderately high temperatures materially affect both the carbohydrate composition and physical appearance of rice grains.

Exposure to moderately high temperature during grain filling of rice induces a chalky appearance in 10 to 15 percent of kernels (Tashiro and Wardlaw, 1991b), which reduces their physical attractiveness to consumers and, hence, their market value (Mohapatra, Patel, and Sahu, 1993). Although there are different manifestations of chalkiness in rice kernels (opacity, milky-white, white-back, and white-core kernels), each is a symptom of the same cause—irregular packing of starch within the endosperm results from a stress-induced change in the size and shape of starch granules. This permits air spaces into what would otherwise be a densely packed continuum of starch and protein (Tashiro and Wardlaw, 1991b) and changes the refraction of light within the kernel, inducing a "chalky" appearance at variance with the preferred translucence.

Moderately high postanthesis temperature also causes chemical changes in the rice kernel. The most important of these is the decreased proportion of amylose in rice starch (Asaoka et al., 1984, 1985). Amylose content is probably the most important determinant of the cooking and processing characteristics of rice, and it is responsible for the commercially important distinction between *indica* ("high" amylose) and *japonica* ("low" amylose) types of grain. *Indica* rice is long grained and, when cooked, tends to be dry and easily separated, whereas *japonica* rice is short grained and moist, firm, and sticky when cooked. These properties are related to the amount of amylose in the grain, which directly affects water absorption and expansion during cooking, and which is inversely related to the stickiness, glossiness and tenderness that Japanese consumers prefer in their rice.

Starch structure is most sensitive to temperature modification in the period five to fifteen days after anthesis, when starch accumulation is most active. It has been estimated that each 1°C rise in temperature during this period reduces starch amylose content by about 0.37 percent (Okuno, 1985). Temperature-induced reductions in the proportion of amylose in starch are accompanied by an increase in amylopectin content. Furthermore, the fine structure of amylopectin is altered such that the number of large chains increases and the number of small chains decreases (Asaoka et al., 1984). The net effect of these changes in the gross and fine structure of rice starch is that its average molecular weight increases with postanthesis temperature. This, in turn, increases the temperature of gelatinization, which helps to explain the changes in cooking properties and mouth-feel that are evident in rice that has matured under elevated temperatures.

Grain protein percentage of rice increases with temperature via the same mechanism as in wheat (Tashiro and Wardlaw, 1991a). This is not usually considered to be an important determinant of rice quality.

Maize. In many respects, the response of maize to elevated temperature is similar to that of rice. As temperatures during the grain-filling period increase, starch and amylose contents decline, starch granule size is reduced, and although the effects of temperature on amylopectin branching appear to vary with genotype, the average molecular mass of starch decreases (Lu et al., 1996). As a result of these changes, starch gelatinization temperature increases (White, Polak, and Burkhart, 1991; Lu et al., 1996).

This has a dramatic effect on at least one of the end uses of maize: expanded snack and breakfast foods. Expanded maize products are esteemed for their crispness, a property conferred on them by their rapid expansion during cooking. The extrusion cooking process has evolved into a complex and highly controlled art, which is perturbed somewhat by unpredictable variations in input quality. In addition to the effect of input variability itself, the temperature-induced changes in maize quality cited previously act to significantly reduce the expansion ratio and increase the bulk density of maize products (Chinnaswamy and Hanna, 1988). In short, high postanthesis temperatures significantly reduce the quality of maize-based expanded foods.

Very High Temperature

The effect of temperature on grain protein percentage is similar for both the moderately high and very high temperature ranges. Field (Blumenthal, Bekes, et al., 1991; Borghi et al., 1995) and controlled-environment studies (Bhullar and Jenner, 1985; Stone and Nicolas, 1995c) generally confirm that wheat protein percentage increases in response to exposure to very high temperature, although this does not always occur (Graybosch et al., 1995). As in the moderately high temperature range, the increase in protein percentage is primarily due to a reduction in grain starch content, as nitrogen accumulation is comparatively unresponsive to brief episodes of very high temperature (Bhullar and Jenner, 1986; Blumenthal, Gras, et al., 1995), although significant reductions in protein accumulation are common (Stone and Nicolas, 1996b; Stone, Nicolas, and Wardlaw, 1996). The response of grain protein percentage to heat stress varies among wheat varieties (Blumenthal, Gras, et al., 1995; Stone and Nicolas, 1995c), although under field conditions, other environmental factors may exert a greater influence on grain protein content (Graybosch et al., 1995).

As discussed previously, increases in grain protein percentage caused by moderately high temperature often result in an increase in bread dough quality. By contrast, where grain protein percentage increases as a result of exposure to very high temperature, bread-making quality frequently declines (Randall and Moss, 1990; Blumenthal, Bekes, et al., 1991; Borghi et al., 1995).

The early fieldwork of Finney and Fryer (1958) showed that temperatures above a certain (32°C) threshold during the grain-filling period of wheat resulted in a breakdown in the usually positive relationship between flour protein percentage and grain quality. Specifically, as the number of hours above 32°C in the last fifteen days of the grain-filling period increased, loaf volume and dough mixing time fell below that which would have been predicted on the basis of cultivar and the protein percentage of unstressed grain. For many years, there was no further examination of the effects of heat stress on grain quality, until a controlled-environment study by Randall and Moss (1990) showed that exposure to as little as three days of very high temperature (35/30°C) during grain filling substantially reduced dough strength, although loaf volume was not responsive to heat stress. Subsequent field-based studies by Blumenthal and colleagues (Blumenthal, Bekes, et al., 1991; Blumenthal, Batey, et al., 1991) showed that maximum dough resistance and loaf volume declined as the number of hours above 35°C increased, while dough extensibility tended to increase with heat stress. Clearly, very high temperature exerts a significant influence on grain quality, and as this appears to be independent of the effects of heat stress on protein percentage (Randall and Moss, 1990; Graybosch et al., 1995), it suggests that protein composition is affected by heat stress.

It has been hypothesised that very high temperatures (> 35°C) during grain filling reduce dough strength by activating heat shock elements in gliadin genes, thereby increasing gliadin synthesis during heat stress and decreasing the ratio of polymer:monomer (or glutenin:gliadin) in mature grain (Blumenthal, Batey, Bekes, et al., 1990; Blumenthal, Batey, et al., 1991). The proposed mechanism by which this occurs has been schematized by Blumenthal and colleagues (1993) (Figure 8.11). To explain this hypothesis and its implications for wheat quality more fully, it is necessary to develop an understanding of the heat shock response.

FIGURE 8.11. Schematized Hypothesis of the Mechanism by Which Very High Temperature Reduces Dough Strength

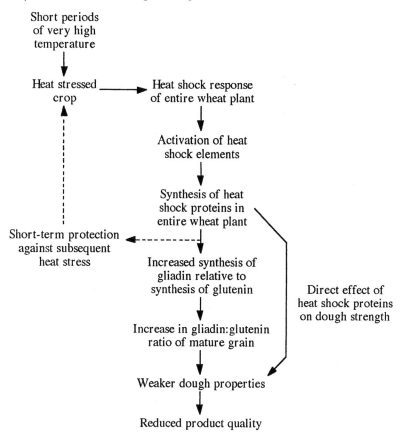

Source: Adapted from Blumenthal, Barlow, and Wrigley, 1993.

The Heat Shock Response

Exposure to very high temperature induces the heat shock response in all known living organisms (Barnett et al., 1980; Ashburner, 1982). This response to heat stress is complex but is generally characterized by (1) selective transcription and translation of heat shock mRNA (messenger ribonucleic acid), at the expense of "normal" mRNA (Ballinger and Pardue, 1982; Hickey and Weber, 1982); leading to (2) a simultaneous reduction of "normal" protein synthesis, and the stimulation of synthesis of heat shock proteins (Hendershot, Weng, and Nguyen, 1992); which putatively (3) induces acquired thermotolerance or enhances the ability of heat-shocked organisms to tolerate and recover from heat stress (Henle and Dethlefsen, 1978; Lindquist et al., 1982; Howarth and Skøt, 1994).

The acquired thermotolerance afforded by the heat shock response has been shown to be a potentially important means of reducing the effects of heat stress on plants, in general (Sachs and Ho, 1986; Vierling, 1991), and wheat, in particular (Blumenthal, Bekes, et al., 1990; Vierling and Nguyen, 1992). Unfortunately, most of the work on acquired thermotolerance in wheat has been performed using either seedlings, excised leaves, or cell suspensions, and, as mentioned previously, there is not strong evidence to show that acquired thermotolerance is important in alleviating the effects of heat stress on yield.

Paradoxically, while heat shock proteins appear to confer benefit by reducing the effects of heat stress on plant function, it is believed that they indirectly and directly contribute to reduced wheat quality. The indirect influence of the heat shock response on grain quality is shown by the vertical pathway in Figure 8.11. Blumenthal and colleagues (Blumenthal, Batey, Bekes, et al., 1990; Blumenthal, Batey, Wrigley, and Barlow, 1990) have shown that in wheat coleoptiles sudden heat stress induces the synthesis of a heat shock protein that shares considerable homology with certain gliadin genes. This prompted the hypothesis that gliadin synthesis may be enhanced by heat shock conditions, a supposition that was subsequently shown by the enhanced accumulation of ^{14}C-labeled amino acids into gliadin protein during heat stress (Blumenthal, Batey, et al., 1991). The preferential synthesis of gliadin during heat stress results in an increase in the gliadin:glutenin ratio, which has been shown to result in reduced dough strength (MacRitchie, 1984; Wrigley et al., 1994) and bread-making quality (Simmonds, 1989). Thus, there is good evidence for an indirect effect of heat shock protein synthesis on grain quality.

Evidence for a direct effect of heat shock proteins on grain quality is scarce and is complicated by the fact that the composition of constitutive (or "normal") proteins is simultaneously altered by heat stress. Nevertheless, it has been proposed that a (approximately) 70 kDa protein detectable only in heat-stressed grain may directly reduce dough strength by inhibiting the development of normal protein-protein interactions in the gluten matrix (Bernardin, 1994). This hypothesis was tested by adding purified heat shock protein to flour and testing its effects on mixing properties

(Blumenthal et al., 1998). The effects were minimal, even with very large additions of heat shock protein. This strongly suggests that heat shock proteins do not interfere with protein interactions after proteins have attained their "normal" conformation, but it does not rule out the possibility that heat shock proteins modify the folding and aggregation of gluten proteins in situ during grain filling, thereby altering their dough-forming potential (Blumenthal et al., 1998).

What is clear is that the proportions of different classes of wheat protein are significantly altered by very high temperature. Analysis of protein composition on a mass (milligram per kernel) rather than a proportion (percent of flour protein) basis has indicated that protein composition may be altered less by enhanced synthesis of certain proteins than by a *less reduced* synthesis of some proteins under stress. As for the moderately high temperatures, evidence suggests that the synthesis of dough-weakening proteins is reduced less under very high temperature conditions than that of dough-strengthening proteins (Stone and Nicolas, 1996a; Stone, Nicolas, and Wardlaw, 1996). This naturally leads to a balance of proteins that reduces dough strength. As yet, little evidence helps to explain this differential sensitivity of different protein fractions to heat stress, although variation in their time of synthesis during grain filling has been suggested (Stone and Nicolas, 1996b).

The hypotheses for the mechanisms by which very high temperature reduces dough quality have provided a firm basis for further research. It is clear that wheat genotypes differ in their response of grain quality to heat stress (Blumenthal, Bekes, et al., 1995; Stone and Nicolas, 1995c) and that the timing of stress influences the degree of response (Randall and Moss, 1990; Blumenthal, Bekes, et al., 1991). Furthermore, genotype and timing of stress interact, which currently makes it difficult to predict the effects of stress on either protein composition or dough properties (Stone and Nicolas, 1996a).

STRATEGIES FOR COPING WITH HEAT STRESS

The main aim of this chapter has been to examine some of the effects of heat stress that limit yield and quality of the world's three main crops. This provided an overview of some of the problems posed by heat stress and their physiological bases. To offer some relief from this picture of seemingly inescapable gloom, the discussion now briefly turns to mechanisms by which the effects of heat stress on crop yield and quality can be minimized.

Stress Avoidance

Avoiding high temperature events is the simplest and most obvious means of reducing the effects of heat stress on crop yield and quality. In many cases, this involves selection of genotypes that mature or pass critical stages of growth or development before the onset of stress conditions. As this frequently requires a truncation of the growing season for the crop in question, the gains made through stress avoidance must be carefully balanced against reduction in yield potential, if yield is to be maintained overall. Similarly, altering sowing time to avoid heat stress at establishment or maturity frequently involves a trade-off between favorable conditions at one end of the season or the other. The complexities of these apparently simple means of avoiding heat stress are probably best investigated using mechanistic crop models, a method that is becoming increasingly reliable and efficient. The effects on quality of temporally avoiding stress are probably more difficult to ascertain, especially given our rudimentary understanding of the effects of preanthesis growth on many aspects of grain composition.

The temperature conditions that give rise to heat stress may also be avoided by directly altering the cropping environment. As outlined previously, mulches have proved an effective and economical means of reducing heat stress during seedling emergence for many summer crops. By reducing the radiation load at the soil surface, judicious mulching can reduce soil temperatures by as much as 20°C (Townend et al., 1996), which can markedly increase yields by improving crop establishment and decelerating crop development (Lal, 1974; Abrecht and Bristow, 1996). The crop microclimate is also readily altered by irrigation, which cools both plant organs and the surrounding air (Saadia, Huber, and Lacroix, 1996), particularly in low-humidity environments.

Novel approaches to reducing tissue temperature such as increased glaucousness or increased reflectivity of leaf cuticular waxes do not appear to have been extensively studied in modern cereals, despite the fact that they are under simple genetic control (Waines, 1994).

Stress Tolerance

Improving the ability of crop genotypes to minimize the injury arising from heat stress has been explored by many workers. A key to success is the ability to quantify the damage arising from heat stress, so that useful traits or genotypes can be selected. A number of general approaches have been used, each with particular advantages and disadvantages, as summarized in Table 8.2.

TABLE 8.2. Advantages and Disadvantages Associated with Different Methods of Assessing Heat Tolerance of Crop Plants

Method	Advantages	Disadvantages
Field trial	• Direct measurement of important traits • Large sample size enables many traits to be measured	• Difficulty in controlling the timing and intensity of stress • Confounding influence of other factors (photoperiod, disease, etc.) • Time-consuming and costly
Controlled environment	• High degree of control over the timing and intensity of stress • Can usually be performed more than once per year in each facility	• Results may not relate adequately to field responses • Small sample size often restricts the number and reliability of trait measurements • Time-consuming and costly
Laboratory	• Direct measurement of some traits • Rapid • Economical	• Provides indirect measure of trait • Usually related to only one trait • Result may relate to only one stage of crop growth

The most appropriate method of selection will clearly vary with factors such as the trait or traits of interest, the stage at which selection is occurring (screening wild germplasm, as opposed to elite breeding lines), and whether a known indirect indicator of heat tolerance exists for a given trait. Laboratory assessment of indirect measures of heat tolerance is becoming increasingly common as an adjunct to more traditional field assessment of performance under stress. This has been made possible by the establishment of marker traits and means of rapidly and inexpensively assessing them. Some of the more popular methods are (1) measurement of cellular integrity following heat stress, either via electrical conductivity (Blum and Ebercon, 1981) or by reduction of TTC (2,3,5-triphenyltetrazolium chloride) (Krishnan, Nguyen, and Burke, 1989; Porter, Nguyen, and Burke 1994), both of which have been related to field performance under certain conditions (Porter and Nguyen, 1989; Shanahan et al., 1990); (2) measurement of photosynthetic function, usually via chlorophyll fluorescence (Schreiber and Berry, 1977; Moffatt,

Sears, and Paulsen, 1990); and (3) assessment of heat shock protein synthesis, usually in seedling (Blumenthal, Bekes, et al., 1990) or mature leaf tissue (Hendershot, Weng, and Nguyen, 1992).

Many of the major crop species do not show evidence of conscious selection for heat tolerance, although distinct genotypic differences in heat tolerance exist. Wardlaw and colleagues (Wardlaw et al., 1989) reported that there does not appear to have been selection of wheat cultivars for high and stable yield under elevated temperatures. They surveyed sixty-six cultivars of wheat and found that individual kernel mass was reduced by 10 to 60 percent in response to high temperature throughout grain filling, and that varietal differences appeared to be related to country of origin; that is, tropical lines tended to be better adapted to high temperature than European wheats (Wardlaw, Dawson, and Munibi, 1989). Similarly, the japonica-type rices found in cooler climates tend to be more sensitive to heat than the warmer climate indica types (Chowdhury and Wardlaw, 1978), and maize sourced from warm tropical lowlands is more heat tolerant than that from cooler highlands (Eagles and Lothrop, 1994; Lafitte and Edmeades, 1997).

The response to elevated temperature is similar in genetically related lines, and this, coupled with significant differences in response among unrelated lines, indicates that it may be possible to breed for heat tolerance. Subsequent experiments (examining the progeny of crosses between cultivars contrasting in heat tolerance) have shown that, in wheat, heat tolerance has a low but significant heritability (Wardlaw, 1994), and that breeding for specific adaptation is beneficial in maize (Lafitte, Edmeades, and Johnson, 1997).

CONCLUSIONS

Clearly, heat stress is a common occurrence in many of the world's cereal-growing regions. Recognition of this fact is a necessary first step in countering the deleterious effects of heat stress on crop yield and quality, and, ultimately, of increasing the yield and quality potential of many of the world's crops. The second step in reducing the impact of heat stress on crop production is to identify the timing and intensity of stress and, thereby, the stages of crop growth that are most likely to be affected. This is a precondition for understanding the mechanisms by which high temperature reduces yield and alters quality, a greater appreciation of which would facilitate more effective research aimed at reducing the effects of heat stress on cereal yield and quality.

REFERENCES

Abernethy, R.H., D.S. Thiel, N.S. Petersen, and K. Helm (1989). Thermotolerance is developmentally dependent in germinating wheat seed. *Plant Physiology* 89: 569-576.

Abrecht, D.G. and K.L. Bristow (1996). Coping with soil and climatic hazards during crop establishment in the semi-arid tropics. *Australian Journal of Experimental Agriculture* 36: 971-983.

Aldrich, S.R., W.O. Scott, and R.G. Hoeft (1986). How the corn plant grows. In *Modern Corn Production,* eds. S.R. Aldrich, W.O. Scott, and R.G. Hoeft. Champaign, IL: A & L Publications, pp. 1-15.

Altschuler, M. and J.P. Mascarenhas (1982). Heat shock proteins and the effect of heat shock in plants. *Plant Molecular Biology* 1: 103-115.

Asaoka, M., K. Okuno, K. Hara, and H. Fuwa (1985). Effect of environmental temperature at the milky stage on amylose content and fine structure of amylopectin of waxy and nonwaxy endosperm starches of rice (*Oryza sativa* L.). *Agricultural and Biological Chemistry* 49: 373-379.

Asaoka, M., K. Okuno, Y. Sugimoto, J. Kawakami, and H. Fuwa (1984). Effect of environmental temperature during development of rice plants on some properties of endosperm starch. *Starch/Stärke* 36: 189-193.

Ashburner, M. (1982). The effects of heat shock and other stress on gene activity: An introduction. In *Heat Shock from Bacteria to Man,* eds. M.J. Schlesinger, M. Ashburner, and A. Tissieres. New York: Cold Spring Harbor, pp. 1-9.

Bagga, A.K. and H.M. Rawson (1977). Contrasting responses of morphologically similar wheat cultivars to temperatures appropriate to warm climates with hot summers: A study in controlled environment. *Australian Journal of Plant Physiology* 4: 877-887.

Ballinger, D.G. and M.L. Pardue (1982). The subcellular compartmentalization of mRNAs in heat-shocked *Drosophila* cells. In *Heat Shock from Bacteria to Man,* eds. M.J. Schlesinger, M. Ashburner, and A. Tissieres. New York: Cold Spring Harbor, pp. 183-190.

Barnett, T., M. Altschuler, D.N. McDaniel, and J.P. Mascarenhas (1980). Heat shock induced proteins in plant cells. *Developmental Genetics* 1: 331-340.

Bernardin, J.E. (1994). Genetic modification of the wheat genome to allow growth in hot climates. In *Proceedings 44th Australian Cereal Chemistry Conference,* eds. J.F. Panozzo and P.G. Downie. Melbourne: Royal Australian Chemical Institute, pp. 60-61.

Bhullar, S.S. (1984). Physiological and biochemical responses to elevated temperatures influencing grain weight in wheat. PhD Thesis, The University of Adelaide, Australia.

Bhullar, S.S. and C.F. Jenner (1983). Responses to brief periods of elevated temperatures in ears and grains of wheat. *Australian Journal of Plant Physiology* 10: 549-560.

Bhullar, S.S. and C.F. Jenner (1985). Differential responses to high temperature of starch and nitrogen accumulation in the grain of four cultivars of wheat. *Australian Journal of Plant Physiology* 12: 363-375.

Bhullar, S.S. and C.F. Jenner (1986). Effects of a brief episode of elevated temperatures on grain filling in wheat ears cultured on solutions of sucrose. *Australian Journal of Plant Physiology* 13: 617-626.

Björkman, O., H.A. Mooney, and J. Ehleringer (1975). Photosynthetic responses of plants from habitats with contrasting photosynthetic characteristics of intact plants. *Carnegie Institute of Washington Yearbook* 74: 743-748.

Blum, A. and A. Ebercon (1981). Cell membrane stability as a measure of drought and heat tolerance in wheat. *Crop Science* 21: 43-47.

Blum, A. and B. Sinmena (1994). Wheat seed endosperm utilization under heat stress and its relation to thermotolerance in the autotrophic plant. *Field Crops Research* 37: 185-191.

Blumenthal, C.S., E.W.R. Barlow, and C.W. Wrigley (1993). Growth environment and wheat quality: The effect of heat stress on dough properties and gluten proteins. *Journal of Cereal Science* 18: 3-21.

Blumenthal, C.S., I.L. Batey, F. Bekes, C.W. Wrigley, and E.W.R. Barlow (1990). Gliadin genes contain heat-shock elements: Possible relevance to heat-induced changes in grain quality. *Journal of Cereal Science* 11: 185-187.

Blumenthal, C.S., I.L. Batey, F. Bekes, C.W. Wrigley, and E.W.R. Barlow (1991). Seasonal changes in wheat-grain quality associated with high temperatures during grain filling. *Australian Journal of Agricultural Research* 42: 21-30.

Blumenthal, C.S., I.L. Batey, C.W. Wrigley, and E.W.R. Barlow (1990). Involvement of a novel peptide in the heat shock response of Australian wheats. *Australian Journal of Plant Physiology* 17: 441-449.

Blumenthal, C.S., F. Bekes, I.L. Batey, C.W. Wrigley, H.J. Moss, D.J. Mares, and E.W.R. Barlow (1991). Interpretation of grain quality results from wheat variety trials with reference to high temperature stress. *Australian Journal of Agricultural Research* 42: 325-334.

Blumenthal, C., F. Bekes, P.W. Gras, E.W.R. Barlow, and C.W. Wrigley (1995). Identification of wheat genotypes tolerant to the effects of heat stress on grain quality. *Cereal Chemistry* 72: 539-544.

Blumenthal, C., F. Bekes, C.W. Wrigley, and E.W.R. Barlow (1990). The acquisition and maintenance of thermotolerance in Australian wheats. *Australian Journal of Plant Physiology* 17: 37-47.

Blumenthal, C., P.W. Gras, F. Bekes, E.W.R. Barlow, and C.W. Wrigley (1995). Possible role for the Glu-D1 locus with respect to tolerance to dough-quality change after heat stress. *Cereal Chemistry* 72: 135-136.

Blumenthal, C., P.J. Stone, P.W. Gras, F. Bekes, B. Clarke, E.W.R. Barlow, R. Appels, and C.W. Wrigley (1998). Heat-shock protein 70 and dough-quality changes resulting from heat stress during grain filling in wheat. *Cereal Chemistry* 75: 43-50.

Borghi, B., M. Corbellini, M. Ciaffi, D. Lafiandra, E. De Stefanis, D. Sgrulletta, G. Boggini, and N. Di Fonzo (1995). Effect of heat shock during grain filling on grain quality of bread and durum wheats. *Australian Journal of Agricultural Research* 46: 1365-1380.

Burke, J.J. (1990). Variation among species in the temperature dependence of the reappearance of variable fluorescence following illumination. *Plant Physiology* 93: 652-656.

Cheikh, N. and R.J. Jones (1995). Heat stress effects on sink activity of developing maize kernels grown *in vitro*. *Physiologia Plantarum* 95: 59-66.

Chinnaswamy, R. and M.A. Hanna (1988). Relationship between amylose content and extrusion-expansion properties of corn starches. *Cereal Chemistry* 65: 138-143.

Chowdhury, S.I. and I.F. Wardlaw (1978). The effect of temperature on kernel development in cereals. *Australian Journal of Agricultural Research* 29: 205-223.

Christy, A.L., D.R. Williamson, and A.S. Wideman (1985). Maize source development and activity. In *Regulation of Carbon and Nitrogen Reduction and Utilization in Maize,* eds. J.C. Shannon and C.D. Boyer. Rockville, MD: American Society of Plant Physiologists, pp. 11-20.

Cooper, P. and T.H.D. Ho (1984). Heat shock proteins in maize. *Plant Physiology* 71: 215-222.

Cooper, P., T.H.D. Ho, and R.M. Hauptmann (1984). Tissue specificity of the heat shock response in maize. *Plant Physiology* 75: 431-441.

Dawson, I.A. and I.F. Wardlaw (1989). The tolerance of wheat to high temperatures during reproductive growth. III. Booting to anthesis. *Australian Journal of Agricultural Research* 40: 965-980.

Doekes, G.J. and L.M.J. Wennekes (1982). Effect of nitrogen fertilization on quantity and composition of wheat flour proteins. *Cereal Chemistry* 59: 276-278.

Eagles, H.A. and J.E. Lothrop (1994). Highland maize from Central México—Its origin, characteristics and use in breeding programs. *Crop Science* 34: 11-19.

Felton, W.L., D.M. Freebairn, N.A. Fettell, and J.B. Thomas (1987). Crop residue management. In *Tillage: New Directions in Australian Agriculture,* eds. P.S. Cornish and J.E. Pratley. Melbourne: Inkata, pp. 171-193.

Finney, K.F. and H.C. Fryer (1958). Effect on loaf volume of high temperatures during the fruiting period of wheat. *Agronomy Journal* 50: 28-34.

Fischer, R.A. and O.R. Maurer (1976). Crop temperature modification and yield potential in a dwarf spring wheat. *Crop Science* 16: 855-859.

Ford, M.A., I. Pearman, and G.M. Thorne (1976). Effects of variation in ear temperature in growth and yield of spring wheat. *Annals of Applied Biology* 82: 317-333.

Graybosch, R.A., C.J. Peterson, P.S. Baenziger, and D.R. Shelton (1995). Environmental modification of hard red winter wheat flour protein composition. *Journal of Cereal Science* 22: 45-51.

Hall, A.E. (1992). Breeding for heat tolerance. *Plant Breeding Reviews* 10: 129-168.

Hawker, J.S. and C.F. Jenner (1993). High temperature affects the activity of enzymes in the committed pathway of starch synthesis in developing wheat endosperm. *Australian Journal of Plant Physiology* 20: 197-209.

Hawker, J.S., C.F. Jenner, and C.M. Niemietz (1991). Sugar metabolism and compartmentation. *Australian Journal of Plant Physiology* 18: 227-237.

Hendershot, K.L., J. Weng, and H.T. Nguyen (1992). Induction temperature of heat-shock protein synthesis in wheat. *Crop Science* 32: 256-261.

Henle, K.J. and L.A. Dethlefsen (1978). Heat fractionation and thermotolerance: A review. *Cancer Research* 38: 1843-1851.

Herrero, M.P. and R.R. Johnson (1980). High temperature stress and pollen viability of maize. *Crop Science* 20: 796-800.

Hickey, E. and L.A. Weber (1982). Preferential translation of heat-shock mRNAs in HeLa cells. In *Heat Shock from Bacteria to Man*, eds. M.J. Schlesinger, M. Ashburner, and A. Tissieres. New York: Cold Spring Harbor, pp. 199-206.

Howarth, C.J. and K.P. Skøt (1994). Detailed characterization of heat shock protein synthesis and induced thermotolerance in seedlings of *Sorghum bicolor* L. *Journal of Experimental Botany* 45: 1353-1363.

Hunt, L.A. and S. Pararajasingham (1995). Cropsim-wheat—A model describing the growth and development of wheat. *Canadian Journal of Plant Science* 75: 619-632.

Inaba, K. and K. Sato (1976). High temperature injury of ripening in rice plant. VI. Enzyme activities of kernel as influenced by high temperature. *Proceedings of the Crop Science Society of Japan* 45: 162-167.

Jenner, C.F. (1991a). Effects of exposure of wheat ears to high temperature on dry matter accumulation and carbohydrate metabolism in the grain of two cultivars. I. Immediate responses. *Australian Journal of Plant Physiology* 18: 165-177.

Jenner, C.F. (1991b). Effects of exposure of wheat ears to high temperature on dry matter accumulation and carbohydrate metabolism in the grain of two cultivars. II. Carry-over effects. *Australian Journal of Plant Physiology* 18: 179-190.

Jenner, C.F., T.D. Ugalde, and D. Aspinall (1991). The physiology of starch and protein deposition in the endosperm of wheat. *Australian Journal of Plant Physiology* 18: 211-226.

Jones, R.J., S. Quattar, and R.K. Crookston (1984). Thermal environment during endosperm cell division and grain filling in maize: Effects on kernel development and growth in vitro. *Crop Science* 24: 133-137.

Jones, R.J., J.A. Roessler, and S. Quattar (1985). Thermal environment during endosperm cell division in maize: Effect on number of endosperm cells and starch granules. *Crop Science* 25: 830-834.

Keeling, P.L., R. Banisadr, L. Barone, B.P. Wasserman, and G.W. Singeltary (1994). Effect of temperature on enzymes in the pathway of starch biosynthesis in developing wheat and maize grain. *Australian Journal of Plant Physiology* 21: 807-827.

Kolderup, F. (1975). Effects of temperature, photoperiod, and light quality on protein production in wheat grains. *Journal of the Science of Food and Agriculture* 26: 583-592.

Kolderup, F. (1979a). Application of different temperatures in three growth phases of wheat. I. Effects on grain and straw yields. *Acta Agriculturae Scandinavia* 29: 6-10.

Kolderup, F. (1979b). Application of different temperatures in three growth phases in wheat. III. Effects on protein content and composition. *Acta Agriculturae Scandinavia* 29: 379-384.

Krishnan, M., H.T. Nguyen, and J.J. Burke (1989). Heat shock protein synthesis and thermal tolerance in wheat. *Plant Physiology* 90: 140-145.

Kuroyanagi, T. and G.M. Paulsen (1985). Mode of high temperature injury to wheat. II. Comparisons of wheat and rice with and without inflorescences. *Physiologia Plantarum* 65: 203-208.

Lafitte, H.R. and G.O. Edmeades (1997). Temperature effects on radiation use and biomass partitioning in diverse tropical maize cultivars. *Field Crops Research* 49: 231-247.

Lafitte, H.R., G.O. Edmeades, and E.C. Johnson (1997). Temperature responses of tropical maize cultivars selected for broad adaptation. *Field Crops Research* 49: 215-229.

Lal, R. (1974). Soil temperature, soil moisture and maize yield from mulched and unmulched soils. *Plant and Soil* 40: 129-143.

Langer, R.H.M. and L.B. Olugbemi (1970). A study of New Zealand wheats. IV. Effects of extreme temperature at different stages of development. *New Zealand Journal of Agricultural Research* 13: 878-886.

Lindquist, S., B. DiDomenico, G. Bugaisky, S. Kurtz, L. Petko, and S. Sonoda (1982). Regulation of the heat-shock response in *Drosophila* and yeast. In *Heat Shock from Bacteria to Man,* eds. M.J. Schlesinger, M. Ashburner, and A. Tissieres. New York: Cold Spring Harbor, pp. 167-175.

Loomis, R.S. and W.A. Williams (1963). Maximum crop productivity: An estimate. *Crop Science* 3: 67-72.

Lösch, R (1979). Responses of stomata to environmental factors—Experiments with isolated epidermal strips of *Polypodiom vulgare*. II. Leaf bulk water potential, air humidity, and temperature. *Oecologia* 39: 229-238.

Lu, T., J.L. Jane, P.L. Keeling, and G.W. Singletary (1996). Maize starch fine structures affected by ear developmental temperature. *Carbohydrate Research* 282: 157-170.

MacRitchie, F. (1984). Baking quality of wheat flours. *Advances in Food Research* 29: 201-277.

Marcellos, H. and W.V. Single (1972). The influence of cultivar, temperature, and photoperiod on post-flowering development of wheat. *Australian Journal of Agricultural Research* 23: 533-540.

Matsui, T., O.S. Namuco, L.H. Ziska, and T. Horie (1997). Effects of high temperature and CO_2 concentration on spikelet sterility in indica rice. *Field Crops Research* 51: 213-219.

Moffatt, J.M., R.G. Sears, and G.M. Paulsen (1990). Wheat high temperature tolerance during reproductive growth. I. Evaluation by chlorophyll fluorescence. *Crop Science* 30: 881-885.

Mohapatra, P.K., R. Patel, and S.K. Sahu (1993). Time of flowering affects grain quality and spikelet partitioning within the rice panicle. *Australian Journal of Plant Physiology* 20: 231-241.

Monteith, J.L. (1972). Solar radiation and productivity in tropical ecosystems. *Journal of Applied Ecology* 9: 747-766.

Monteith, J.L. and R.K. Scott (1982). Weather and yield variation of crops. In *Food, Nutrition and Climate*, eds. K. Blaxter and L. Fowden. Englewood, NJ: Applied Science Publishing, pp. 127-149.

Moonen, J.H.E., A. Scheepstra, and A. Graveland (1982). Use of the SDS-sedimentation test and SDS-polyacrylamide gel electrophoresis for screening breeder's samples of wheat for bread-making quality. *Euphytica* 31: 677-690.

Morris, M.L., A. Belaid, and D. Byerlee (1991). Wheat and barley production in rainfed marginal environments of the developing world. In *1990-91 CIMMYT World Wheat Facts and Trends: Wheat and Barley Production in Rainfed Marginal Environments of the Developing World, Part 1.* Mexico, D.F.: CIMMYT, pp. 19-20.

Muchow, R.C. (1990). Effect of high temperature on grain-growth in field-grown maize. *Field Crops Research* 23: 145-158.

Muchow, R.C., T.R. Sinclair, and J.M. Bennett (1990). Temperature and solar radiation effects on potential maize yield across locations. *Agronomy Journal* 82: 338-343.

Nicolas, M.E., R.M. Gleadow, and M.J. Dalling (1984). Effects of drought and high temperature on grain growth in wheat. *Australian Journal of Plant Physiology* 11: 553-566.

Nielsen, K.F. and E.C. Humphries (1966). Effects of root temperature on plant growth. *Soils and Fertilisers* 29: 1-7.

Oda, M., Y. Yasuda, S. Okazaki, Y. Yamauchi, and Y. Yokoyama (1980). A method of flour quality assessment for Japanese noodles. *Cereal Chemistry* 57: 253-254.

Okuno, K. (1985). Environmental control of gene expression in mutants of rice. *Rice Genetics Newsletter* 2: 65-66.

Ougham, H.J., J.M. Peacock, J.L. Stoddart, and P. Soman (1988). High temperature effects of seedling emergence and embryo protein synthesis of sorghum. *Crop Science* 28: 251-253.

Paulsen, G.M. (1994). High temperature responses in crop plants. In *Physiology and Determination of Crop Yield*, eds. K.J. Boote, J.M. Bennett, T.R. Sinclair, and G.M. Paulsen. Madison, WI: American Society of Agronomy, pp. 365-394.

Petersen, N.S. and H.K. Mitchell (1981). Recovery of protein synthesis after heat shock: Prior heat-treatment affects the ability of cells to translate mRNAs. *Proceedings of the National Academy of Sciences, USA* 78: 1708-1711.

Porter, D.R. and H.T. Nguyen (1989). Chromosomal location of genes controlling heat shock proteins in hexaploid wheats. *Theoretical and Applied Genetics* 78: 873-878.

Porter, D.R., H.T. Nguyen, and J.J. Burke (1994). Quantifying acquired thermal tolerance in winter wheat. *Crop Science* 34: 1686-1689.

Preston, K.R., O.M. Lukow, and B. Morgan (1992). Analysis of relationships between flour quality properties and protein fractions in a world wheat collection. *Cereal Chemistry* 69: 560-567.

Randall, P.J. and H.J. Moss (1990). Some effects of temperature regime during grain filling on wheat quality. *Australian Journal of Agricultural Research* 41: 603-617.

Rawson, H.M. (1986). High-temperature-tolerant wheat: A description of variation and a search for some limitations to productivity. *Field Crops Research* 14: 197-212.

Rawson, H.M. and L.T. Evans (1970). The pattern of grain growth within the ear of wheat. *Australian Journal of Biological Science* 23: 753-764.

Rawson, H.M. and M. Zajac (1993). Effects of higher temperatures, photoperiod and seed vernalisation on development in two spring wheats. *Australian Journal of Plant Physiology* 20: 211-222.

Reynolds, M.P., M. Balota, M.I.B. Delgado, I. Amani, and R.A. Fischer (1994). Physiological and morphological traits associated with spring wheat yield under hot, irrigated conditions. *Australian Journal of Plant Physiology* 21: 717-730.

Rijven, A.H.G.C. (1986). Heat inactivation of starch synthase in wheat endosperm tissue. *Plant Physiology* 81: 448-453.

Ristic, Z., D.J. Gifford, and D.D. Cass (1991). Heat shock proteins in two lines of *Zea mays* L. that differ in drought and heat resistance. *Plant Physiology* 97: 1430-1434.

Ritchie, R.T. and D.S. NeSmith (1991). Temperature and crop development. In *Modeling Plant and Soil Systems,* eds. R.J. Hanks and R.T. Ritchie. Madison, WI: American Society of Agronomy, Monograph 31, pp. 5-29.

Russell, E. and S. Coupe, eds. (1994). *The Macquarie World Atlas.* Sydney, Australia: Macquarie Library.

Saadia, R., L. Huber, and B. Lacroix (1996). Using evaporative cooling to fight heat stress in corn—The potential of sprinkler irrigation to reduce air and reproductive organ temperature. *Agronomie* 16: 465-477.

Sachs, M.M. and T.H.D. Ho (1986). Alteration of gene expression during environmental stress in plants. *Annual Review of Plant Physiology* 37: 363-376.

Saini, H.S. and D. Aspinall (1982). Abnormal sporogenesis in wheat (*Triticum aestivum* L.) induced by short periods of high temperature. *Annals of Botany* 49: 835-846.

Saini, H.S., M. Sedgley, and D. Aspinall (1983). Effect of heat stress during floral development on pollen tube growth and ovary anatomy in wheat (Triticum aestivum L.). *Australian Journal of Plant Physiology* 10: 137-144.

Saini, H.S., M. Sedgley, and D. Aspinall (1984). Developmental anatomy in wheat of male sterility induced by heat stress, water deficit or abscisic acid. *Australian Journal of Plant Physiology* 11: 243-253.

Satake, T. and S. Yoshida (1978). High temperature-induced sterility in indica rice at flowering. *Japanese Journal of Crop Science* 47: 6-17.

Sato, K. and K. Inaba (1976). High temperature injury of ripening in rice plant. V. On the early decline of assimilate storing ability of grains at high temperature. *Proceedings of the Crop Science Society of Japan* 45: 156-161.

Savin, R., P.J. Stone, and M.E. Nicolas (1996). Responses of grain growth and malting quality of barley to short periods of high temperature in field studies using portable chambers. *Australian Journal of Agricultural Research* 47: 465-477.

Savin, R., P.J. Stone, M.E. Nicolas, and I.F. Wardlaw (1997). Grain growth and malting quality of barley. 2. Effects of temperature regime before heat stress. *Australian Journal of Agricultural Research* 48: 625-634.

Schipper, A., W. Jahn-Deesbach, and D. Weipert (1986). Untersuchungen zum Klimateinfluss auf die Weizenqualitat. *Getreide, Mehl und Brot* 40: 99-103.

Schreiber, U. and J.A. Berry (1977). Heat-induced changes of chlorophyll fluorescence in intact leaves, correlated with damage of the photosynthetic apparatus. *Planta* 136: 233-238.

Shanahan, J.F., I.B. Edwards, J.S. Quick, and J.R. Fenwick (1990). Membrane thermostability and heat tolerance of spring wheat. *Crop Science* 30: 247-251.

Shi, Y.-C., P.A. Seib, and J.E. Bernardin (1994). Effects of temperature during grain-filling on starches from six wheat cultivars. *Cereal Chemistry* 71: 369-383.

Shpiler, L. and A. Blum (1986). Differential reaction of wheat cultivars to hot environments. *Euphytica* 35: 483-492.

Simmonds, D.H. (1989). *Wheat and Wheat Quality in Australia*. Melbourne: CSIRO.

Singletary, G.W., R. Banisadr, and P.L. Keeling (1994). Heat stress during grain filling in maize: Effects on carbohydrate storage and metabolism. *Australian Journal of Plant Physiology* 21: 829-841.

Skogqvist, I. and N. Fries (1970). Induction of thermosensitivity and salt sensitivity in wheat roots (*Triticum aestivum*) and the effects of kinetin. *Experimentia* 26: 1160-1162.

Slafer, G.A. and H.M. Rawson (1994). Sensitivity of wheat phasic development to major environmental factors: A re-examination of some assumptions by physiologists and modellers. *Australian Journal of Plant Physiology* 21: 393-426.

Slatyer, R.O. (1977). Altitudinal variation in photosynthetic characteristics of snow gum, *Eucalyptus pauciflora* Sieb. ex Spreng. III. Temperature response of four populations grown at different temperatures. *Australian Journal of Plant Physiology* 4: 301-312.

Smillie, R.M. and G.C. Gibbons (1981). Heat tolerance and heat hardness in crop plants measured by chlorophyll fluorescence. *Carlsberg Research Communications* 46: 395-403.

Smyth, D.R. (1997). Plant development—Attractive ovules. *Current Biology* 7: R64-R66.

Sofield, I., L.T. Evans, M.G. Cook, and I.F. Wardlaw (1977). Factors influencing the rate and duration of grain filling in wheat. *Australian Journal of Plant Physiology* 4: 785-797.

Sofield, I., I.F. Wardlaw, L.T. Evans, and S.Y. Zee (1977). Nitrogen, phosphorus and water contents during grain development and maturation in wheat. *Australian Journal of Plant Physiology* 4: 799-810.

Sosulski, F.W., E.A. Paul, and W.L. Hutcheon (1963). The influence of soil moisture, nitrogen fertilization, and temperature on quality and amino acid composition of Thatcher wheat. *Canadian Journal of Soil Science* 43: 219-228.

Spiertz, J.H.J. (1974). Grain growth and distribution of dry matter in the wheat plant as influenced by temperature, light energy and ear size. *Netherlands Journal of Agricultural Science* 22: 207-220.

Stone, P.J., P.W. Gras, and M.E. Nicolas (1997). The influence of recovery temperature on the effects of a brief heat shock on wheat. III. Grain protein composition and dough properties. *Journal of Cereal Science* 25: 129-141.

Stone, P.J. and M.E. Nicolas (1994). Wheat cultivars vary widely in their responses of grain yield and quality to short periods of post-anthesis heat stress. *Australian Journal of Plant Physiology* 21: 887-900.

Stone, P.J. and M.E. Nicolas (1995a). Effect of timing of heat stress during grain filling on two wheat varieties differing in heat tolerance. I. Grain growth. *Australian Journal of Plant Physiology* 22: 927-934.

Stone, P.J. and M.E. Nicolas (1995b). Comparison of sudden heat stress with gradual exposure to high temperature during grain filling in two wheat varieties differing in heat tolerance. I. Grain growth. *Australian Journal of Plant Physiology* 22: 935-944.

Stone, P.J. and M.E. Nicolas (1995c). A survey of the effects of high temperature during grain filling on yield and quality of 75 wheat cultivars. *Australian Journal of Agricultural Research* 46: 475-492.

Stone, P.J. and M.E. Nicolas (1996a). Effect of timing of heat stress during grain filling on two wheat varieties differing in heat tolerance. II. Fractional protein accumulation. *Australian Journal of Plant Physiology* 23: 739-749.

Stone, P.J. and M.E. Nicolas (1996b). Varietal differences in mature protein composition of wheat resulted from different rates of polymer accumulation during grain filling. *Australian Journal of Plant Physiology* 23: 727-737.

Stone, P.J., M.E. Nicolas, and I.F. Wardlaw (1996). The influence of recovery temperature on the effects of a brief heat shock on wheat. II. Fractional protein accumulation during grain growth. *Australian Journal of Plant Physiology* 23: 605-616.

Stone, P.J. and M.E. Nicolas (1998). The effect of duration of heat stress during grain filling on two wheat varieties differing in heat tolerance: Grain growth and fractional protein accumulation. *Australian Journal of Plant Physiology* 25: 13-20.

Stone, P.J., R. Savin, I.F. Wardlaw, and M.E. Nicolas (1995). The influence of recovery temperature on the effects of a brief heat shock on wheat. I. Grain growth. *Australian Journal of Plant Physiology* 22: 945-954.

Subjeck, J.R., J.J. Sciandra, and R.J. Johnson (1982). Heat shock proteins and thermotolerance. *British Journal of Radiology* 55: 579-584.

Sutton, K.H., R.L. Hay, and W.B. Griffin (1989). Assessment of the potential bread baking quality of New Zealand wheats by RP-HPLC of glutenins. *Journal of Cereal Science* 10: 113-121.

Tashiro, T. and I.F. Wardlaw (1989). A comparison of the effects of high temperature on grain development in wheat and rice. *Annals of Botany* 64: 59-65.

Tashiro, T. and I.F. Wardlaw (1990a). The response to high temperature shock and humidity changes prior to and during the early stages of grain development in wheat. *Australian Journal of Plant Physiology* 17: 551-561.

Tashiro, T. and I.F. Wardlaw (1990b). The effect of high temperature at different stages of ripening on grain set, grain weight and grain dimensions in the semi-dwarf wheat 'Banks'. *Annals of Botany* 65: 51-61.

Tashiro, T. and I.F. Wardlaw (1991a). The effect of high temperature on the accumulation of dry matter, carbon and nitrogen in the kernel of rice. *Australian Journal of Plant Physiology* 18: 259-265.

Tashiro, T. and I.F. Wardlaw (1991b). The effect of high temperature on kernel dimensions and the type and occurrence of kernel damage in rice. *Australian Journal of Agricultural Research* 42: 485-496.

Tester, R.F., W.R. Morrison, R.H. Ellis, J.R. Piggott, G.R. Batts, T.R. Wheeler, J.I.L. Morison, P. Hadley, and D.A. Ledward (1995). Effects of elevated growth temperature and carbon dioxide levels on some physicochemical properties of wheat starch. *Journal of Cereal Science* 22: 63-71.

Tollenaar, M. and T.W. Bruulsema (1988a). Efficiency of maize dry matter production during periods of complete leaf area expansion. *Agronomy Journal* 80: 580-585.

Tollenaar, M. and T.W. Bruulsema (1988b). Effects of temperature on rate and duration of kernel dry matter accumulation of maize. *Canadian Journal of Plant Science* 68: 935-940.

Townend, J., P.W. Mtakwa, C.E. Mullins, and L.P. Simmonds (1996). Soil physical factors limiting establishment of sorghum and cowpea in two contrasting soil types in the semi-arid tropics. *Soil and Tillage Research* 40: 89-106.

Toyokawa, H., G.L. Rubenthaler, J.R. Powers, and E.G. Schanus (1989). Japanese noodle qualities. II. Starch components. *Cereal Chemistry* 66: 387-391.

Vierling, E. (1991). The roles of heat shock proteins in plants. *Annual Review of Plant Physiology and Molecular Biology* 42: 579-620.

Vierling, R.A. and H.T. Nguyen (1992). Heat-shock protein gene expression in diploid wheat genotypes differing in thermal tolerance. *Crop Science* 32: 370-377.

Vose, R.S., R.L. Schmoyer, P.M. Steurer, T.C. Peterson, R. Heim, T.R. Karl, and J. Eischeid (1992). *The Global Historical Climatology Network: Long Term Monthly Temperature, Precipitation, Sea Level Pressure, and Station Pressure Data*. Oak Ridge, TN: Carbon Dioxide Information Analysis Center, Oak Ridge National Laboratory.

Waines, J.G. (1994). High temperature stress in wild wheats and spring wheats. *Australian Journal of Plant Physiology* 21: 705-715.

Walker, J.M. (1969). One-degree increments in soil temperature affect maize seedling behavior. *Soil Science Society of America Proceedings* 33: 729-736.

Wardlaw, I.F. (1970). The early stages of grain development in wheat: Response to light and temperature in a single variety. *Australian Journal of Biological Science* 23: 765-774.

Wardlaw, I.F. (1994). The effect of high temperature on kernel development in wheat: Variability related to pre-heading and postanthesis temperatures. *Australian Journal of Plant Physiology* 21: 731-739.

Wardlaw, I.F., I.A. Dawson, and P. Munibi (1989). The tolerance of wheat to high temperatures during reproductive growth. II. Grain development. *Australian Journal of Agricultural Research* 40: 15-24.

Wardlaw, I.F., I.A. Dawson, P. Munibi, and R. Fewster (1989). The tolerance of wheat to high temperatures during reproductive growth. I. Survey procedures and general response patterns. *Australian Journal of Agricultural Research* 40: 1-13.

Wardlaw, I.F. and C.W. Wrigley (1994). Heat tolerance in temperate cereals: An overview. *Australian Journal of Plant Physiology* 21: 695-703.

Warrington, I.J. and E.T. Kanemasu (1983). Corn growth response to temperature and photoperiod. III. Leaf number. *Agronomy Journal* 75: 762-766.

Weaich, K., K.L. Bristow, and A. Cass (1996). Modeling pre-emergent maize shoot growth. II. High temperature stress conditions. *Agronomy Journal* 88: 398-403.

White, P., L. Polak, and S. Burkhart (1991). Thermal properties of starches from corn grown in temperate and tropical environments. *Cereal Foods World* 36: 704.

Wiegand, C.L. and J.A. Cuellar (1981). Duration of grain filling and kernel weight of wheat as affected by temperature. *Crop Science* 21: 95-101.

Williams, M., P.R. Shewry, and J.L. Harwood (1994). The influence of the "greenhouse effect" on wheat (*Triticum aestivum* L.) grain lipids. *Journal of Experimental Botany* 45: 1379-1385.

Wilson, D.R., R.C. Muchow, and C.J. Murgatroyd (1995). Model analysis of temperature and solar radiation limitations to maize potential productivity in a cool climate. *Field Crops Research* 43: 1-18.

Wrigley, C.W. (1996). Biopolymers—Giant proteins with flour power. *Nature* 381: 539-540.

Wrigley, C.W., C. Blumenthal, P.W. Gras, and E.W.R. Barlow (1994). Temperature variation during grain filling and changes in wheat-grain quality. *Australian Journal of Agricultural Research* 21: 875-885.

Xiao, C.-M. and J.P. Mascarenhas (1985). High temperature-induced thermotolerance in pollen tubes of *Tradescantia* and heat shock proteins. *Plant Physiology* 78: 887-890.

Index

Page numbers followed by the letter "f" indicate figures; those followed by the letter "t" indicate tables.

Abscisic acid (ABA)
    calcium mediation, 158, 159
    and chilling tolerance, 89, 134,
        135-136, 152, 153-154, 155f
    and stomatal closure, 85
Acclimation. *See also* Acquired
        thermotolerance; Cold
        acclimation
    in chilled root systems, 90-98
    to chilling, 3-5, 4f, 7f, 8fi, 28, 63
    and chilling-induced water deficit,
        83, 84f
    to lipid peroxidation, 10-11
    mineral uptake, 90-91
    in thermotolerance, 184
    water uptake, 91-93
Acquired thermotolerance, 184-186
    cereal grain crops, 276
Active oxygen species. *See also*
        Reactive oxygen species
        (ROS)
    antioxidant biosynthesis effects,
        66
    excess symptoms, 53-54
    presence in light, 62, 63
    scavenging of, 55f
Adenylate energy charge (AEC), 179
Alpha-tocopherol, 5, 55, 56, 60
    in chilling response studies, 6
Alternate respiratory pathway, 12, 13
Alternative oxidase, 13
Amino acid accumulation, in cold
        acclimation, 118-119

Anthocyanins, 59
    accumulation in *sfr* mutants, 166
    and environmental stress, 66
Antifreeze proteins, 112, 113
Antioxidant defense mechanism
    chilling stress, 30
    interactivity of, 68
Antioxidant systems, classes, 58
Antioxidants, 5, 10, 28
    capacity in different plant species,
        67-68
    and chilling tolerance, 16-19,
        29-30
    classes of, 58
    heat shock response, 179, 195
    optimal configuration, 68
Antiparallel spin, singlet oxygen, 57
Apoplast, 22
Aquaporins, and chilling
        acclimation, 93, 98
Arabidopsis thaliana
    ABA levels, 154, 161
    calcium channel activity, 156
    cold-inducible gene promoters,
        135-136
    cold-inducible genes, 112, 123,
        124t
    fatty acid desaturation, 96, 97
    heat stress induced calmodulin,
        189, 196
    HSP100 in, 194
    HSP70 in, 191

Arabidopsis thaliana (continued)
   lipid unsaturation in heat
      tolerance, 197
   proline accumulation, 119
   protein kinases, 160, 162
   sfr mutants, 166
   smHSPs in, 192-193
Ascorbate (AsA), 5, 18, 55, 55f, 59
   in chilling response studies, 65
   H_2O_2 removal, 56
   OH reactions, 57
Ascorbate peroxidase (APX), 5, 17,
      18, 20, 31-32, 56, 58
   in chilling response studies, 64, 65
Ascorbate-glutathione cycle, 18-19,
      55f, 59
ATP depletion, in heat stress, 179
ATP/ADP activation, in chilling
      stress, 9-10, 14
Autotrophic, early seedlings, 3, 247
Autooxidation, 54
Avoidance mechanisms, heat stress,
      182

Beta-carotene, 5, 58, 60
   in chilling response studies, 65-66
BN115
   cold-inducible gene, 123, 124t
   low-temperature response, 164
BnPRP (cold-induced mRNA), cell
      wall effects, 121
Brassica napus
   cold-inducible gene promoters,
      135-136
   cold-inducible genes, 123, 124t
   heat shock proteins, 130
Bread dough strength, high
      temperature effects, 269,
      270-271, 274, 275f

Calcium channel activity, and
      membrane fluidity, 156
Calcium ion (Ca2+), 22, 23, 35

Calcium ion (Ca2+) *(continued)*
   CDPRs regulation, 161-162
   cell wall storage of, 27
   heat stress signaling, 195-196
   multiple stress signaling, 203
   second messenger stress response,
      25-26, 157-158
   signal specificity, 159-160
Calcium-dependent protein kinases
      (CDPKs), and stress
      signaling. 161-162
Calmodulin, 195, 196
Calvin cycle, H_2O_2 inhibition, 56
Cap function, heat shock effects,
      226, 227
Caratenoids, in chilling response
      studies, 65
Carbohydrate accumulation
   in cold-acclimated plants, 114-118
   low-temperature effects, 82
Carbohydrate composition, rice
      grains, 272
Carbohydrate metabolism, related
      enzymes, 114, 115t
Carbon dioxide (CO_2), high-
      temperature effects, 248,
      249-250
Carbon partitioning, low-temperature
      effects, 81
Carbonyl content, chilled seedlings,
      15
Cat genes, 18, 23
Catalase (CAT), 5, 7f, 8f, 13, 17, 18,
      33, 55f, 58
   in chilling response studies, 63-64
   in heat stress response, 180, 181,
      195
   H_2O_2 removal, 56
Cell membrane stability (CMS), heat
      stress test, 185, 186, 199, 206
Cell wall rigidity, and cold tolerance,
      119-120
Cereal grain crops
   genotype assessment methods,
      278, 279t

Cereal grain crops *(continued)*
  high-temperature effects on
      quality, 267
  quality definition , 266-267
Chaperones, 15, 130-131
  HSPs as, 187t-188t, 190, 222
Chilling acclimation-responsive
      genes, 16
Chilling stress, 2
  light/dark effects, 62, 87-89
  mechanisms, 6-15, 61
  oxidative stress from, 1, 28
  symptoms, 4, 61
Chilling tolerance, 78, 92
Chilling-induced water deficit, 83
  hydraulic effects, 84
  stomatal effects, 85
Chilling-sensitive crops, 2
Chlorophyll, singlet oxygen
      production, 58
Chloroplasts, antioxidants in, 58, 59,
      60
Chronic high temperature, 244
Circadian rhythms, and low
      temperature, 87-88
Cis element
  identification, 27, 35
  low-temperature mediation, 163,
      164
Cold acclimation, 109, 151-152
  and amino acid and amine
      accumulation, 118-119
  and carbohydrate accumulation,
      115-118
  and cell wall changes, 119-121
  genetic engineering potential,
      167-168
  multiple regulatory pathways,
      134-135
Cold stress, subdivisions, 2
Cold-induced genes *(cor)*, 16, 122,
      124t, 126t-128t, 133t
  specificity of, 135
Cold-tolerant hybrid selection, 1

Cor15, cold-inducible *arabidopsis*
      gene, 123, 124t, 153, 168
Cross-tolerance, multiple stresses,
      36, 203
Cryoprotection
  proline and polyamines, 118
  soluble sugar role in, 117
CuZnSOD, 16, 20, 25
Cyclic adenosine diphosphate
      (ADP)-ribose (cADPR),
      calcium release, 158, 159
Cytosol, antioxidants in, 58, 59
Cytosolic (cyt) oxidase, 12, 13

Deep supercooling, 109, 113-114
Dehydration
  cell wall responses, 119-120
  and cold-stress differences, 163
  and deep supercooling activity,
      114
Dehydrin-like genes/proteins,
      stress-inducible, 125,
      126t-128t, 129
Dehydroascorbate reductase
      (DHAR), 19, 56, 59
  in chilling response studies, 64, 65
Desaturases, temperature effects,
      96-98
Desiccation tolerance, 114, 168
Developmental stages
  and chilling sensitivity, 2, 63, 68
  elevated temperature effects,
      250-251
  and heat sensitivity, 178, 179, 206
  and HSP expression, 221, 222
Differential thermal analysis (DTA),
      plant water status, 111
Disease resistance genes, 21
Dismutation, superoxide ion, 55, 58
Diurnal rhythms, and low
      temperature, 87-88
DNA chip/microarray technology,
      207

DNA marker technology, stress
 resistance, 205, 206
Drought response element (DRE),
 Arabidopsis, 164
Drought stress, and heat stress, 201
Dry matter accumulation, low-
 temperature effects, 80, 81-82

Electron transport, photosystem I
 (PSI) and II (PSII), 62, 181,
 197, 201
End uses, affected by grain quality,
 266, 268
Enzymatic activity increase, and
 carbohydrate metabolism,
 114
Ethylene, and chilling tolerance, 89
Eukaryotic initiation factors (eIFs),
 and thermal stress, 226, 227,
 228
Evaporation, and soil temperatures,
 245
Excitation states, oxygen species, 57
Expansion-induced lysis, plasma
 membrane, 121
Expressed sequence tags (EST)
 sequencing, 207
Extracellular freezing, 109, 111-113,
 151

Fatty acid
 root system composition, 95t
 synthesis, 94
Fatty acid desaturase *(fad)* genes, 96,
 97
Fertilization, and temperature
 sensitivity, 253-254
Flavonoids, 59
Fluidity, plant membrane
 and calcium channel activity, 156
 and chilling acclimation, 93-94

Fluidity, plant membrane *(continued)*
 and chilling injury, 7, 9, 10, 80
 and high temperature, 189
Fluorescence, radiative decay, 58
Food crops, tropical/subtropical
 origins of, 1
Fos, mammalian cell transcription
 factor, 24
Fracture-jump lesions, plasma
 membrane, 122
Free radical, definition, 57
Freezing stress. *See* Chilling stress

Gene expression
 antioxidant increase signals, 66-67
 Arabidopsis CDPK effects,
  161-162
 cluster manipulation, 168
 desaturase in chilling acclimation,
  96
 heat shock effects, 186, 189, 221
 low-temperature activation,
  163-165
Gene transfer technology, 2, 35, 36
Genetic engineering, chilling
 tolerance, 30
Germination, maize chilling
 sensitivity during, 2
Glutathione (GSH), 5, 19, 23, 25,
 55f, 59
 in chilling response studies, 65
 in heat stress response, 195
 H_2O_2 removal, 56
 OH reactions, 57
Glutathione reductase (GR), 5, 17,
 18, 32-33, 55f, 56, 59, 64
Glycerolipid production, root system,
 94
Grain filling, high-temperature
 effects, 256, 257, 259f, 260f,
 261
Growth, in chilled root system, 78,
 79f, 80-82
 acclimation, 90, 93

Haber-Weiss reaction, 55, 56
  formula, 57
Heat shock elements (HSE), 223,
  224
  wheat gliadin genes, 274
Heat shock factor (HSF), 223, 224
Heat shock granules (HSGs), 192
Heat shock proteins (HSPs), 130,
  179, 185, 186, 187t-188t,
  189-193, 220
  membrane effects, 197
  in multiple abiotic stresses,
  201-203
  pollen effects, 253-254
  protein/mRNA stability, 198-199
  species variation in synthesis,
  200-201
Heat shock (HS) response
  cereal grain quality, 275-277
  temperature range, 219
  transcription effects, 222-224, 225,
  229
  translation effects, 225, 226-229
  and water status, 221
Heat stress
  antioxidant activity in, 195
  association with other stresses,
  201-203
  avoidance mechanisms, 182
  cereal crop coping strategies,
  277-280
  cereal grain filling effects,
  257-266
  definition, 243, 244
  gene expression induced by, 186,
  189
  grain protein response, 274
  HSP100 role in, 193-195
  HSP70 role in, 190-191
  photosynthesis effects, 249-250
  plant injury, 180, 181
  plant organ effects, 251-252
  seedlings, 248
  signaling pathway model, 196
  small HSPs role in, 191-193

Heat tolerance
  assessment methods, 278-280,
  279t
  and transpiration, 249
Heterotrophic preemergent seedlings,
  2
High temperature (HT)
  classes of heat stress inducing, 244
  crop damage, 178
  photosynthesis effects, 248-250
  tolerance, 182, 183t-184t
HSP100 (heat shock protein family),
  187t, 193-195
  splicing efficiency, 198
HSP70 (heat shock protein family),
  187t, 190-191, 206
HVA1, cold-inducible rice gene, 129
Hydraulic conductance, and root
  chilling, 83-84, 92, 98
Hydrogen peroxide (H_2O_2), 5, 6, 7f,
  8f, 13, 15, 18, 19, 54, 56
  activation Oxy R, 23
  Ca2+ increase, 26, 56
  as defense mechanism, 20
  gene expression induction, 21
  NF-kB activation, 24
  scavenging of, 31, 55f
  signaling role, 19-23, 28, 34-35
  in vacuole, 60
Hydroxycinnamic acid derivatives,
  59
Hydroxyl radical (OH), 5, 10, 54, 55,
  57

Ice formation, determination
  technologies, 111
Ice nucleators, cacti, 112
Indica rice, amylose content, 272
Inherent thermotolerance, 184
Inositol (1,4,5)-triphosphate (IP_3),
  calcium release, 159
Intracellular ice formation, 109

Iron-induced free-radical reactions, 59
Isoprene, and thermotolerance, 197-198

Japonica rice, amylose content, 272
Jun, mammalian cell transcription factor, 24

Killing point (LT50), freezing, 110
Kinase cascade
    effects of HSP70 on, 191
    in heat stress signaling model, 196
Krebs cycle enzymes, chilling deactivation, 12-14

Late-embryogenesis-abundant (LEA) proteins, 125, 153
Light inception, heat stress effects, 252-253
Light/dark
    chilling effects in, 62
    and oxidative stress, 5
Linolenic acid, root system membranes, 94, 96
Lipid membranes
    and chilling tolerance, 9, 33-34
    composition change plasma, 122, 152
    composition in roots, 94
    and water flux, 83
Lipid peroxidation, 10-11, 11f, 19, 28, 53, 58, 60
Lipid peroxides, signaling, 27-28
Lipid-soluble antioxidants, 60
Liquid-crystalline phase, plant membrane, 7
Low molecular weight (LMW) HSP, 220. *See also* SmHSP
Low-temperature induced (*lti*) genes, 153
    ABA response, 154, 155, 155f

Low-temperature response element (LTRE), *Arabidopsis,* 164
Low-temperature signaling, mutant studies, 165-167
LT50 (lowest temperature 50% die), 109-110

Maize
    chilling sensitivity, 2, 3
    expanded products temperature effects, 273
    high-temperature effects, 254, 255, 256f, 259f, 273
    lipid peroxidation, 10, 11f
    model of chilling damage/tolerance, 28, 29f
    optimum growth temperature, 3, 247
Manganese ion (Mn^2), 55
Marker-assisted selection (MAS), stress resistance, 205
Membrane fluidity
    and calcium channel activity, 156-157
    and chilling acclimation, 93-94
    and chilling injury, 7, 9, 10, 80
    and high temperature, 189
    impact of sterols on, 98
Membrane lipids
    chilling injury, 78, 79f, 80
    root system, 94-95
Membrane-associated antioxidants, 60
Membranes *See also* Lipid membranes; Plasma membrane, freezing-induced injury
    in heat stress response, 180, 181
    phase transition and chilling injury, 7-8, 61, 62, 121
    thermotolerance role, 196-198
Metal ions/chelates, in active oxygen species reactions, 57
Methyl jasmonate, 27, 28

Michaelis constant (K_m), plant enzyme, 178
Micronutrient uptake, in chilled root system, 87
Mineral uptake, in chilled root system, 78, 79f, 85-87
acclimation, 90-91
Mitochondria
antioxidants in, 58, 59
function, 11-14
Mitogen-activated protein (MAP) kinases, stress signaling role, 162
MnSOD, and Sox R regulation, 23
Moderately high temperature effects
grain quality, 269-273
postanthesis, 255-262
Molecular mapping, stress resistance, 206
Molecular oxygen (O_2), by-product toxicity, 53
Monodehydroascorbate reductase (MDHAR), 17, 56, 59
in chilling response studies, 64
mRNA stability, heat shock effects, 225-226, 230-232
MsaCiA (cold-induced mRNA), cell wall effects, 120-121
Mulches, soil temperature effects, 247
Mutant studies, low-temperature signaling, 165-167

NADP+ (nicotinamide adenine dinucleotide phosphate, oxidized), 55f, 62
NADPH (nicotinamide adenine dinucleotide phosphate), 55f
NADPH (nicotinamide adenine dinucleotide phosphate)-dependent superoxide synthase, 20, 22
NF-kB, transcription factor, 14, 24, 26, 34

Nitrogen uptake, in chilled root system, 86
Nutritional deficits, and chilling stress, 87

Optimal temperature, maize seedling growth, 3, 247
Optimum temperature range, definition, 178
Osmolytes, osmotic stress tolerance, 197
Osmoregulation, soluble sugar role, in, 117
Osmotic stress, and heat stress, 202, 203
"Overacclimation" phase, 110
Overexpression studies, 167, 168
Over-wintering, 109, 151
plant antifreeze activity during, 113
Ovules, heat stress effects, 254
Oxidative stress
anthocyanin defense, 66
in chilling injury, 1, 28, 30
in chilling response studies, 64, 65
and heat stress, 181
light/dark effects, 5
in mitochondria, 11-14
protein degradation, 14
response variations to, 34
Oxidized glutathione (GSSG), 19, 23, 55f, 59
Ozone (O_3), 18

Paclobutrazol, induced thermotolerance, 194
Pathogenesis-related proteins, antifreeze activity, 112
Peroxidase (POX), 20
Peroxidation of lipids, 10-11, 11f, 19, 28
Phase transition, plant membranes and chilling, 7-8, 9, 61, 62
Phenylalanine ammonia-lyase (PAL), 21

Phospholipid synthesis, during chilling, 98
Phosphorus, and low soil temperature, 87
Phosphorylation, 23, 27, 157
 Ca2+ effects, 26, 45
 inhibition, 14
 and thermal stress, 226, 227, 228-229
Photoinhibition
 chilling induced, 3, 62, 63
 heat stress induced, 181
Photosynthesis
 and chilling effects, 61-62
 high temperature effects, 248, 249, 249f
 and oxygen species, 60
Plant lipid synthetic pathways, 94
Plant organs, heat stress effects, 251-252
Plant species
 and chilling sensitivity, 2, 63, 68
 and heat sensitivity, 178
Plasma membrane, freezing-induced injury, 121, 122
Pollen, and HSPs, 253-254
Poly (A) tail function, heat shock effects, 226, 230
Polyadenylation, mRA by high temperatures, 199
Polyamines, accumulation in cold acclimation, 119
Postanthesis temperature effects, 254-255
Potassium, and low soil temperature, 87
Preemergent seedlings, chilling sensitivity, 2
Proline, accumulation in cold acclimation, 118, 119
Protease activity, chilled seedlings, 14, 15
Protein degradation, 14-15
Protein denaturation, 19, 53
Protein oxidation, 28

Protein phosphorylation
 cold-induced gene role, 132
 gene expression regulation, 165
 signal transduction, 160-161
Protein transport, blockage in chilling, 9-10

Quality, grain, definition, 266-267
Quantitative trait loci (QTLs), thermotolerance, 200, 206, 207

Radicle emergence, and temperature, 80
Reactive oxygen species (ROS), 5, 6, 10, 14, 16, 34. *See also* Active oxygen species
 antioxidant scavenging of, 30
 dual role in chilling, 28
 heat stress generation of, 181, 195
 in multiple stresses, 201
 production of, 54
 as second messengers, 19, 35
Receptor-like protein kinase (RPK1), stress signaling role, 162-163
Redox-sensitive protein kinase signaling, 26
Reproductive growth, high-temperature effects, 253
Respiration, and chilling injury, 11-12
Rice, high-temperature effects, 254, 255, 256f, 257, 259f, 272-273
RNA-binding proteins, 131-132
Root cell plasma membrane, chilling acclimation, 93
Root plastids, 94
Root system
 diurnal sensitivity to cold, 88
 fatty acid composition, 95t
 functions affected by chilling, 78, 79f

Root system *(continued)*
    membrane lipids, 94-95
    soil temperature tolerance, 77
Root tip elongation, and temperature,
    80, 81
Root zone temperature, 77, 82
Root-shoot temperature differentials,
    82

Scavenging potential, antioxidant,
    53, 67
Seedling growth, and soil
    temperatures, 247
Sensitivity to freezing *(sfr)* mutants,
    Arabidopsis thaliana, 166
Signal transduction
    calcium role in, 157-160
    H_2O_2 role in, 21, 22f
    low-temperature stress, 154, 156f
    mutant studies of pathways,
        165-167
    oxidative stress response, 34
    photosynthesis in, 157
    protein kinases role in, 161-163
    protein phosphorylation role in,
        160-161
Singlet oxygen ($^1O_2$), 54, 57-58
Sink demand, low-temperature
    effects, 82
SmHSP (heat shock protein family),
    188t, 191-193. *See also* Low
    molecular weight (LMW)
    HSP
Soil
    high-temperature effects, 246, 247
    low-temperature effects, 77, 86,
        87, 88
Solid-gel phase, plant membrane, 7
Soluble sugars, and freezing
    tolerance, 116, 117, 118
SoxR and *SoxS* genes, 23
Splicing stability, in thermotolerance,
    198-199

Sterols, impact on membrane
    fluidity, 98
Stigma and style, heat stress effects,
    254
Stomatal closure, 85, 248
Stomatal conductance, 248-249
Stress avoidance, high temperature,
    278
Stress signaling triggers, 157
Stress tolerance, high temperature,
    278-280
Subtropical origins, food crops, 1, 61
Sucrose levels, grain high-
    temperature effects, 259
Summer crops, temperature
    sensitivity, 255
Superoxide anion (O_2^-), 54
Superoxide dismutase (SOD), 5, 16,
    18, 20, 30-31, 55f, 58
    in chilling response studies, 64, 65
    in heat stress response, 180, 195
Superoxides, 54-55
    in chilling stress, 5, 6, 10, 12
Synchronization, and low
    temperature, 88
Systemic acquired resistance (SAR),
    21, 24

Thermal kinetic window (TKW),
    178, 179, 198
    modification, 204
Thermostability, enzyme and protein,
    198-199
Thermotolerance
    acquisition of, 184-186
    genetic variation in, 199-201
    group ratings, 182
    heat shock proteins role in,
        189-193, 276
    membrane role in, 196-198
    and multiple stresses, 201-203
    protein/mRNA stability, 198-199
    transgenic manipulation, 203-205
Thiol groups, H_2O_2 oxidation, 56

Thylakoid proteins, heat stress, 181,197

Trans-acting factors
    identification, 27
    low-temperature mediation, 163

Transcription, heat shock effects, 222-224, 229

Transgenic manipulation, thermotolerance, 203-205

Translation, heat shock effects, 226-229

Translational efficiency, heat shock effects, 225, 226, 229-230

Transpiration, and leaf temperature, 220, 221, 249

Trehalose
    and freezing tolerance, 168
    and high-temperature tolerance, 197

Triphenyl tetrazolium chloride (TTC) cell viability assay, heat stress, 185, 186, 199, 200, 279

Tropical origins, food crops, 1, 61

Ultraviolet (UV) irradiation, and heat stress, 201, 203

Unsatuaration
    membrane lipids, 9, 197
    root lipids, 95-96
    root system membranes, 93-94, 96, 97, 98

Vacuoles
    antioxidants in, 59, 60
    low-temperature sensors in, 157

Vapor pressure deficit, in stomatal closure, 248

Very high temperature effects
    grain quality, 273-275
    postanthesis, 262-266

Vivo generated ROS, 6, 15

Water channels, temperature dependence of, 119

Water uptake, in chilled root system, 78, 79f, 80, 83-85
    acclimation, 91-93, 98

Water-selective channel proteins (aquaporins), and chilling acclimation, 93

Water-soluble antioxidants, 59

Wheat, high temperature effects, 254, 255, 256f, 257, 258f, 259f, 269-272

Wild plants, heat tolerance of, 182

Winter crops, temperature sensitivity, 255

Winter rye, 112, 113, 116, 117

Winterhardiness, soluble sugar role in, 116, 118

Woody plant cold acclimation, 152

Xanthophyll cycle carotenoids, 60, 66

Zeaxanthin, 60, 65

Order Your Own Copy of
This Important Book for Your Personal Library!

CROP RESPONSES AND ADAPTATIONS
TO TEMPERATURE STRESS

_____ in hardbound at $94.95 (ISBN: 1-56022-890-3)

_____ in softbound at $49.95 (ISBN: 1-56022-906-3)

COST OF BOOKS_____

OUTSIDE USA/CANADA/
MEXICO: ADD 20%_____

POSTAGE & HANDLING_____
(US: $4.00 for first book & $1.50
for each additional book
Outside US: $5.00 for first book
& $2.00 for each additional book)

SUBTOTAL_____

IN CANADA: ADD 7% GST_____

STATE TAX_____
(NY, OH & MN residents, please
add appropriate local sales tax)

FINAL TOTAL_____
(If paying in Canadian funds,
convert using the current
exchange rate. UNESCO
coupons welcome.)

☐ **BILL ME LATER:** ($5 service charge will be added)
(Bill-me option is good on US/Canada/Mexico orders only;
not good to jobbers, wholesalers, or subscription agencies.)

☐ Check here if billing address is different from
shipping address and attach purchase order and
billing address information.

Signature_____

☐ **PAYMENT ENCLOSED: $**_____

☐ **PLEASE CHARGE TO MY CREDIT CARD.**

☐ Visa ☐ MasterCard ☐ AmEx ☐ Discover
☐ Diner's Club ☐ Eurocard ☐ JCB

Account #_____

Exp. Date_____

Signature_____

Prices in US dollars and subject to change without notice.

NAME_____

INSTITUTION_____

ADDRESS_____

CITY_____

STATE/ZIP_____

COUNTRY_____ COUNTY (NY residents only)_____

TEL_____ FAX_____

E-MAIL_____

May we use your e-mail address for confirmations and other types of information? ☐ Yes ☐ No
We appreciate receiving your e-mail address and fax number. Haworth would like to e-mail or fax special
discount offers to you, as a preferred customer. **We will never share, rent, or exchange your e-mail
address or fax number.** We regard such actions as an invasion of your privacy.

Order From Your Local Bookstore or Directly From
The Haworth Press, Inc.
10 Alice Street, Binghamton, New York 13904-1580 • USA
TELEPHONE: 1-800-HAWORTH (1-800-429-6784) / Outside US/Canada: (607) 722-5857
FAX: 1-800-895-0582 / Outside US/Canada: (607) 772-6362
E-mail: getinfo@haworthpressinc.com
PLEASE PHOTOCOPY THIS FORM FOR YOUR PERSONAL USE.
www.HaworthPress.com

BOF00